AROMATICITY

AROMATICITY

PETER J. GARRATT

Department of Chemistry
University College London

A Wiley–Interscience Publication

JOHN WILEY & SONS

New York Chichester Brisbane Toronto Singapore

Library of Congress Cataloging in Publication Data:

Garratt, P. J. (Peter Joseph), 1934–
Aromaticity.

"A Wiley-Interscience publication."
Bibliography: p.
Includes index.
1. Aromatic compounds. 2. Chemical bonds.
I. Title.

QD331.G28 1986 547′.6 85-17816
ISBN 0-471-80703-6

Printed in the United States of America

10 9 8 7 6 5 4 3 2 1

E. HEILBRONNER: *On Mr. Binsch's slide the following definition is proposed: "A conjugated π-electron system is called aromatic if it shows neither strong first-order nor second-order double bond fixation." Now, could you point out a molecule, except benzene, which classifies as "aromatic."*

B. BINSCH: *Benzene is a perfect example!*

E. HEILBRONNER: *Name a second one.*

Jerusalem Symposia on Quantum Chemistry and Biochemistry, Vol. 3, *Aromaticity, Pseudoaromaticity, Antiaromaticity*, Israel Academy of Science and Humanities, 1971.

PREFACE

The classification of organic substances into aliphatic and aromatic compounds goes back to the early part of the last century. Its continued scientific justification arose from the finding that certain unsaturated radicals, such as C_6H_5, unexpectedly passed unchanged through a series of reactions carried out on these substances. Attempts to explain this behavior have provided one of the most fruitful areas for the interaction of theory and experiment known to organic chemistry, and this interaction continues to the present day. One of the initial major achievements of the application of quantum mechanics to large molecules, carried out by Hückel, was to provide not only an explanation for the stability of benzene but to predict other molecules which would be similarly stabilized. How these predictions were fulfilled and extended and how theory and experiment continue to interact in this area are the subjects of this text.

The text adopts a similar format to that I had used for a previous text in this area. It is aimed at final-year undergraduates and first-year graduate students and attempts to integrate theory and experiment. In order to improve readability, references have not been given except in the form of further reading at the end of each chapter. I hope that such provision, largely to reviews, will provide an easy and sufficient access to the primary literature for anyone seeking to clarify or develop the arguments used in the book.

The initial chapter provides an introduction both to the problem of aromaticity and to quantum theory; it may be omitted by those cognizant with the latter. The second chapter is pivotal to the rest of the text, expounding the theoretical and experimental differences between cyclobutadiene, benzene, and cyclooctatetraene. The following seven chapters describe the various types of compounds to which the attribute aromatic has been given; these chapters are each self-contained and, if the book is used as a teaching text, can be used or omitted to suit the instructor's taste. The penultimate chapter discusses the concept of homoaromaticity and related topological possibilities, an area which continues to provoke disputation, and the final chapter, after a cursory examination of the importance of aromaticity in the transition state, summarizes the conclusions of the preceding ten.

A number of people read the manuscript in whole or part during the course of its preparation. I would like to thank my students, Vahid Alikhani, Kate Baldwin, Alka Khurana, Stewart Mitchell, David Payne, John Porter, and Andrew Tsotinis, our visitors from Hannover, Uwe Matthies and Stefan Wolff, and my colleague Chris Cooksey, for reading and commenting on many or all of the chapters. Professor John Ridd read and gave me his expert advice on Chapter 3. I would also like to thank all of those who read and commented on the previous text and particularly Professor Peter Vollhardt, who translated and revised the previous German text with me. I am again indebted to Dr. Jim Parkin, but now also for his expert advice on word processing, a technique that allowed me to prepare my own typescript.

My friend and mentor, Professor Thomas Katz, initiated me into this area during a very happy period spent at Columbia University. It was sustained by the late Professor Franz Sondheimer, to whose memory I dedicate this book.

Acknowledgments. I thank Messrs Butterworths for permission to publish Figures 2.16 and 9.7; Professor R. McWeeny and Messrs Taylor and Francis for permission to publish Figure 2.14; and Pergamon Press for permission to publish Figure 4.19.

PETER GARRATT

CONTENTS

AROMATICITY

1

THE AROMATICITY PROBLEM

1.1. INTRODUCTION

One of the major achievements in chemistry during the nineteenth century was the formulation of an empirical theory of valency. This theory resulted largely from the study of organic compounds, the systematic investigation of which was begun in the early part of the century by Wohler and Leibig. Whereas the ionic theory of Berzelius explained much of the inorganic chemistry that was known at the time, Dumas and others clearly demonstrated that this theory was not successful in accounting for the behavior of organic compounds. Thus, although chlorine and hydrogen were of opposite nature in the system of Berzelius, chlorine could replace hydrogen in organic systems. Furthermore, as was pointed out by Laurent, in most cases such a *substitution* did not greatly affect the properties of the compound. A second important observation was that groups of atoms in organic molecules appeared to pass unchanged through complex series of reactions. These groups of atoms, such as CH_3 and CH_3CH_2, were called *radicals.* After a long and acrimonious argument between the proponents of the substitution theory and those of the radical theory, these theories were combined by Gerhardt, using the experimental results of Hofmann and of Williamson, into his "theory of types." In this theory, the hydrogen atoms of the parent system of the type were successively replaced by organic "radicals." Thus, methanol, CH_3OH, was of the water type, one of the hydrogens of H_2O having been replaced by the CH_3 radical. Dimethyl ether, CH_3OCH_3, was also of the water type, but in this case both of the hydrogens had been replaced by CH_3 radicals.

A second problem, which was resolved at about the same time, was the question whether atomic or equivalent weights should be used in the formulation of organic molecules. Berzelius had established excellent values for the atomic weights of many of the elements early in the century, but unfortunately his atomic weight for carbon was most seriously in error. Because of this, some chemists had come to favor equivalent weights, with carbon having the value of 6, for the composition of organic compounds. Complete confusion resulted, since, for example, H_2O_2 could be the formula of either water or hydrogen peroxide! However, at the Karlsruhe Conference in 1860, Cannizarro circulated a pamphlet that, although it did not lead to immediate agreement on a single system, had a profound effect on the conference participants. In this pamphlet, the Italian chemist summarized his previously reported results and explained how, by the use of Avogadro's hypothesis, a unique system of atomic weights could be obtained that was applicable to both organic and inorganic compounds. Most chemists accepted these findings, and molecular formulas were able to be established on which all could agree.

In 1857, just before the Karlsruhe Conference, Kekulé had recognized that by using the new atomic weights of Cannizarro, carbon appeared to be tetravalent in a number of its compounds. In his celebrated paper of 1858, Kekulé extended this view to all carbon compounds and also introduced the concept that the carbon atoms could be linked to one another. Similar views were put forward independently by Couper, who used a dotted line to represent a valency bond. These ideas were developed for other atoms, and it was concluded that each atom had one or more valencies. The belief that the formula now represented an arrangement of atoms began to be more commonly accepted, and structural formulas were introduced by Crum Brown. These structural formulas were then shown to require modification to express the three-dimensional nature of the molecules, and the concept of the tetrahedral arrangement of the carbon valencies was independently advanced by van't Hoff and Le Bel.

A number of carbon compounds, such as carbon monoxide, did not conform to the Kekulé-Couper theory of tetravalent carbon. The unsaturated hydrocarbons ethylene, C_2H_4, and acetylene, C_2H_2, were two further exceptions to the rule, and three serious explanations were put forward to account for the apparent lack of tetravalency of carbon in these systems. Couper proposed that carbon could be divalent as well as tetravalent, Kekulé suggested that some of the carbon valencies were unsatisfied, and Erlenmeyer proposed that the carbons were linked to each other with more than one valency. Benzene and the aromatic compounds raised further complications, since these molecules, unlike ethylene and acetylene, do not readily undergo addition reactions to give derivatives in which carbon is tetravalent. The paradox of benzene, unsaturated and yet inert, is the central theme of this book. The ramifications of this problem lead us far away from benzene, but it remains the aromatic compound par excellence and will be the subject of discussion in the rest of this and much of the two next chapters.

1.2. EARLY INVESTIGATIONS INTO THE STRUCTURE OF BENZENE

Benzene was discovered by Faraday in 1825 in the condensate obtained by compression of the gas generated by pyrolysis of whale oil. Faraday determined the composition, vapor pressure, and melting point (42°F). He deduced the correct molecular formula, which he expressed as C_6H_3, using the equivalent weights current at that time. The first synthesis of benzene was accomplished by Mitscherlich in 1833 by the decarboxylation of benzoic acid. Mitscherlich confirmed the correct molecular formula and subsequently synthesized a number of derivatives.

Benzene was recognized as the parent of a number of compounds, all of which contained the C_6H_5 radical, and this radical was shown to be inert, remaining intact throughout series of chemical reactions. Benzene did not fit into the Kekulé–Couper tetravalent theory of carbon and joined the unsaturated hydrocarbons ethylene and acetylene in providing difficulties for the theory.

The first satisfactory formula for benzene was put forward by Kekulé in 1865. In this formulation, the six carbon atoms of benzene were considered to be linked alternately by one or two valencies, which left eight valencies unsatisfied. Two of the unsatisfied valencies were then used to link the terminal carbon atoms to form a *cyclic system.* The six carbon atoms then had six unused valencies to which six monovalent hydrogens could be attached. This was originally illustrated by Kekulé as the structure shown in Figure 1.1*a* in which the ellipses represent the carbon atoms, the lines depict the bonds between the carbon atoms, the dots represent the points of attachment of the hydrogen atoms, and the arrows indicate the junction of the terminal carbon atoms. Subsequently, Kekulé modified these concepts and introduced the now familiar hexagon structure, with alternate double and single bonds (Fig. 1.1*b*).

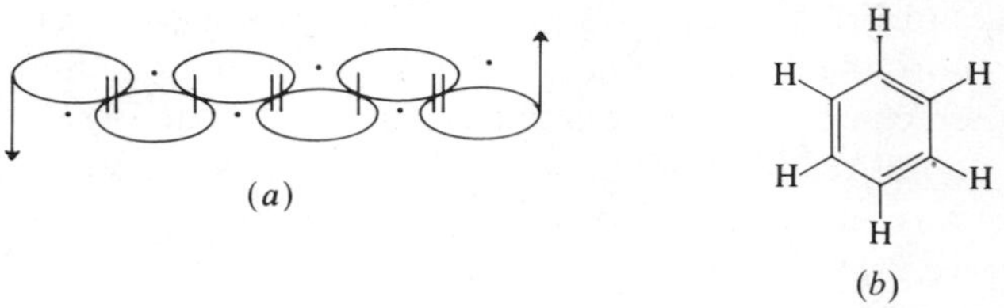

(a)

(b)

Figure 1.1. Structures for benzene suggested by Kekulé. For a description of structure (a) see text.

The hexagonal formula predicts that all six carbon atoms, but not the six bonds, are equivalent, and this prediction was confirmed by investigations carried out independently by Ladenburg and Wroblewsky. Wroblewsky prepared the five possible monobromobenzoic acids, using suitable blocking groups that could be subsequently removed, and he found that the 1,2 and 1,6, and the 1,3 and 1,5 derivatives were identical (Fig. 1.2). Thus, carbons 2 and 6 and carbons 3 and 5 are identical. Ladenburg had previously found that the three isomeric hydroxybenzoic acids gave the same phenol on decarboxylation and the same benzoic acid on reduction. The phenol was then converted to benzoic acid, which was shown to be identical with benzoic acid obtained by reduction. Thus, the carbon atoms at positions 1, 2, 3, and 4 are equivalent. The combination of these data with Wroblewsky's demonstrates that the six carbon atoms are equivalent to each other (Fig. 1.2).

These experiments now placed the requirement on any suggested structural formula of benzene that all the carbon atoms must be equivalent.

Ladenburg now protested that Kekulé's hexagon formula required *four*, rather than the *three* disubstituted products that had been prepared (Fig. 1.3). To counter this objection, Kekulé introduced the concept that there were two equivalent structures for benzene, Figure 1.4, which rapidly interconverted by a "mechanical motion." This mechanical motion is equivalent to the oscillation of the double bonds around the ring and the interchange between **1***a* and *b* renders the two 1,2-disubstituted derivatives in Figure 1.3 equivalent.

Although the "resonating" structures for benzene now appear to us as an example of Kekulé's intuitive genius, at the time it was introduced, the idea was considered a mere device to save the hexagon theory. Many other structures for benzene were proposed, such as the "para bonded" formula **2** by Dewar and Wislicenus and the "diagonal" formulas **3** and **4** by Claus. We now recognize that such formulas have stereochemical implications as three-dimensional structures, but this was not completely recognized at the

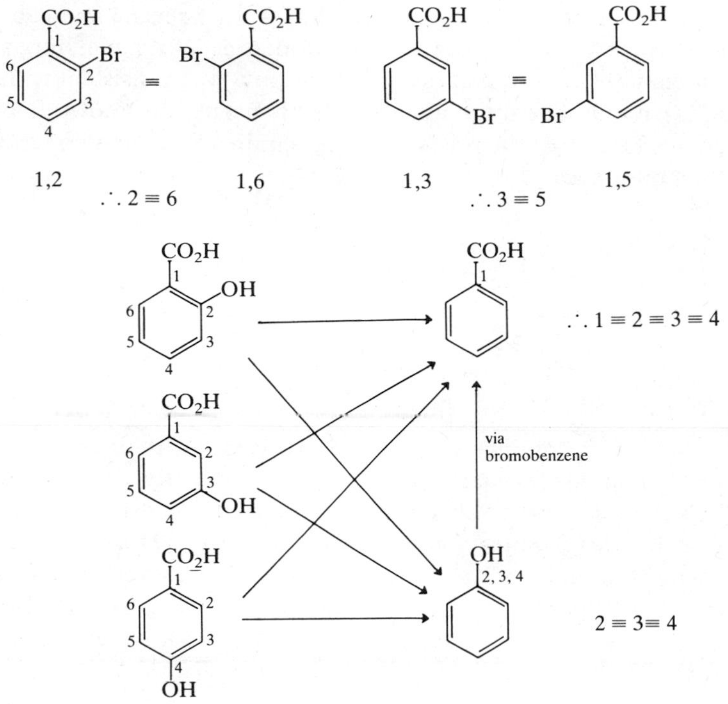

CO_2H
Br
1,2
∴ 2 ≡ 6
1,6
1,3
∴ 3 ≡ 5
1,5
1,4
OH
∴ 1 ≡ 2 ≡ 3 ≡ 4
via
bromobenzene
2, 3, 4
2 ≡ 3 ≡ 4

Figure 1.2

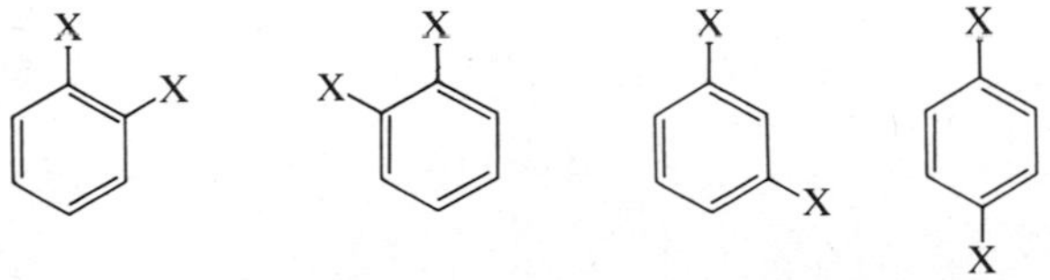

X

Figure 1.3. The four possible disubstituted benzenes assuming the static Kekulé hexagonal structure.

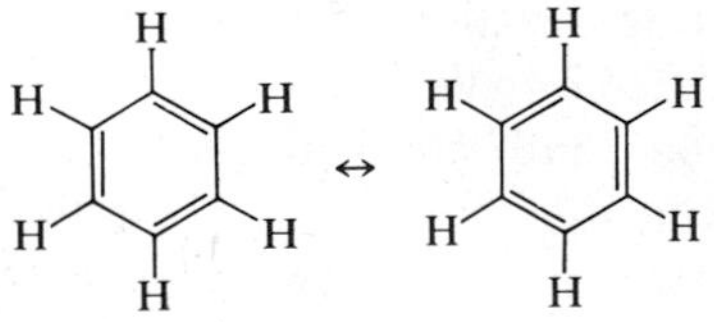

H
↔

a **1** *b*

Figure 1.4

time. Ladenburg, however, also proposed formula **4** for benzene and subsequently recognized the three-dimensional properties of "Ladenburg's prism" (**5**). Ladenburg believed that the prism formula was consistent with the observation that only one ortho disubstituted benzene was known, but van't Hoff demonstrated that the prism formula requires that enantiomeric "ortho" forms should exist.

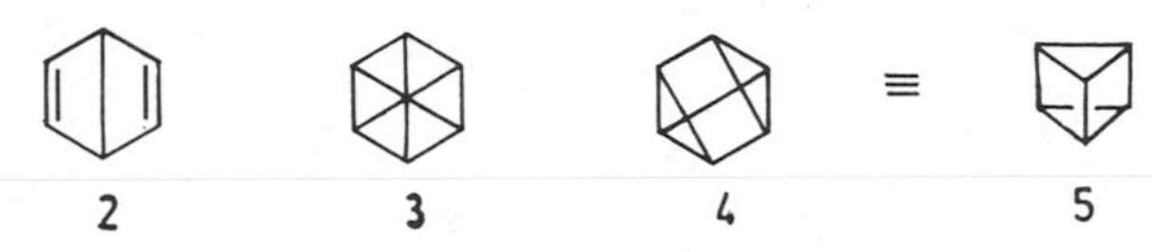

Figure 1.5

Kekulé's formulation continued to provide conceptual difficulties for many workers. Lothar Meyer and Armstrong independently proposed a model in which the six unused valencies were directed toward the center of the hexagon (**6**); this formula was adopted by Baeyer (**7**). Bamberger extended this type of formula to naphthalene (**8**) and to heterocyclic systems, such as pyrrole (**9**). Bamberger recognized that in these systems *six* unsatisfied valencies are required for each ring, and he clearly considered that this arrangement must provide a stabilized inert system.

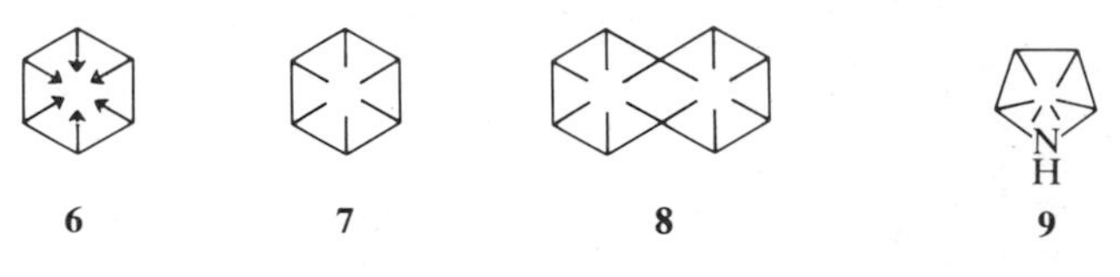

Figure 1.6

These structural theories for benzene are a remarkable achievement of nineteenth-century chemistry, but they do not explain the lack of reactivity of benzene and related aromatic systems. It is difficult to understand why either the Kekulé structure or those with unsaturated valencies should not readily undergo addition reactions. The Dewar, Claus, and Ladenburg structures, when translated into three-dimensional forms, are unacceptable. Dewar's structure **2** would indicate that two carbons are different from the other four, besides having the stereochemical implications shown in formula **10**. The diagonal formula **3** of Claus is not translatable into three-dimensional terms with reasonable bond angles (Fig. 1.8). A number of these structures will be discussed further in Chapter 3 as valence isomers of benzene.

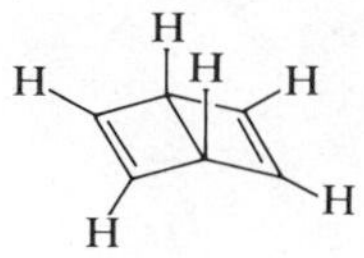

H H H H H H

10

Figure 1.7

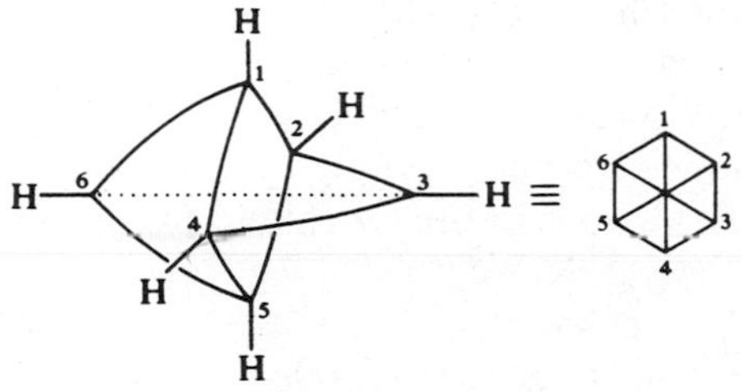

H H H H H H

Figure 1.8. The spacial implications of the Claus benzene formula.

Bamberger had realized the importance of the sextet of affinities in benzene and had, as we have previously seen, used this concept to explain the aromaticity of other systems. The first major attempt to account for the lack of reactivity of benzene was, however, provided by Thiele within the framework of his theory of partial valency. To account for the 1,4-addition of reagents to butadiene, Thiele suggested that the diene had partial unsatisfied valencies at the terminal atoms and a partial double bond between the central carbon atoms. When this concept is applied to benzene, the terminal unsatisfied valencies are now shared by the 1,6 carbon atoms, and a structure is obtained in which all of the bonds have a character similar to the central bond of butadiene (Fig. 1.9).

Thiele's structure for benzene suggests that all cyclic polyenes should have similar properties to benzene, and the synthesis of cyclooctatetraene (**11**) by Willstätter and co-workers, which had the properties of an olefin rather than benzene, caused Thiele's view to fall out of favor.

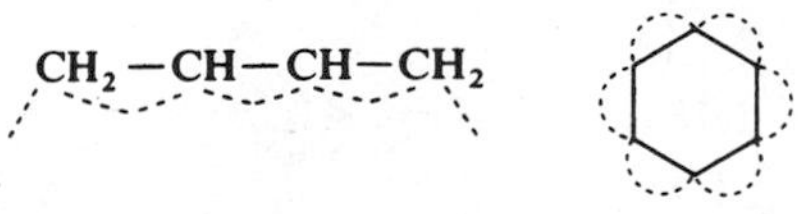

$CH_2-CH-CH-CH_2$

(a) *(b)*

Figure 1.9. The structures of butadiene (*a*) and benzene (*b*) according to Thiele's theory of partial valency.

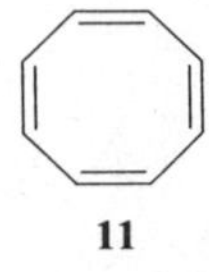

11

Figure 1.10

During the period between Kekulé's formulation for benzene and Willstätter's synthesis of cyclooctatetraene, a revolution in our understanding of the nature of matter had occurred. The electron had been discovered by Thomson, and this had led to attempts to correlate affinity with electron availability. In 1916, Kossel and Lewis gave an interpretation of atomic structure in which the electrons occupied shells, with two electrons in the first shell and eight electrons in the two subsequent shells. When the shells are filled, the inert gas structure results. Thus, with the inner two-electron shell filled, the helium configuration is reached; with the two-electron and first eight-electron shells filled, the ten-electron neon structure is attained. These ideas were significantly extended by Lewis in his suggestion that the atoms forming molecules can *share* electrons so that each may attain the inert gas configuration. This proposal embodies the concept of the "localized bond," formed by the sharing of two electrons.

The electronic theory of valency was applied to benzene by Armit and Robinson and by Ingold. Armit and Robinson supposed the six free affinities to be six electrons, and they reinterpreted Bamberger's theory in terms of this "aromatic sextet" of electrons which, like the octet, was presumed to have a closed configuration. Ingold extended the Kekulé model to include the three Dewar structures with *para*-bonds between the 1,4, 2,5, and 3,6 atoms, respectively, and formulated benzene as a combination of all five structures.

Neither the Armit–Robinson nor the Ingold theory accounts for the stability of benzene or why, in the Armit–Robinson theory, six rather than eight electrons form a stable shell. However, during the 1920s quantum mechanics was invented and, arising from this, theories of chemical bonding were developed by which it is possible to account for the stability of the aromatic sextet. These theories are discussed in the next section.

1.3. QUANTUM MECHANICS AND THE STRUCTURE OF COMPLEX MOLECULES

The state of a molecular system may be represented by a function ψ, called the wave function. The wave function ψ is a solution of the Schrödinger wave equation

$$H\psi = E\psi \tag{1.1}$$

in which H is the Hamiltonian operator that is associated with the observable energy of the system, E. This equation has in general discrete solutions ψ_i called *eigenfunctions*, and the corresponding energies E_i of these eigenfunctions are *eigenvalues*. The solution ψ_0 corresponding to the lowest eigenvalue E_0 represents the ground state of the system, and the other solutions of ψ_i corresponding to higher eigenvalues represent excited states. In general, the Schrödinger equation is only analytically soluble for one-electron systems such as the hydrogen atom or the hydrogen molecular ion H_2^+. For complex systems, approximate methods must be used, and the most useful of these is the *variation method*. The *variation principle* states that the mean value of the energy calculated using the function ψ_i is never less than the energy of the lowest eigenstate E_0. Thus,

$$E_0 \leqslant E = \frac{\int \psi_i H \psi_i \, d\tau}{\int \psi_i \psi_i \, d\tau} \tag{1.2}$$

The value of the trial function ψ_i is then varied so that the value of E more closely approaches E_0. The usual method of obtaining trial functions ψ_i is to take linear combinations of the atomic orbitals (LCAO) of the atoms involved, such that

$$\psi_i = \sum_r c_{ir} \phi_r \tag{1.3}$$

where ϕ_r is the atomic orbital of the rth atom, and c_{ir} is the coefficient of that orbital.

If equation (1.3) is substituted into equation (1.2), then equations (1.4) and (1.5) can be obtained:

$$E = \frac{\int \left(\sum_r c_r \phi_r\right) H \left(\sum_r c_r \phi_r\right) d\tau}{\int \left(\sum_r c_r \phi_r\right)^2 d\tau} \tag{1.4}$$

$$= \frac{\sum_r \sum_s c_r c_s \int \phi_r H \phi_s \, d\tau}{\sum_r \sum_s c_r c_s \int \phi_r \phi_s \, d\tau} \tag{1.5}$$

Denoting $\int \phi_r H \phi_s \, d\tau$ by H_{rs}, and $\int \phi_r \phi_s \, d\tau$ by S_{rs}, equation (1.5) can be expressed as

$$E = \frac{\sum_r \sum_s c_r c_s H_{rs}}{\sum_r \sum_s c_r c_s S_{rs}} \tag{1.6}$$

The coefficients c_r and c_s can now be independently varied so that E tends toward zero. This is accomplished by setting each of the partial derivatives $\partial E/\partial c_i$ equal to zero, which then provides a set of secular equations

$$\begin{aligned}
c_1(H_{11} - S_{11}E) + c_2(H_{12} - ES_{12}) \cdots c_n(H_{1n} - ES_{1n}) &= 0 \\
c_1(H_{12} - S_{12}E) + c_2(H_{22} - ES_{22}) \cdots c_n(H_{2n} - ES_{2n}) &= 0 \\
\vdots \qquad\qquad\qquad \vdots \qquad\qquad\qquad \vdots \quad & \\
c_1(H_{1n} - S_{1n}E) + c_2(H_{2n} - ES_{2n}) \cdots c_n(H_{nn} - ES_{nn}) &= 0
\end{aligned} \tag{1.7}$$

These may be expressed in more compact form as

$$\sum_r c_r(H_{rs} - ES_{rs}) = 0 \tag{1.8}$$

There is an algebraic theorem that states that a set of simultaneous equations have nontrivial solutions (i.e., not solutions in which $c_1 = c_2 = c_n = 0$) only when the determinant formed from the coefficient c_1 vanishes. The secular determinant corresponding to equation (1.7), which must be equated to zero, is equation (1.9):

$$\begin{vmatrix}
H_{11} - ES_{11} & H_{12} - ES_{12} \cdots & H_{1n} - ES_{1n} \\
H_{12} - ES_{12} & H_{22} - ES_{22} \cdots & H_{2n} - ES_{2n} \\
\vdots & \vdots & \vdots \\
H_{1n} - ES_{1n} & H_{2n} - ES_{2n} \cdots & H_{nn} - ES_{nn}
\end{vmatrix} = 0 \tag{1.9}$$

Equation (1.9) can be solved to give values of the energy of the system. It is, however, often convenient at this stage to make certain approximations to simplify the problem. These approximations, introduced by Hückel, are discussed in the next section, with particular reference to benzene.

1.4. THE HÜCKEL METHOD AND ITS APPLICATION TO BENZENE

The benzene molecule contains 12 nuclei and 42 electrons, and it would thus be an enormous task to calculate the wave function by the method described in the previous section. Indeed, if such a calculation were carried out, the resulting wave functions would bear no relationship to concepts such as the chemical bond. Major approximations are therefore made that substantially simplify the problem.

The carbon and hydrogen nuclei, together with the carbon $1s^2$ electrons, are assumed not to take part in the bonding and are neglected. The problem is thus reduced to the remaining 30 electrons, the valence electrons, of which 4 are contributed by each carbon and 1 by each hydrogen. The configuration of the carbon atoms is assumed to be $s\,p_x\,p_y\,p_z$, and the s, p_x, p_y orbitals are combined together to form three hybrid orbitals arranged at 120° from each other in a plane. Two of these hybrid orbitals are used to bond with two other carbons and the third to bond to an hydrogen atom. The six carbons and six hydrogens thus form a hexagonal structure, and the orbitals concerned with this bonding, which are symmetric with regard to the molecular plane, are called σ bonds. The hexagonal structure is thus formed from σ bonds, which is shown in Figure 1.11. The σ framework of

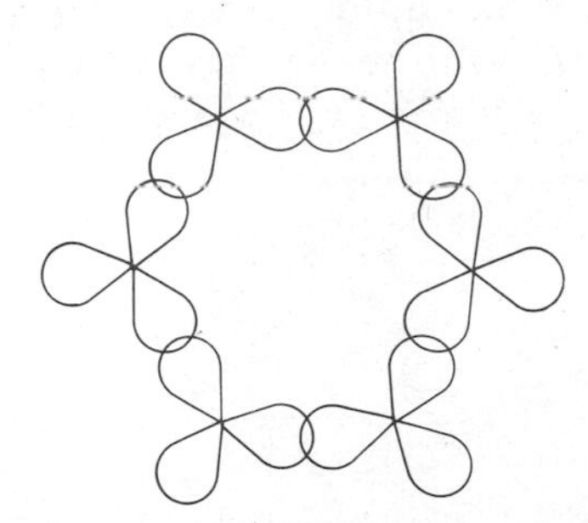

Figure 1.11. The σ framework of benzene.

benzene has required us to use 24 of the 30 valence electrons; the remaining 6 electrons are in p_z orbitals. These latter orbitals are antisymmetric with regard to the ring plane of the hexagon and are termed π orbitals. It is the six electrons in the π orbitals that are treated as though they are responsible for the properties of benzene, and these six π electrons can now be considered separately. The separation of the π electrons from the σ electrons is known as the *Hückel Approximation.*

Hückel then developed a method, now called the Hückel molecular orbital (HMO) method, in which the π electrons can be treated by equation (1.9). In the HMO method, several simplifications are introduced into this equation.

The integral H_{rr}, termed the *coulombic integral*, is considered to have a value that depends only on the characteristic atomic orbital ϕ_r of the atom on which it is centered. The coulomb integral is the approximate measure of the electron attracting power of the atom involved, and it is assumed to be independent of the rest of the system. The coulomb integral for carbon, in the HMO method, is given by

$$H_{rr} = \alpha \tag{1.10}$$

The integral H_{rs} is called the *resonance integral* and is a measure of the binding power of the bond *rs*. H_{rs} is assumed to become vanishingly small except when r and s are nearest neighbors. The Hückel method assumes $H_{rs} = 0$ when r and s are *not* joined by a σ bond. For carbons that are nearest neighbors,

$$H_{rs} = H_{sr} = \beta \tag{1.11}$$

The integral S_{rs} is called the *overlap integral* and has the value 0.25 for nearest neighbors. This value rapidly decreases for nonnearest neighbors, and as a consequence H_{rs} vanishes in these cases (see above). The HMO method assumes that $S_{rs} = 0$ when $r \neq s$, and it is self-evident that $S_{rr} = 1$.

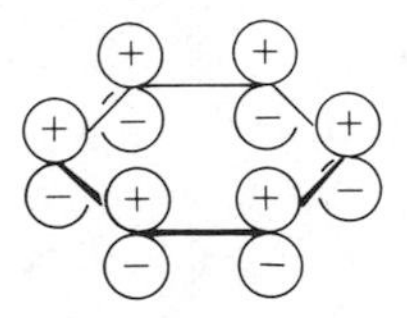

Figure 1.12

Using these approximations, equation 1.9 can be applied to the six atomic π orbitals shown in Figure 1.12. The six atomic π orbitals combine to give a six-term secular equation that generates six molecular orbitals. Using the integrals defined above, then

$$H_{11} = H_{22} = H_{33} = H_{44} = H_{55} = H_{66} = \alpha$$

$$H_{12} = H_{23} = H_{34} = H_{45} = H_{56} = H_{61} = \beta$$

and

$$H_{13} = H_{14} = H_{15} = H_{24} = \text{etc.} = 0$$

The secular equation (1.9) now has the form of equation (1.12):

$$\begin{vmatrix} \alpha - E & \beta & 0 & 0 & 0 & \beta \\ \beta & \alpha - E & \beta & 0 & 0 & 0 \\ 0 & \beta & \alpha - E & \beta & 0 & 0 \\ 0 & 0 & \beta & \alpha - E & \beta & 0 \\ 0 & 0 & 0 & \beta & \alpha - E & \beta \\ \beta & 0 & 0 & 0 & \beta & \alpha - E \end{vmatrix} = 0 \qquad (1.12)$$

The determinant can be simplified by dividing through by β and setting $(\alpha - E)/\beta = x$, which gives equation (1.13):

$$\begin{vmatrix} x & 1 & 0 & 0 & 0 & 1 \\ 1 & x & 1 & 0 & 0 & 0 \\ 0 & 1 & x & 1 & 0 & 0 \\ 0 & 0 & 1 & x & 1 & 0 \\ 0 & 0 & 0 & 1 & x & 1 \\ 1 & 0 & 0 & 0 & 1 & x \end{vmatrix} = 0 \qquad (1.13)$$

Equation (1.13) can then be solved and the six roots determined. These give six solutions for the energy: $E = \alpha - 2\beta$, $E = \alpha - \beta$ (two roots), $E = \alpha + \beta$ (two roots), and $E = \alpha + 2\beta$. Since β is a negative quantity, positive coefficients of β represent more stable energy levels. The energy levels of benzene are shown in Figure 1.13.

From the earlier discussion, since H_{rr} measures the electron affinity of the carbon atom and is equal to α, a molecular orbital with an energy α

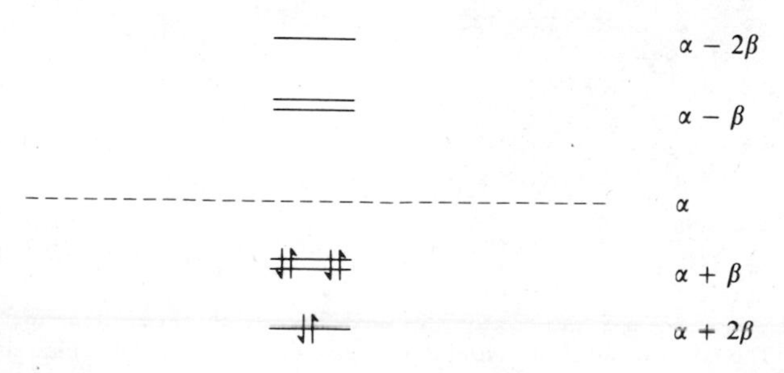

Figure 1.13

has the same energy as the atomic orbital from which it was derived. Molecular orbitals of energy α are therefore termed *nonbonding molecular orbitals* (NBMO). Orbitals that have energies in which the coefficient of β is positive have lower energies than the $2p$ atomic orbitals from which they were derived and are therefore *bonding orbitals*, whereas those with negative coefficients of β have higher energies than the $2p$ atomic orbitals and are *antibonding orbitals*. Each molecular orbital can accommodate two electrons of opposite spin quantum number, and the six π electrons of benzene can thus enter three bonding orbitals (Fig. 1.13).

The HMO method applied to benzene thus gives an explanation for its stability in that all six π electrons are bonding. The reason for the "magic" number six—the aromatic sextet—also becomes apparent, since there are only three bonding molecular orbitals available from the cyclic combination of six atomic orbitals and thus only six π electrons can be accommodated. The total energy for the six π electrons is $6\alpha + 8\beta$. It would be of considerable interest to know the value of β, but this value is difficult to determine, as will be discussed in Chapter 2.

The six π-molecular orbitals of benzene are represented in Figure 1.14. Of the bonding orbitals, the lowest energy $\alpha + 2\beta$ orbital has no nodes, and the two $\alpha + \beta$ orbitals have one node each. In the case of the antibond-

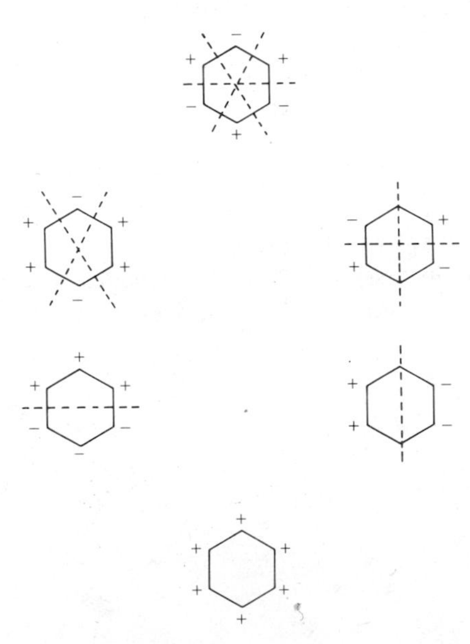

Figure 1.14. The six π orbitals of benzene, showing the phase of the orbital at each carbon atom and the nodal planes.

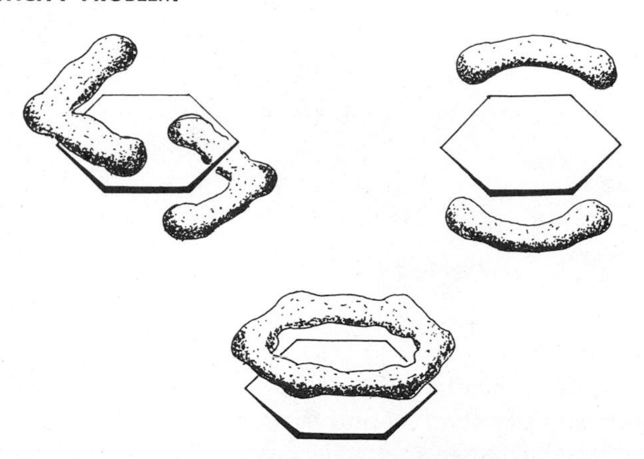

Figure 1.15. A diagrammatic representation of the three bonding orbitals of benzene. (For clarity, only the positive phase of the orbital is shown, the negative phase being the mirror image of the positive phase in the ring plane.)

ing orbitals, the $\alpha - \beta$ orbitals have two nodes and the $\alpha - 2\beta$ orbital has three nodes. The phases of the six π orbitals are indicated in Figure 1.14. The positive phases of the three bonding orbitals are shown pictorially in Figure 1.15; the negative phase, which has been omitted for clarity, is the mirror image of the positive phase reflected in the ring plane. When the six π electrons are introduced into the three bonding orbitals, the plot of the π-electron density (ψ^2) has the form shown in Figure 1.16, with a node in the ring plane and two circular regions of high density above and below the σ framework.

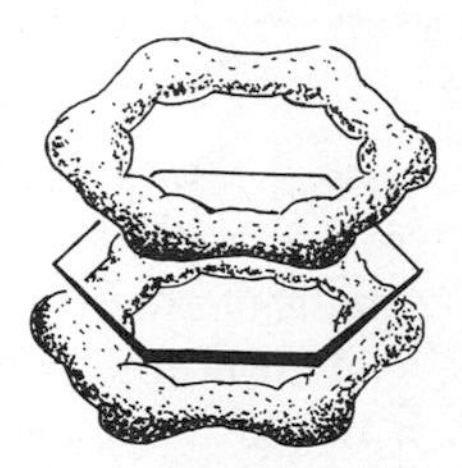

Figure 1.16. The π-electron density (ψ^2) resulting from the introduction of the six electrons into the three bonding orbitals.

Benzene is a special representative of the group of n-membered monocyclic conjugated systems, and in general the π-electron molecular orbital levels in these systems can be shown to be given by the expression

$$E = \alpha + 2\beta \cos \frac{2\pi r}{n} \tag{1.14}$$

where

$$r = 0, \pm 1, \pm 2 \cdots 2/n$$

if n is even and

$$r = 0, \pm 1, \pm 2 \cdots \pm \frac{n-1}{2}$$

if n is odd.

When n is even, it can be easily shown that the molecular orbitals are arranged symmetrically about α and that the lowest orbital has an energy of $\alpha + 2\beta$ and the highest an energy of $\alpha - 2\beta$. *All* of the remaining orbitals are *doubly degenerate.* In the case in which n is odd, the lowest orbital is of energy $\alpha + 2\beta$ and the remaining orbitals are *doubly degenerate* (Fig. 1.17).

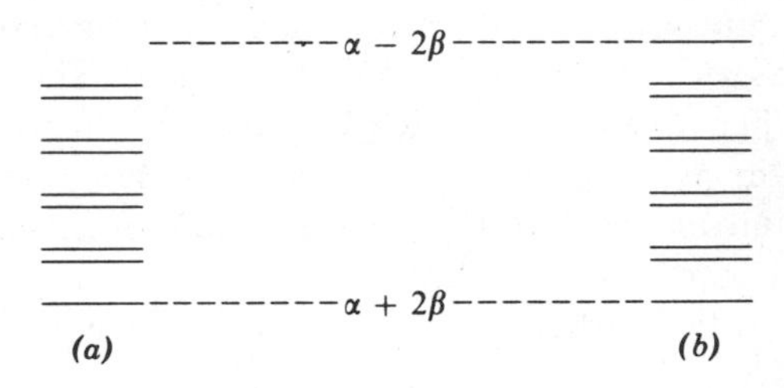

Figure 1.17. HMO π-orbital pattern for odd (a) and even (b) monocyclic systems.

Since a doubly degenerate orbital can hold four electrons and the lowest molecular orbital can hold two electrons, then if the orbitals are filled on the aufbau principle, complete shells of electrons can *only* occur in *mono*cyclic conjugated systems if the system contains $4n + 2\pi$ electrons, where n is an integer. This observation forms the basis of the *Hückel Rule,* which will be discussed further in Chapter 2. These findings can be expressed in a simple geometrical form due to Frost and Musulin. In this method, the system is inscribed in a circle of diameter 2β so that one atom is at the bottom of the vertical axis (Fig. 1.18).

The horizontal axis now corresponds to the energy level α, and the energies of the orbitals correspond to the positions of the atoms on the circle. Benzene with six atoms has n even and, as can be seen in Figure 1.18, has two nondegenerate orbitals of energy $\alpha + 2\beta$ and $\alpha - 2\beta$. Of the four remaining orbitals, the degenerate bonding orbitals are of energy

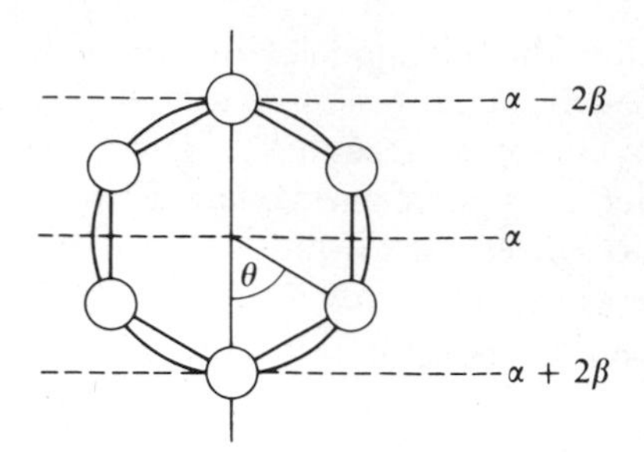

Figure 1.18

$\alpha + 2\beta \cos \theta$, where $\theta = 60°$, and thus $E = \alpha + \beta$, while the degenerate antibonding orbitals are of energy $\alpha - 2\beta \cos 60°$, and thus $E = \alpha - \beta$. These values, of course, correspond to those obtained from equation (1.14). Frost and Musulin diagrams will be used on numerous occasions throughout the rest of this book.

One other consequence of equation (1.14) should be noted here. Since the lowest molecular orbital is always $\alpha + 2\beta$ and the highest is never more than $\alpha - 2\beta$ in energy, as the value of n increases the degenerate levels are forced closer and closer together. The properties of a series of monocyclic systems should reflect the result of this orbital packing.

The implications of the assumptions made in the Hückel method have been discussed in considerable detail by theoretical chemists, and the interested reader is referred to the references at the end of this chapter. It is generally agreed that this method only gives satisfactory results for groups of closely related molecules, such as the benzenoid hydrocarbons, and that it predicts the order of orbital energies much better than either the total π energy of the system or the energy between orbitals. A further discussion of the shortcomings of the Hückel theory and some efforts toward its rehabilitation will be found in Chapter 2.

1.5. THE VALENCE BOND METHOD

Besides the molecular orbital (MO) method, which has been described above in its simplest form, a second approximate method is available, the valence bond (VB) method. Both the MO and VB methods start by taking atomic orbitals, but whereas in the MO model a linear combination of these orbitals is first formed, in the VB model wave functions similar to Heitler–London two-electron bond functions are constructed. The VB model thus gives a picture that is much more similar to the chemist's conception of a

chemical bond than does the MO model. However, in the case of conjugated molecules, the simple VB model has to be modified by the introduction of a further concept, *resonance*, to account for delocalization. In this respect, the MO method, which uses polycentric one-electron functions, is superior.

The VB model for benzene takes into account the two Kekulé (**1**) (**2**) and three Dewar (**3**) (**4**) (**5**) structures. These are shown in Figure 1.19, and the wave functions can be set up in the form

$$\psi = c_1\Theta_1 + c_2\Theta_2 + c_3\Theta_3 + c_4\Theta_4 + c_5\Theta_5 \qquad (1.15)$$

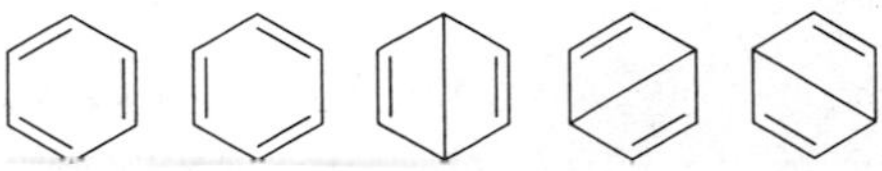

(1) (2) (3) (4) (5)

Figure 1.19

where c_1 is the coefficient and Θ_1 the wave function of structure (**1**). The Kekulé structures are more important than the Dewar structures, and this will be expressed in the value of the coefficients. Other structures, such as dipolar structures, may also be added, but the coefficients will be small.

The wave function ψ can be evaluated in terms of the coulombic integral Q, which closely corresponds to H_{rr} in the HMO treatment, and the exchange integral J, which represents the interchange of two electrons between *one* pair of carbon atoms. The π-electron energies of benzene can be evaluated in these terms, a 5×5 secular determinant being formed, the roots of which have the values $E = Q + 2.16J$, $Q - 4.16J$, Q, $Q - 2J$, and $Q - 2J$. The lowest energy has the value $E = Q + 2.16J$, whereas the energy of a single Kekulé structure is $Q + 1.5J$. The difference in energy, $0.66J$, represents the increase in stability of benzene over a single Kekulé structure. Thus the VB method, like the MO method, accounts for the stability of benzene. The VB method can be applied to other aromatic systems, but the application becomes more difficult as the systems become more complex and the number of contributing structures increases.

1.6. CONCLUSIONS

By the end of the 1930s an explanation of the stability of benzene was thus available in quantum mechanical terms. The conclusions of both the MO and VB methods with regard to benzene were the same; but in the simple

form in which we have just described the predictions of the MO method regarding systems homologous to benzene, they are different from the predictions of the VB method. The nature of these predictions and how they have been fulfilled will be the subject of succeeding chapters.

FURTHER READING

For more detailed discussion of the historical aspects of the benzene problem see: J. R. Partington, *A History of Chemistry*, Vol. 4, Macmillan, New York, 1964; C. K. Ingold, *Structure and Mechanism in Organic Chemistry*, Ed. 2, Cornell University Press, Ithaca, N.Y., 1969; J. P. Snyder in J. P. Snyder (Ed.), *Nonbenzenoid Aromatics*, Vol. 1, Academic Press, New York, 1969; K. Hafner, *Angew. Chem. Int. Ed.*, 1979, **18**, 641.

For general accounts of the MO and VB theories see C. A. Coulson, *Valence*, Ed. 2, Oxford University Press, New York, 1962; A. Streitwieser, *Molecular Orbital Theory for Organic Chemists*, Wiley, New York, 1961; T. E. Peacock, *Electronic Properties of Aromatic and Heterocyclic Molecules*, Academic Press, New York, 1965; J. N. Murrell, S. F. A. Kettle, and J. M. Tedder, *Valence Theory*, Ed. 2, Wiley, New York, 1970; J. N. Murrell, S. F. A. Kettle, and J. M. Tedder, *The Chemical Bond*, Wiley, New York, 1978; L. Salem, *The Molecular Orbital Theory of Conjugated Systems*, W. A. Benjamin, New York, 1966; M. J. S. Dewar, *The Molecular Orbital Theory of Organic Chemistry*, McGraw-Hill, New York, 1969; W. T. Borden, *Modern Molecular Orbital Theory for Organic Chemists*, Prentice-Hall, Englewood Cliffs, N.J., 1975; K. Yates, *Hückel Molecular Orbital Theory*, Academic Press, New York, 1978.

For a critical account of the concept of aromaticity developed here and in Chapter 2, see D. Lewis and D. Peters, *Facts and Theories of Aromaticity*, Macmillan, New York, 1975.

2

CYCLOBUTADIENE, BENZENE, AND CYCLOOCTATETRAENE

2.1. APPLICATION OF THE HMO THEORY

In Chapter 1 we examined the history of the benzene problem and presented a simple description of benzene in quantum mechanical terms. The HMO theory satisfactorily accounts for the stability of benzene since the 6 π electrons enter bonding orbitals and form a closed electronic shell. We will now apply the HMO theory to the two nearest homologues of benzene—cyclobutadiene (**1**) and cyclooctatetraene (**8**)—using the method of Frost and Musulin, assuming that the systems are planar, with equal bond lengths and bond angles. The diagrams are shown in Figure 2.1 together with that of benzene.

A difference in the pattern of orbitals between benzene on the one hand and cyclobutadiene and cyclooctatetraene on the other can immediately be discerned. Whereas both cyclobutadiene and cyclooctatetraene have two NBMOs of energy α, orbitals of this energy are not present in benzene. Furthermore, 2 of the 4 π electrons of cyclobutadiene and of the 8 π

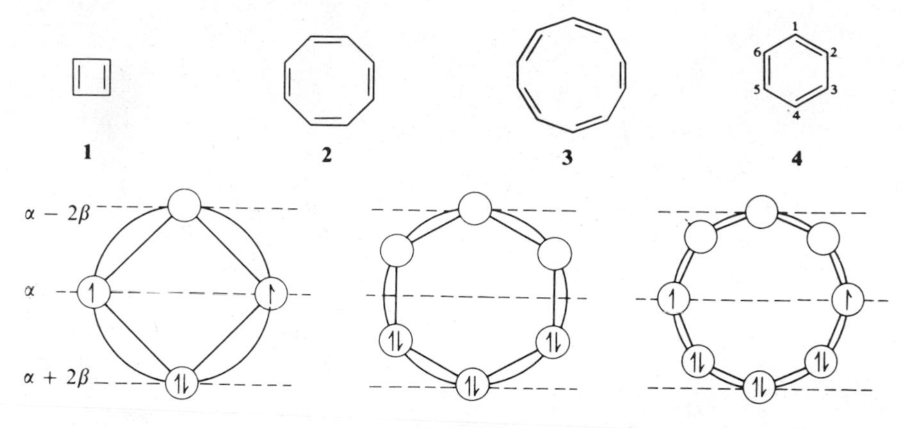

Figure 2.1

electrons of cyclooctatetraene must enter the NBMOs, which are then only half filled since these orbitals can contain four electrons. The closed shell structure is thus not duplicated in these structures for cyclobutadiene and cyclooctatetraene. Two possible arrangements of the electrons in the NBMO could be made: The electrons could be paired in one orbital with opposite spin, or each could occupy a separate orbital with parallel or antiparallel spins. Hund's Rule suggests that the latter arrangement with parallel spins will be of lower energy, but the differences in energy between the two states is likely to be small. The state with the unpaired electron will be a triplet diradical, whereas that with paired electrons will be a singlet. Application of the HMO method to planar cyclodecapentaene (**3**) shows that this system has a closed electronic shell similar to benzene. Hückel distinguished between these two types of systems; those with $4n$ π electrons, where n is an integer, have open configurations with electrons in NBMOs, whereas those with $(4n + 2)$ π electrons will have closed electronic configurations. This is the basis for *Hückel's Rule*, which states that planar, monocyclic systems with $(4n + 2)$ π electrons will be aromatic, whereas those with $4n$ will not.

What conclusions does the HMO method come to with regard to the relative stabilities of the localized and delocalized systems? A calculation of a single Kekulé structure for benzene, **4**, can be made, taking the same values for the coulombic and resonance integrals, α and β, as were taken for benzene. In this model, the π-electron interactions between C2–C3, C4–C5, and C1–C6 are ignored, and the system is treated as a set of three noninteracting ethene bonds. The determinant then has the form of equation (2.1), which can be readily reduced to the three determinants of the form of equation (2.2).

$$\begin{vmatrix} \alpha - E & \beta & 0 & 0 & 0 & 0 \\ \beta & \alpha - E & 0 & 0 & 0 & 0 \\ 0 & 0 & \alpha - E & \beta & 0 & 0 \\ 0 & 0 & \beta & \alpha - E & 0 & 0 \\ 0 & 0 & 0 & 0 & \alpha - E & \beta \\ 0 & 0 & 0 & 0 & \beta & \alpha - E \end{vmatrix} = 0 \qquad (2.1)$$

$$\begin{vmatrix} \alpha - E & \beta \\ \beta & \alpha - E \end{vmatrix} = 0 \qquad (2.2)$$

The three determinants of type of equation (2.2) give a total of six solutions, three of the form $E = \alpha + \beta$ and three of the form $E = \alpha - \beta$. The six π electrons can then be put in the three $\alpha + \beta$ orbitals of lower energy, and the total π energy of **4** is $6\alpha + 6\beta$. The difference in energy between benzene and the Kekulé structure **4** is thus 2β. Similar calculations can be carried out for the planar delocalized forms of cyclobutadiene, cyclooctatetraene, and higher homologues of the series. The energy of the localized systems in each case is equal to the product of the number of π electrons and $(\alpha + \beta)$. The calculated π energies of the localized and delocalized forms of the first four members of the series are shown in Table 2.1, and the results for the series are expressed graphically in Figure 2.2.

TABLE 2.1

System	No. of π Electrons	E Delocalized	E_0 Localized	$E - E_0$
C_4H_4	4	$4\alpha + 4\beta$	$4\alpha + 4\beta$	0
C_6H_6	6	$6\alpha + 8\beta$	$6\alpha + 6\beta$	2β
C_8H_8	8	$8\alpha + 9.66\beta$	$8\alpha + 8\beta$	1.66β
$C_{10}H_{10}$	10	$10\alpha + 12.95\beta$	$10\alpha + 10\beta$	2.95β

The HMO theory thus predicts that the $(4n + 2)$ π-electron systems will have closed electron shells, whereas those with $4n$ will not, and that there will be a gradual increase in the difference in energy between the localized and delocalized forms with increasing ring size. The smaller $(4n + 2)$ π-electron systems are predicted to be more stable than the neighboring $4n$ π-electron systems, but this difference in stability virtually disappears for the larger rings. The theory specifically predicts that cyclobutadiene will be a square planar, triplet diradical with zero resonance energy, that benzene

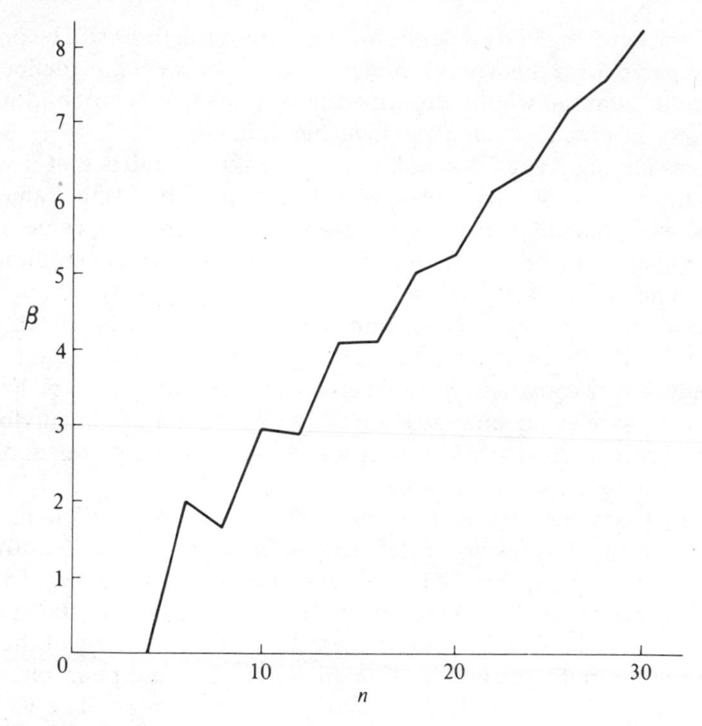

Figure 2.2. The delocalization energy (in β) calculated by the HMO method for monocyclic conjugated systems.

will be a symmetrical hexagon with considerable resonance energy, and that cyclooctatetraene will be a planar, octagonal triplet diradical, again with considerable resonance energy.

So far a value has not been assigned to β, and a discussion of its magnitude will be left to later in this chapter. In the next section, the experimental evidence for the actual structures of cyclobutadiene, benzene, and cyclooctatetraene will be discussed, and it will be decided how far these structures conform to those predicted by the HMO theory.

2.2. EXPERIMENTAL EVIDENCE FOR THE STRUCTURES OF CYCLOBUTADIENE, BENZENE, AND CYCLOOCTATETRAENE

In the preceding discussion of cyclobutadiene, benzene, and cyclooctatetraene, the theoretical models have assumed these systems to be planar, symmetric systems with all of the C—C bonds equivalent. In the present section, the experimental evidence for the actual structures will be presented and compared with theoretical models. The experimental evidence for the structure of benzene will be described first, since this compound has been the subject of the most extensive investigation.

The resonating Kekulé formulation of benzene predicts that it will be a planar, hexagon of D_{6h} symmetry, with all the C—C bonds the same length. The first experimental evidence in support of this structure came not from benzene itself but from a number of crystalline aromatic compounds (e.g., naphthalene and anthracene), which were investigated by x-ray crystallographic methods by Robertson, Lonsdale, Pinney, and others in the late 1920s and early 1930s. It was found that the x-ray diffraction data on these compounds were consistent with structures made up of symmetric hexagons. A particularly relevant study was that of Lonsdale on hexamethylbenzene, in which a planar hexagonal structure was deduced without prior assumption that the structure was of this type.

The first conclusive evidence for the D_6h symmetry of benzene was adduced by Ingold and his collaborators from an extensive study of the infrared and Raman spectra of benzene and deuterated benzenes. The planar hexagonal model for benzene requires it to have 20 fundamental vibrations, which can be divided into 10 classes in terms of the symmetry of the system. Of these vibrations, 7 should be Raman active, 4 should be infrared active, and the remaining 9 should be inactive in either mode. The vibrational spectrum of benzene showed the expected 7 fundamental lines in the Raman and 4 fundamental lines in the infrared spectrum. The Raman spectrum of hexadeuterobenzene, in which the hydrogens are replaced by deuterium, allows the correlation of these bands with the vibrational mode from which they arise. Figure 2.3 shows, diagrammatically, the Raman spectrum of benzene and hexadeuterobenzene. It is readily seen that there are two types of lines: those that are shifted to much longer wavelength in hexadeuterobenzene than in benzene and those that show much smaller shifts. The bands with large shifts are those in which the C and H atoms

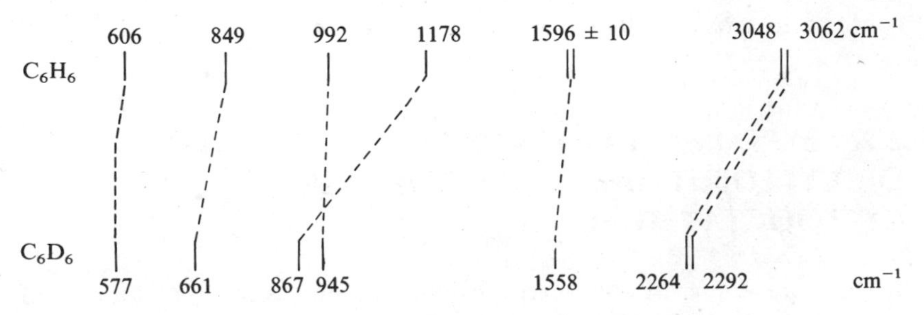

Figure 2.3

of each C—H group move *independently* of each other during the vibration, whereas the bands with small shifts are those in which the C—H group moves as a unit in the vibration. In the first case, the substitution of deuterium for hydrogen causes a change in the mass ratio of 2:1, whereas in the second case the mass ratio change is 14:13. The band at 3062 cm^{-1} in benzene appears at 2292 cm^{-1} in C_6D_6 and is assigned to the C—H breathing motion, whereas that at 3048 cm^{-1} is found at 2264 cm^{-1}, and is assigned to a vibrational mode in which the hexagon flattens (Fig. 2.4). In both of these vibrational modes, the C—H bond length changes. The band at 606 cm^{-1} in benzene appears at 577 cm^{-1} in C_6D_6 and is assigned to a vibration in which the hexagon elongates, whereas the band at 1596 cm^{-1} appears at 1558 cm^{-1} and is assigned to a vibration in which the atoms at the opposite ends of a hexagon are compressed together (Fig. 2.4). In both of these modes the C—H groups move as units.

The four principal bands in the infrared spectrum (IR) of benzene and hexadeuterobenzene are shown diagrammatically in Figure 2.5. A number

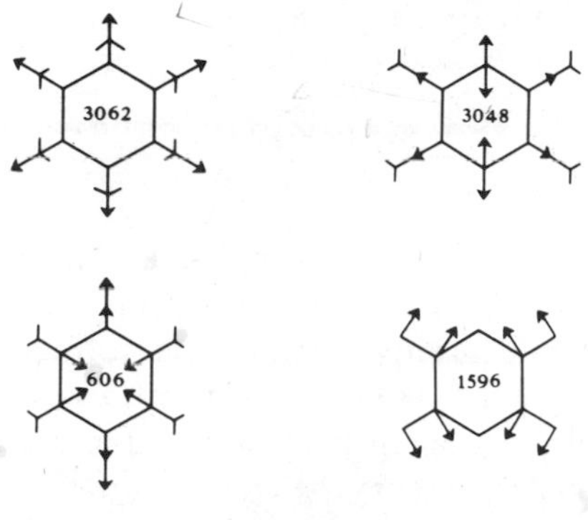

Figure 2.4

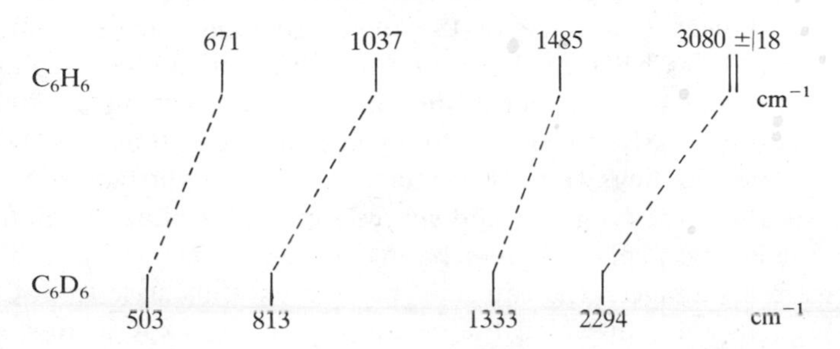

Figure 2.5

of other bands that could be assigned to specific vibrational modes was also observed. The forbidden frequencies were identified by examining the spectra of specifically deuterated benzenes in which the forbidden nature is removed by the lower symmetry of the system.

Values for the bond lengths in benzene were obtained by Stoicheff from a study of the rotational Raman spectrum of the vapor. The C—C bond lengths were 139.7 ± 0.1 pm and the C—H bond lengths were 108.4 ± 0.5 pm. An earlier electron diffraction study by Schomaker and Pauling had given values of 139.3 ± 2 pm (C—C) and 108 ± 4 pm (C—H), and a subsequent x-ray study of crystalline benzene by Cox and co-workers showed that the molecule was centrosymmetric with C—C bond lengths of 139.2 pm. The structure for benzene derived from these and later more refined x-ray data is shown in Figure 2.6.

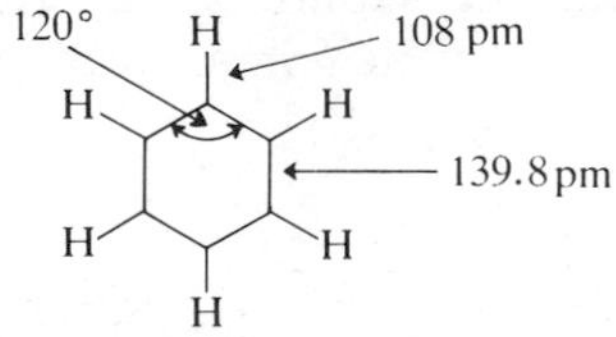

Figure 2.6

The first convincing evidence for the isolation of cyclobutadiene was obtained in the early 1970s by Krantz and Chapman and their respective co-workers from experiments involving the photofragmentation of photo-α-pyrone (**28**) in an argon matrix (see p. 46). Infrared spectra were obtained that were interpreted as arising from cyclobutadiene. The spectral lines were broad, and when other precursors were used some confusion ensued because of discrepancies between the observed spectra, but it was generally concluded that cyclobutadiene had the unexpected square planar ground state. However, Maier and Masamune and their respective co-workers eventually showed that some of the observed bands arose from charge transfer complexes between the cyclobutadiene and the other photofragment, which were held together in the matrix. Two bands below 2000 cm^{-1} could now be clearly attributed to cyclobutadiene, those at 1236 and 573 cm^{-1}. These findings were substantiated by Krantz who labeled α-pyrone with ^{13}C at the carbonyl group and showed that after photoirradiation, the 650 cm^{-1} band was split into two bands, whereas the 1236 and 573 cm^{-1} bands were unchanged. The 650 cm^{-1} band must thus arise from CO_2.

Subsequent work in which the sensitivity of the experiments and the transparency of the medium had been improved has revealed the presence of four bands at 1523, 1240, 723, and 572 cm^{-1}, the two new bands being of very low intensity. The spectrum of tetradeuterocyclobutadiene has also been obtained, and the spectra and spectral frequency shifts are diagrammatically illustrated in Figure 2.7.

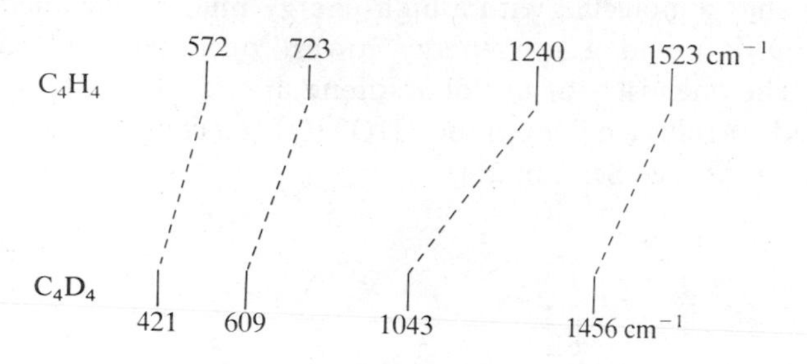

Figure 2.7

The appearance of four bands below 2000 cm^{-1} means that cyclobutadiene must have a lower symmetry than D_{4h}, but whether it is D_{2h} cannot be decided on the available data. Theoretical calculations using ab initio self-consistent-field (SCF) techniques have produced spectra for the rectangular planar structure in which the *relative* intensities of the bands have been very well matched for both the C_4H_4 and C_4D_4 molecules. Relative intensity matching of theoretical and experimental infrared spectra is more meaningful than line frequency matching since current theoretical methods give insufficiently accurate line positions for these to be reliably used.

The IR spectrum of cyclobutadiene is thus in accord with the bond alternated, rectangular structure **1a**. This conclusion is supported by the results obtained from more stable, substituted cyclobutadienes. Thus, the x-ray crystallographic analyses of the derivatives **5a**, **5b**, **6**, and **7** show that all have rectangular structures with long bonds in the range 160.0–150.6 pm and short bonds in the range 133.9–144.1 pm (Fig. 2.8). The photoelectron

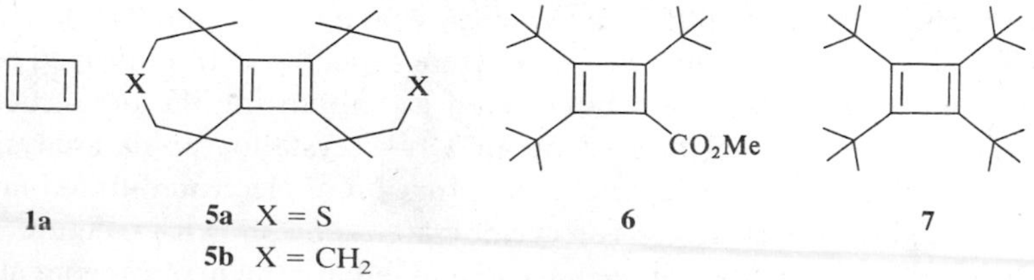

X X CO_2Me

1a **5a** X = S **5b** X = CH_2 **6** **7**

Figure 2.8

(PE) spectra of these compounds are all similar and interpretable in terms of the rectangular, but not the square, structure.

The rectangular planar structure for cyclobutadiene removes the degeneracy of the nonbonding orbitals (Fig. 2.1) and allows electron pairing in a bonding orbital (Fig. 2.9). The operation of this pseudo-Jahn–Teller effect furnishes a molecule with a high-energy highest occupied molecular orbital (HOMO) and a low-energy lowest occupied molecular orbital (LUMO). The chemistry of cyclobutadiene and its derivatives would thus be expected to reflect nucleophilic (HOMO), electrophilic (LUMO), and radical reactions (see Section 2.4).

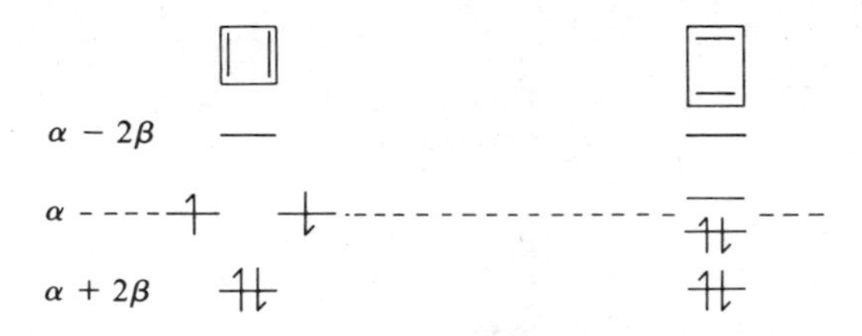

Figure 2.9. Schematic energy levels for square planar and rectangular cyclobutadiene.

Cyclooctatetraene is a pale yellow oil that has an ill-defined electronic spectrum that tails into the visible region. A study of the Raman and IR spectra clearly indicated that cyclooctatetraene is not a planar system of D_{8h} symmetry, since the spectra are too complex with many bands active in both spectral types. On the basis of these spectra, Lippincott and his collaborators suggested that cyclooctatetraene could be in either the "crown" conformation of D_4 symmetry or the "tub" conformation of D_{2d} symmetry, and they suggested that on balance the data favored the crown structure (Fig. 2.10). However, the crown structure has a considerable disadvantage in that each of the double bonds is strained, the hydrogens not being coplanar, a fact at variance with the thermochemical data (see Section 2.3.2). Electron diffraction studies, which had originally been interpreted in favor of the planar structure, were repeated with improved resolution, and the new data were interpreted as supporting the tub structure. The tub structure was confirmed by an x-ray crystallographic analysis of crystalline cyclooctatetraene, and a combination of electron diffraction and x-ray data gives the structure for cyclooctatetraene shown in Figure 2.11.

The three compounds that we have discussed are the first three members of the homologous series $(C_2H_2)_n$, and yet they have completely different

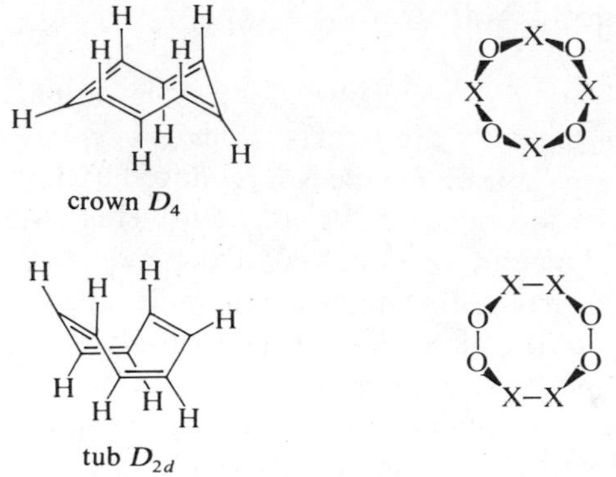

crown D_4

tub D_{2d}

Figure 2.10

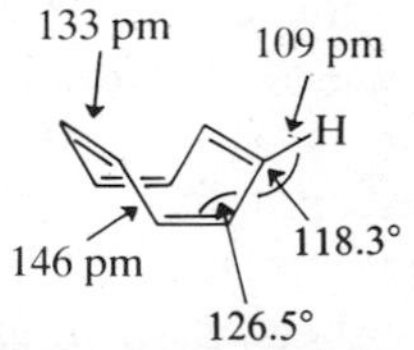

Figure 2.11

structural arrangements. Cyclobutadiene is planar with alternate single and double bonds; benzene is planar with all C—C bonds equal; and cyclooctatetraene is nonplanar with alternate single and double bonds. That these compounds should differ in properties is predicted by Hückel's Rule, but the stabilization of the planar delocalized form of cyclooctatetraene predicted by the HMO method is not observed.

2.3. PHYSICAL PROPERTIES

In this section, the physical properties of cyclobutadiene, benzene, and cyclooctatetraene that have not been discussed in connection with the structure of these molecules will be considered. Such a distinction is artificial, since much of the data given in this section could be, and in fact was, presented as evidence for the structure of these compounds. However, it will now be possible for us to compare the properties of these compounds knowing their difference in structure.

2.3.1. Diamagnetic Anisotropy

The majority of organic molecules do not have permanent magnetic moments and consequently are weakly *diamagnetic*, having negative magnetic susceptibilities. This diamagnetism is caused by Larmor precession of the electrons, which produces small magnetic fields opposing the applied magnetic field. The magnitude of this effect depends on the area of the orbit traversed by the electron. Most diamagnetic molecules are anisotropic; that is, the magnitudes of the diamagnetic susceptibility along the three perpendicular principal magnetic axes are not equal. In the bulk measurement of the diamagnetism of a compound, the average magnetic susceptibility is measured; but in a single crystal, it is possible to determine the magnetic susceptibilities along the crystal axes.

Single crystals of a number of benzenoid hydrocarbons, such as naphthalene, show large diamagnetic anisotropies. Krishnan and his collaborators were able to demonstrate that the relationship between the magnetic and crystal axes in these compounds depended upon the orientation of the molecules in the crystal. Furthermore, they found that the magnitude of the magnetic susceptibilities along these magnetic axes depended upon the magnetic susceptibilities of the principal axes of the molecule. It was observed that for these benzenoid hydrocarbons, the magnetic susceptibility in the axis at right angles (K^3) to the plane of the ring was greater than that along the axes in the plane of the ring (K^1, K^2), which were approximately equal (Fig. 2.12). For the series of compounds benzene, naphthalene, and anthracene, the susceptibilities in the plane of the ring showed only a small increase, whereas that along the normal to the plane increased additively with the increase in number of rings (Table 2.2). A similar sequence of values was found for the series biphenyl, triphenyl, and quaterphenyl.

Raman and Krishnan, following an earlier theory of Ehrenfest, suggested that the large diamagnetic anisotropy of these systems was due to Larmor precession of electrons in orbitals containing many nuclei. This theory was

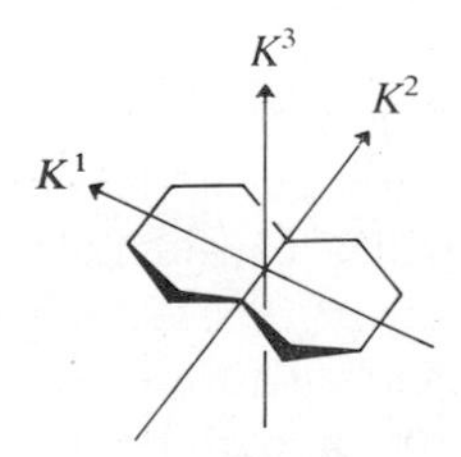

Figure 2.12

put into a quantitative form by Pauling and was subsequently interpreted in quantum mechanical terms by London. Pauling showed that the contribution of the electron in a cylindrically symmetric field about the z axis to the magnetic susceptibility was given by equation (2.3).

$$\chi = -\left(\frac{Ne^2}{4mc^2}\right)(\rho^2)_{av} \tag{2.3}$$

In this equation, χ is the magnetic susceptibility and $(\rho^2)_{av}$ is the mean square of the distance of the electron from the z axis. In the Hückel model for benzene, only the six π electrons need to be considered. If R^2 is taken to be the value for $(\rho^2)_{av}$, where R is the distance (139 pm) from the ring center to the carbon nuclei of benzene, then the calculated value for the anisotropy, ΔK, was 49,* which is in reasonable agreement with that observed (54, Table 2.2).

TABLE 2.2

Compound	$-K^1 \times 10^6$ esu	$-K^2 \times 10^6$ esu	$-K^3 \times 10^6$ esu	$\Delta K \times 10^6$ esu
 (benzene)	37.3	37.3	91.2	53.9
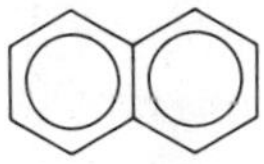 (naphthalene)	39.4	43.0	187.2	146.0
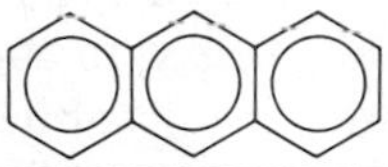 (anthracene)	45.9	52.7	272.5	223.2

This model for benzene, in which the π electrons precess in orbitals extending over the ring, has subsequently been used to explain the deshielding of aromatic protons in the ^{1}H nuclear magnetic resonance (NMR) spectrum. In this model, the applied field, H^0, causes the π electrons to circulate in orbitals extending over the six carbon atoms, and a magnetic field, H^1, is induced, which opposes the applied field. The lines of force resulting from the induced field are shown in Figure 2.13, and the effect of this induced field is that the apparent field inside the ring is decreased, whereas the apparent field outside the ring is increased. Protons *outside* the ring therefore resonate at *lower* field than protons uninfluenced by the induced field, whereas protons *inside* the ring will resonate at *higher* field.

*These, and all subsequent values of magnetic susceptibility, will be in units of 10^{-6} esu.

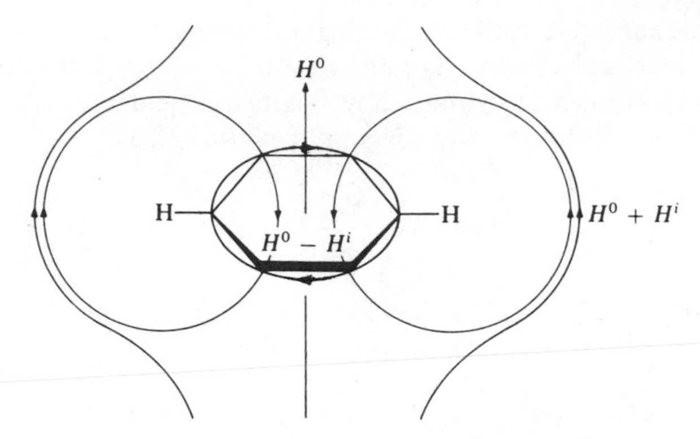

Figure 2.13

The use of this model to account for the position of aromatic protons in the ^{1}H NMR spectrum was originally developed by Pople who, for purposes of calculation, substituted a point dipole at the center of the ring for the induced ring current. Waugh and Fessenden improved this model by using a field in which the π electrons were assumed to be in two identical shells above and below the ring plane. McWeeny developed a quantum mechanical model in which the induced moment arising from the external field was determined by the method of London. A *point dipole*, *m*, was then introduced as a probe at the position of the proton in question, and the coupling energy between the point dipole and the induced current distribution was evaluated. The total magnetic energy differs from that with the dipole absent by the coupling term $-mH'$. For benzene, the resulting equation gave terms proportional to $1/r^3$ and higher powers, where r is the distance in bond lengths of the point dipole from the center of the ring. At large distances from the center, the expression reduces to the Pople approximation, but in the vicinity of the ring these terms introduce a large correction, increasing the estimated field by up to 80%. Both the McWeeny and Waugh–Fessenden calculations predict that the induced field will be of opposite sign inside and outside the ring, whereas the Pople method predicts that the field increases as the center is approached. Figure 2.14 compares the McWeeny and Pople models.

Cyclooctatetraene, unlike benzene, does not exhibit enhanced diamagnetic anisotropy, and the value for the magnetic susceptibility is close to that expected for four double and four single bonds. Pink and Ubbelohde found that the calculated magnetic susceptibility for the structure with alternate single and double bonds was −50 and that for the delocalized structure was −73, whereas the observed value was −52. These authors also

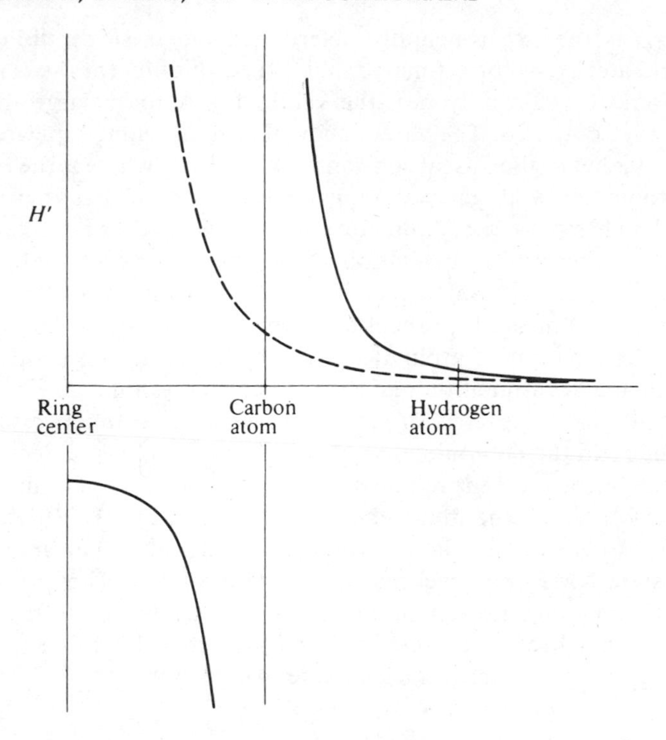

Figure 2.14. Pople (broken line) and McWeeny (solid line) model profiles of the induced magnetic field in benzene. From McWeeny, *Mol. Phys.*, 1958, **1**, 311.

found that the difference in magnetic susceptibility between cyclooctane and cyclooctatetraene was 39.5, which was close to the predicted value of 41 and in contrast to the difference between cyclohexane and benzene, which is 12 instead of the expected 31. In the ^{1}H NMR spectrum, cyclooctatetraene shows a single resonance signal at δ 5.68 in the olefinic region. The magnetic properties of cyclooctatetraene are thus those expected for an olefinic system, and cyclooctatetraene does not show the anomalous magnetic behavior observed in benzene and the benzenoid hydrocarbons.

This method of distinguishing between aromatic and nonaromatic systems has been further developed by Dauben and associates. The original Pascal values used by Pink and Ubbelohde were too inaccurate for a general extension of this method, but more recently more accurate values have become available. Using these and defining the diamagnetic susceptibility exaltation Λ by

$$\Lambda = \chi_M - \chi_{M'} \tag{2.4}$$

(where χ_M is the experimentally determined molar susceptibility of the compound and $\chi_{M'}$ is the estimated molar susceptibility for the corresponding theoretical cyclic polyene), the values for Λ for a large number of systems were obtained. The values show that cyclic nonconjugated olefins have $\Lambda = 0$, that is, the Pascal constants are additive, whereas the benzenoid hydrocarbons show large exaltations. Thus, benzene has $\Lambda = 13.7$ and naphthalene has $\Lambda = 30.5$. Aromatic compounds are thus expected to show diamagnetic susceptibility exaltation. Cyclooctatetraene, as would be expected from the earlier results, has $\Lambda = -0.9$, a value close to zero. The exaltation of diamagnetic susceptibility thus seems to be a valid criterion for the determination of aromaticity in a compound. The main problems are that such a determination requires a calculated estimate of the expected susceptibility in the absence of exaltation, which is a similar disadvantage to that found in the determination of resonance energy (see Section 2.3.2), and that the measurement requires fairly large amounts of compound.

No data on either the diamagnetic exaltation or the ^{1}H NMR spectrum of cyclobutadiene are available, although it is possible that improvements in solid state NMR may eventually provide the latter. The chemical shift (δ 5.38) of the ring proton in the ^{1}H NMR spectrum of tri-*t*-butylcyclobutadiene has been interpreted as indicating that this molecule sustains a paramagnetic ring current (see Chapter 4).

2.3.2. Resonance Energy

When benzene is combusted or hydrogenated, the values obtained from the enthalpy of combustion or hydrogenation do not have the additive values expected for three double bonds of the cyclohexene type. Kistiakowsky and associates determined the heat of hydrogenation of cyclohexene to be -120 kJ mol^{-1}. The actual enthalpy of hydrogenation of benzene was found to be -208 kJ mol^{-1}, and benzene thus has 150 kJ mol^{-1} less energy available than would be expected. The 150 kJ mol^{-1} represents a thermochemical stabilization of benzene, which has been termed the *resonance energy.*

The enthalpies of combustion and hydrogenation of cyclooctatetraene are close to the values expected for a system with four double bonds. Turner and his collaborators found the enthalpy of hydrogenation of cyclooctatetraene to be -410 kJ mol^{-1}, whereas the value calculated from *cis*-cyclooctene (-96 kJ mol^{-1}), would be 384 kJ mol^{-1}. This would give a negative resonance energy of 26 kJ mol^{-1}.

No thermochemical data are available for cyclobutadiene.

Although arguments can be made against both cyclohexene and cyclooctene as suitable models for comparison, these results clearly demonstrate the complete difference in thermochemical behavior between benzene and cyclooctatetraene. Whether the 150 kJ mol^{-1} resonance energy should be mainly attributed to delocalization of the π electrons or to the change in σ-bond properties is difficult to determine. Clearly, cyclohexene is a poor model for benzene, since in the hydrogenation of benzene to cyclohexane six "aromatic" bonds are converted into six "aliphatic" bonds. The bonds change in length from 139 to 154 pm, each carbon rehybridizing from an aromatic $sp^2\pi$ to an aliphatic sp^3 type, whereas in the hydrogenation of cyclohexene the change in bond length is from 133 to 154 pm, and the carbons rehybridize from olefinic $sp^2\pi$ to aliphatic sp^3 type.

In our earlier discussion of the HMO theory, we derived the difference between benzene and a single Kekulé structure as 2β. If we equate this with the resonance energy, then β would have a value of -75 kJ mol^{-1}. However, the models used for the determination of the resonance energy and for the calculation are not the same. The value to be assigned to β depends to a large extent on the assumptions made, and estimated values have ranged from -125 to -40 kJ mol^{-1}. From a consideration that the resonance energies have been overestimated owing to the use of thermochemical data that ignore the difference between the delocalized system and the model compound, Dewar has given the smaller value to β. The problem that is involved is illustrated in Figure 2.15. The value for the enthalpy of hydrogenation of benzene, reaction *a*, is known, and an estimate for the hydrogenation of cyclohexatriene, the reverse of reaction *b*, can be made from the enthalpy of hydrogenation of cyclohexene. The delocalization of a single Kekulé structure to benzene is 2β (equation *d*). The unknown parameter is the energy involved in forming the symmetric Kekulé structure from cyclohexatriene (equation *c*), and most thermochemical estimates in fact equate reaction $c + d$ with 2β.

Although resonance energy can be readily defined in terms of the extra thermodynamic stability of a molecule over that predicted on the basis of

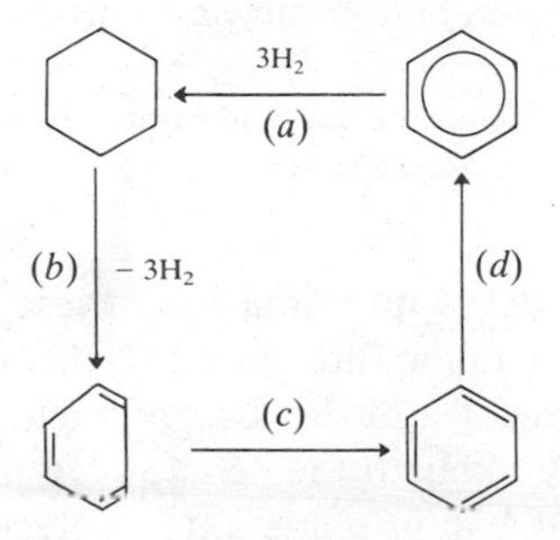

Figure 2.15

a localized bond model, it is a difficult parameter to determine. The best method for estimating the energy of the localized structure would be by the use of bond energies, which would then be treated additively. In the case of benzene, it is necessary to know the energies of the 139 pm "double" and "single" bonds. Unfortunately, such data are not available.

2.3.3. Electronic Spectra

The absorption spectrum of benzene shows three band systems at 255, 205, and 183 nm in the ultraviolet (UV). The long wavelength band at 255 nm is weak and exhibits considerable fine structure, the 205-nm band is of medium intensity, and the 183 nm band is more intense (Fig. 2.16).

The three transitions may be interpreted as electron transitions from the highest occupied e_{1g} orbital to the lowest unoccupied e_{2u} orbital. There are four possible transitions as shown in Figure 2.17, and the symmetries of the states obtained from these transitions are formed from the direct product $e_{1g} \times e_{2u} = B_{1u} + B_{2u} + E_{1u}$. A description of these state symbols is given in Table 2.3, and inspection of Figure 1.14 will confirm that the orbitals of benzene have the characters assigned in Figure 2.17.

TABLE 2.3

Symbol[a]	Description
a, A	Single state, symmetric to rotation around the principal axis
b, B	Single state, antisymmetric to rotation around the principal axis
e, E	Doubly degenerate state
1	Horizontal plane, symmetric to reflection
2	Horizontal plane, antisymmetric to reflection
g	Inversion center, symmetric to inversion
u	Inversion center, antisymmetric to inversion

[a] Lowercase letters are used to describe orbitals, and uppercase letters refer to electronic states. A superscript 1 indicates a singlet state, a superscript 3 a triplet state, e.g., $^1B_{2u}$. $^3B_{2g}$.

In the simple zero-order approximation, these states are degenerate; when electron repulsion is taken into account, this degeneracy is removed, the B_{2u} state being of lowest and the E_{1u} of highest energy. The band at 255 nm was attributed by Sklar to the $^1B_{2u} \leftarrow {}^1A_{1g}$ transition, which since B_{2u} does not transform like an in-plane axis, is electronically forbidden for a $\pi \rightarrow \pi^*$ transition. That the 255-nm band appears in the spectrum is due

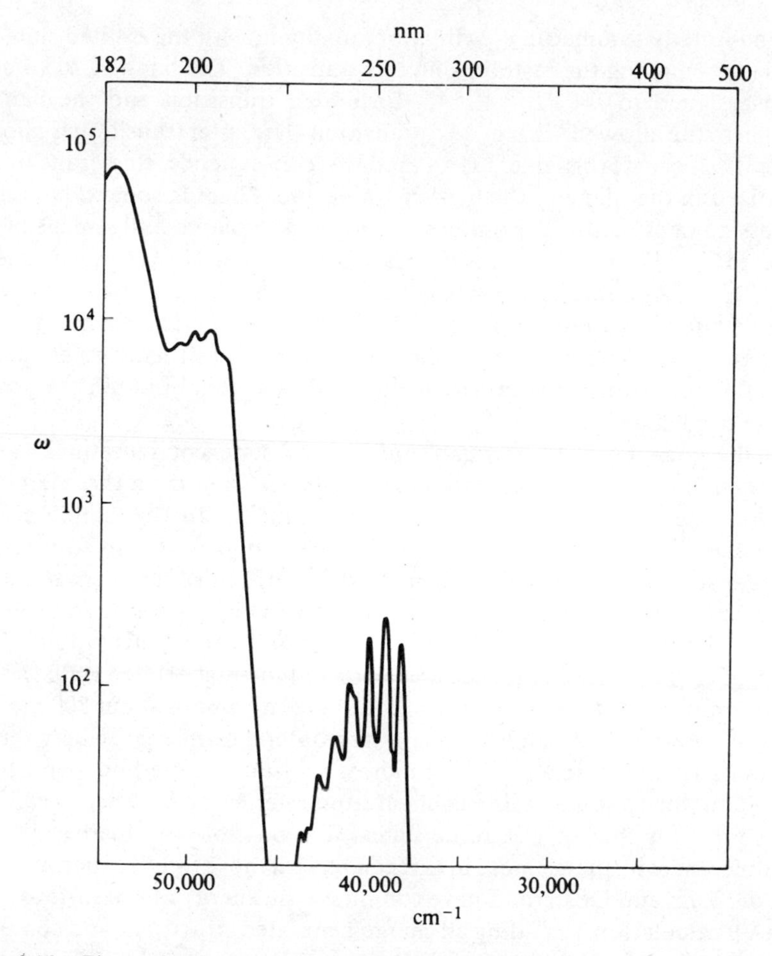

Figure 2.16. Electronic spectrum of benzene in hexane. From Perkampus and Kassebeer, *DMS U.V. Atlas*, Butterworths, Woburn, Mass., 1966.

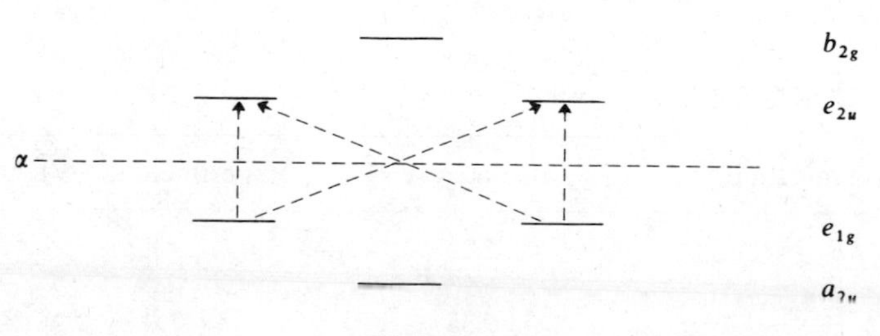

Figure 2.17

to a nontotally symmetric e_{2g} vibration mixing in with the excited state and partially removing the restraints on this transition. The band at 205 nm has been assigned to the $^1B_{1u} \leftarrow {}^1A_{1g}$ forbidden transition and the band at 183 nm to the allowed $^1E_{1u} \leftarrow {}^1A_{1g}$ transition. The latter transition is allowed since E_{1u} transforms like (x, y), and in consequence this transition is polarized in the plane of the benzene molecule. There is some doubt about the assignment of the 205-nm band, and an alternative assignment of this band to an $^1E_{2g} \leftarrow {}^1A_{1g}$ transition has been made. However, on balance the $^1B_{1u} \leftarrow {}^1A_{1g}$ assignment has been more favored.

A detailed, high-resolution rotational analysis of the 255-nm band system of benzene has been made by Callomon, Dunn, and Mills, who concluded from this that benzene is *exactly* planar and hexagonal in both the ground and excited states.

If the energies of the ground and excited state configurations can be determined in terms of the resonance integral β, then from the energies of the transition a value for β could be determined. In the simple Hückel approach, if one electron is promoted from the highest occupied orbital to the lowest unoccupied orbital, then the difference in energy between these, the ground state and the first excited state (which is degenerate in this model), is 2β. Assuming a value of 75 kJ mol^{-1} for β, this would give a value of 1.54 eV between the ground state and the first excited state (1 eV = 96.4 kJ mol^{-1} = 8055 cm^{-1}), which would give a band at about 800 nm. The Hückel theory grossly underestimates the difference in energy between the ground and excited states. Using more sophisticated methods with better wave functions and including configuration interaction between wave functions for a number of electronic states, it is possible to obtain values for the differences in energy more in agreement with those found experimentally. van der Lugt and Oosterhoff have computed the energy of the various states by a VB calculation, including all charge-separated structures, and compared this with the results of MO calculations. The values obtained by these authors, together with the experimental values, are shown in Table 2.4.

There is still some controversy over the electronic spectrum of cyclobutadiene; no strong absorptions are observed above 250 nm. But whereas

TABLE 2.4

Transition	Calculated (eV)	Experimental (eV)
$^1B_{2u} \leftarrow {}^1A_{1g}$	5.22	4.8
$^1B_{1u} \leftarrow {}^1A_{1g}$	6.33	6.0
$^1E_{1u} \leftarrow {}^1A_{1g}$	7.48	7.0

Masamune and co-workers have maintained that a low intensity band can be observed at 300 nm, Maier and co-workers are of the opinion that this is not derived from cyclobutadiene and consider that cyclobutadiene does not absorb above 250 nm. The lack of an intense transition above 200 nm is in accord with the theoretical predictions for the rectangular planar structure but not with the prediction for the square planar structure of a fairly strong absorption at about 370 nm.

The electronic spectrum of cyclooctatetraene shows no distinct bands but exhibits a broad, diffuse absorption from 220 to 380 nm.

2.3.4. Fluxional Properties

The planar symmetric hexagon structure for benzene is the equilibrium configuration of the system, and the vibrational deviations from this structure, which appear in the IR and Raman spectra, were discussed earlier. The force constants for these vibrations show that benzene can be much more easily distorted by out-of-plane vibration than by in-plane vibration. The bending of benzene rings, as in the paracyclophanes, is thus not unexpected (see Chapter 3).

Cyclooctatetraene, besides the vibrational changes found in all molecules, also shows fluxional behavior that occurs at a much slower rate, such that these changes are observed on the NMR time scale (10^{-2}–10^{-4} sec). Two processes have been distinguished, one involving ring inversion and the other the switching of double and single bonds (Fig. 2.18).

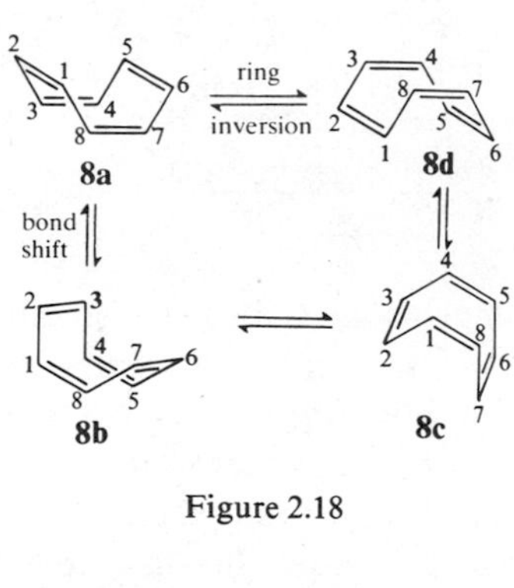

Figure 2.18

The process of bond switching was originally observed by Anet from a study in the NMR spectrum of the dependence of the ^{13}C–proton coupling with temperature. The single broad peak arising from the ^{13}C coupling that is seen at room temperature splits at lower temperatures to a doublet, owing to the different coupling constants across the single and double bonds, which are averaged at room temperature. Thus, in Figure 2.18, if the molecule contained ^{13}C at C-1, then at room temperature the interconversion of **8a** to **8b** occurs at such a rate that during the observation C-1 is joined to C-2 part of the time by a double bond (**8a**) and part of the time by a single bond (**8b**). On cooling the rate is sufficiently slowed that during the observation C-1 is joined to C-2 by either a single or a double bond. In the substituted cyclooctatetraene **9**, both the bond switching and ring inversion processes can be observed. In the course of ring inversion of **9a** to **9b**, the two methyl groups *a*, *b* are interconverted; whereas in the bond shift process, the positions of the ring substituents change (Fig. 2.19). At −35°C, the NMR

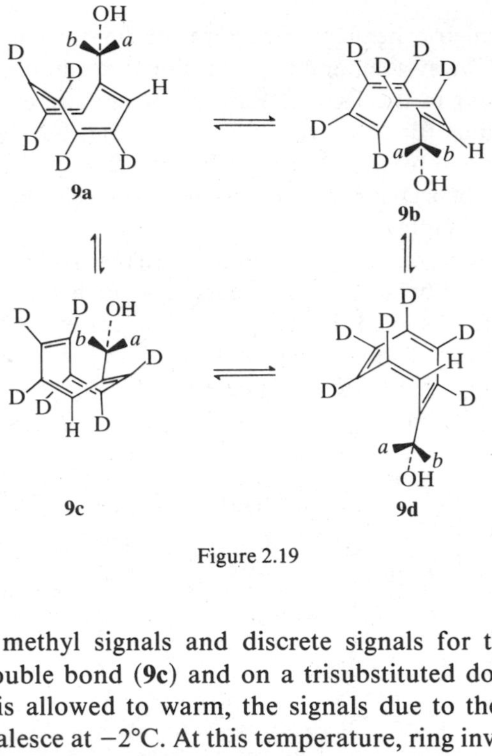

Figure 2.19

shows discrete methyl signals and discrete signals for the proton on a disubstituted double bond (**9c**) and on a trisubstituted double bond (**9a**). As the sample is allowed to warm, the signals due to the methyl groups broaden and coalesce at −2°C. At this temperature, ring inversion is occurring at such a rate that the methyl groups are interchanging environments. Thus, in **9a** the CH_3^b, which is "inside" the ring moves to a position "outside" the ring in **9b**. (Free rotation around the =C1—C-isopropanol bond does not cause the two methyl groups, which are diastereotopic, to become equivalent.) The rate of bond switching is still sufficiently slow at this temperature for discrete ring protons to be observed, but at +41°C the rate of exchange has increased so that the signals due to the ring protons in **9a** (**9b**) and **9c** (**9d**) coalesce. The barrier to ring inversion of **9** has an activation energy at −2°C of $\Delta G = 61$ kJ mol^{-1}, whereas that for bond switching at the same temperature has $\Delta G = 71$ kJ mol^{-1}. The transition states for these processes may be considered to be planar: the structure **10** with localized bonds being the transition state for inversion and structure **11**, with bond delocalization, for bond switching. These results suggest that the localized structure **10** is about 10 kJ mol^{-1} more stable than the delocalized structure **11**. Thus, contrary to the predictions of the HMO theory, the delocalized structure for planar cyclooctatetraene is *less* stable than the localized bond structure.

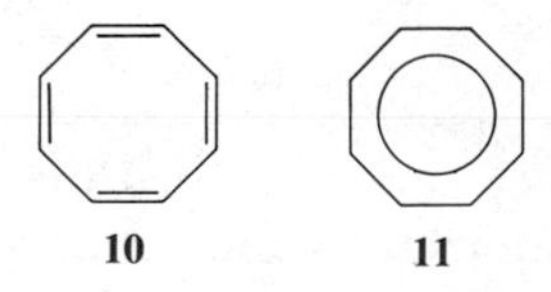

Figure 2.20

The nonplanar bond localized structure of cyclooctatetraene means that suitably substituted derivatives should be chiral and, provided the barriers to ring inversion and bond shift are sufficiently high, resolvable. Paquette and co-workers have successfully resolved a number of substituted, chiral cyclooctatetraenes. Initially, highly substituted derivatives, such as 1,2,3,4-tetramethyl and 1,2,3-trimethylcyclooctatetraene were resolved, but subsequently enantiomers of less highly substituted derivatives, such as 1,3-di-*t*-butylcyclooctatetraene (**12**) were separated (Fig. 2.21). The barriers to ring inversion and bond shift between **12a** and **12b** were found to be $\Delta G =$ 90 kJ mol^{-1} and 99 kJ mol^{-1}, respectively, considerably higher than in monosubstituted derivatives.

12a ⇌ **12b**

Figure 2.21

In cyclobutadiene, the two rectangular structures **1a** and **1a′** are equivalent, but unlike the two Kekulé structures of benzene, they are separated by an energy maximum rather than a minimum (Fig. 2.22).

The rearrangement of **1a** to **1a′** is a valence tautomerism, with square planar cyclobutadiene **1b** as a potential transition state, whereas the benzene Kekulé structures are merely contributing forms of the real delocalized system.

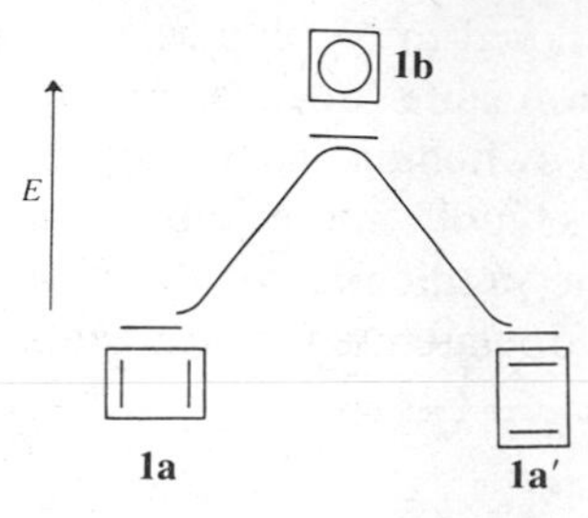

Figure 2.22

Whitman and Carpenter have estimated the barrier to interconversion of the two rectangular tautomers by generating and trapping dideuterocyclobutadiene and measuring the proportions of the 1,2- and 1,4-dideuterocyclobutadienes at different temperatures and with different concentrations of trapping agent. They concluded that ΔH had a lower limit of 6.7 kJ mol^{-1} and an upper limit of 42 kJ mol^{-1}; they also concluded that the entropy of activation for the process was negative and possibly substantially so. Carpenter subsequently suggested that these results, in particular the negative entropy, could be explained if tautomerization takes place largely (97%) by heavy atom tunneling. He also suggested that the same process might contribute to the tautomerization of cyclooctatetraene. Subsequent, more sophisticated calculations support Carpenter's view of the mechanism of interchange in cyclobutadiene.

Maier and co-workers examined the tautomerism of tri-*t*-butylcyclobutadiene (**13**) and found that they were unable to freeze out the tautomers down to −185°C. However, by introducing a $C(CD_3)_3$ group in place of one of the equivalent 1,3-$C(CH_3)$ groups, they were able to observe that the C-1 and C-3 carbon resonance signals in the ^{13}C NMR spectrum were now split, showing the presence of two species **13a** and **13b** with a slight preference for the former (Fig. 2.23). Carpenter has suggested that the

R H ⇌ R H

13a **13b**

R = $C(CH_3)_3$, $C(CD_3)_3$

Figure 2.23

failure to freeze out the tautomers of **13** is in accord with the lower barrier to tunneling enforced by the sterically demanding substituents.

2.4. CHEMICAL PROPERTIES

The chemistry of benzene is extensive and will be treated more comprehensively in Chapter 3. The normal mode of attack on the molecule is by electrophilic reagents, and the system passes through a transition state in which the benzenoid character is removed before the expulsion of a proton to restore the aromatic sextet (Fig. 2.24). Each step of this mechanistic pathway has been investigated in great detail, and the tendency of the benzene nucleus to be retained in this type of substitution reaction was the earliest basis by which aromatic molecules were distinguished from their olefinic counterparts.

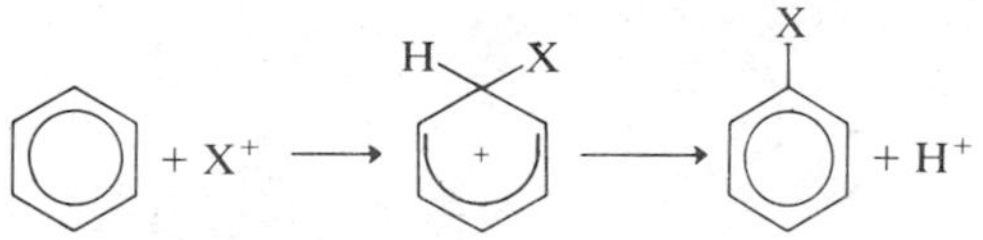

+ X^+ → H X + → X + H^+

Figure 2.24

Cyclooctatetraene was originally synthesized by Willstätter and co-workers by successive Hofmann eliminations on the alkaloid pseudopelletierine, but it is now commercially prepared by the nickel catalyzed tetramerization of acetylene. The discovery of this synthetic method by Reppe and co-workers has led to an extensive chemistry of cyclooctatetrane, which has been comprehensively reviewed. The reactions of cyclooctatetraene are largely those expected for an olefin, except that the products undergo subsequent reactions involving bond reorganization (valence tautomerism). Cyclooctatetraene exists almost exclusively in the monocyclic form **2**, the bicyclic structure **14** contributing less than 0.05% to the equilibrium mixture. The bicyclic isomer has been prepared at low temperature by Vogel and co-workers and is thermodynamically less stable than the monocyclic form by about 29 kJ mol^{-1}.

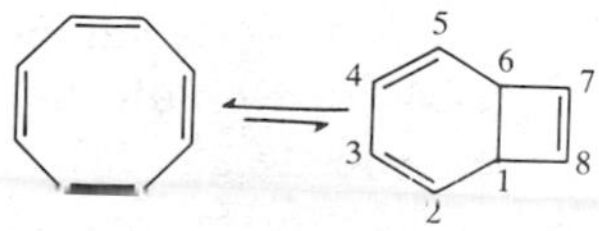

5 4 6 7 3 8 1 2

2 **14**

Figure 2.25

When cyclooctatetraene reacts with halogens, the final products are formally derived from **14** by addition of halogen across the 7,8 double bond. A study of the reaction has shown, however, that electrophilic addition occurs to give the endochlorohomotropylium cation **15**. The cation **15** then reacts with the chloride ions on the endo face to give **16**, which valence tautomerizes by a Cope rearrangement to *cis*-7,8-dichlorobicyclo[4.2.0]octadiene (**17**) (Fig. 2.26). The *cis*-stereochemistry of **17** is controlled by the stereoselectivity of attack on **2** and the intermediate **15**. The exo-chlorohomotropylium cation gives the *trans*-product on subsequent reaction (see Chapter 10).

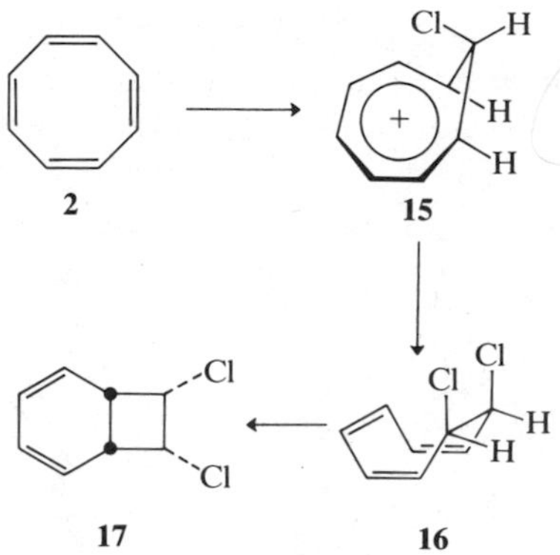

Figure 2.26

Diels–Alder reactions do not occur readily with cyclooctatetraene, and the eventual adducts are derived from **14** (e.g., **18** with maleic anhydride). Epoxidation and carbene addition give products derived from the monocyclic form **2**; both cyclooctatetraene epoxide (**19**) and the bicyclo[6.1.0]nonatrienes (**20**) are useful synthetic intermediates. Cyclooctatetraene is readily reduced to the dianion **21** with alkali metals; this reaction is discussed further in Chapter 5. Hydration with mercury sulphate gives phenylacetaldehyde (**22**). These reactions are summarized in Figure 2.27.

Cyclobutadiene was originally generated as a transient species by Petit and co-workers by the oxidation of the iron tricarbonyl complex **23** with cerium(IV) ions. The cyclobutadiene was trapped with dienophiles to give derivatives of bicyclo[2.2.0]hexadiene (Dewar benzene) (Fig. 2.28). The bicyclo[2.2.0]hexadienes (e.g., **24**) can readily thermally rearrange to the corresponding substituted benzene (e.g., **25**). In the absence of the dienophile, the syn-**26** and anti-**27** dimers were obtained in the ratio 5:1, both of which isomerize thermally to cyclooctatetraene (Fig. 2.29). These

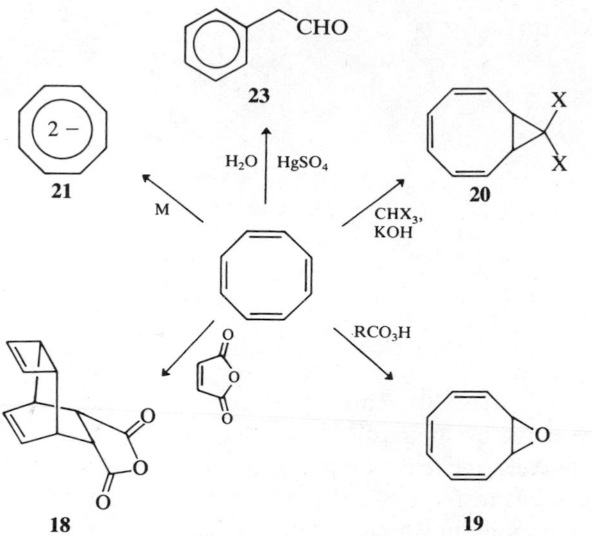

CHO
23
21
2 −
20
X
X
H_2O $HgSO_4$
M
CHX_3, KOH
RCO_3H
18
19

Figure 2.27

results were interpreted by Petit as indicating that cyclobutadiene had a singlet rectangular ground state, since these are reactions of a diene rather than of a diradical. The Diels–Alder adducts from dimethyl fumarate and dimethyl maleate were formed stereospecifically, and no product that would result from bond rotation was observed.

Subsequently, it was shown that when optically active cyclobutadiene iron tricarbonyl derivatives were oxidized, the resultant adducts were not optically active, and the most probable achiral intermediate is the substituted

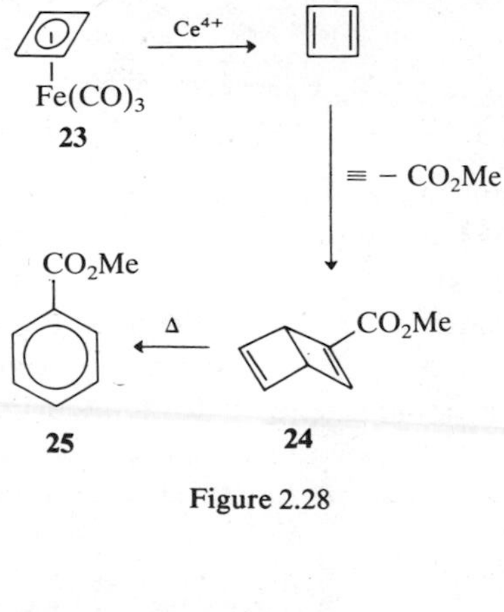

$Fe(CO)_3$
23
Ce^{4+}
$\equiv - CO_2Me$
CO_2Me
Δ
CO_2Me
25
24

Figure 2.28

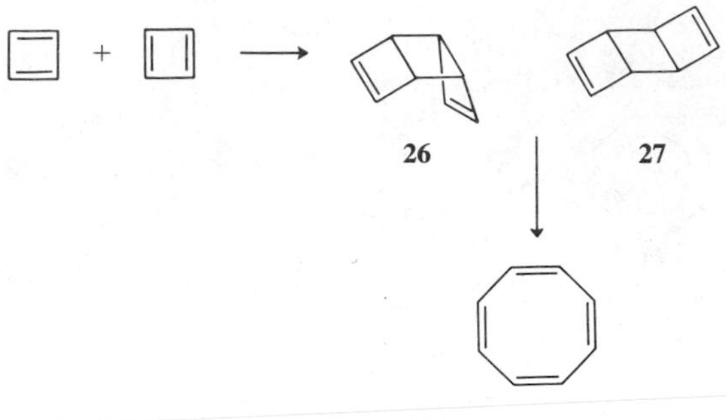

Figure 2.29

cyclobutadiene (Fig. 2.30). This was further substantiated by the reaction of optically active cyclobutadiene complexes with dienophiles; oxidation now gave optically active products via intermediates in which the iron was still complexed. Transfer of cyclobutadiene between an iron complex precursor and a dienophile in which both have been attached to a solid support has also been observed, demanding the transfer of a "free" species.

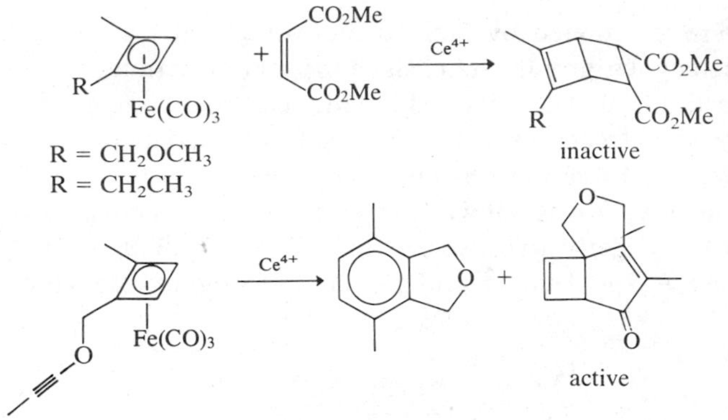

Figure 2.30

The initial formation of cyclobutadiene by the matrix photoirradiation of photo-α-pyrone (**28**) led to the investigation of a large number of possible photolabile precursors, some of which are illustrated in Figure 2.31. Allowing the matrix preparation of cyclobutadiene to thaw leads exclusively to the syn-dimer (**26**), in contrast to the preparation from the metal complexes.

The chemistry of cyclobutadiene is dominated by the ease in which it undergoes self-Diels–Alder reaction to the dimer. Reactions thus have to be conducted at low temperature, but the formation of mixed Diels–Alder

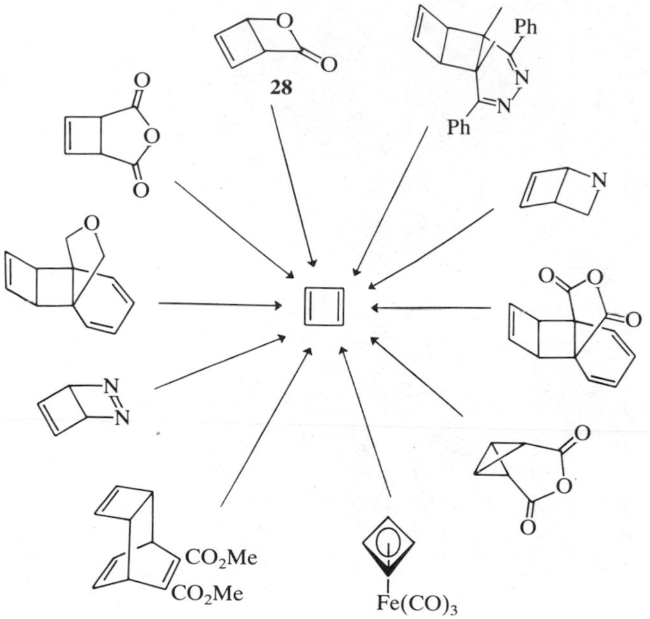

Figure 2.31

adducts, as already indicated, is readily accomplished and cyclobutadiene is now a valuable precursor in a number of syntheses.

The chemistry of stabilized cyclobutadiene derivatives has been much more extensively investigated. Tri-*t*-butylcyclobutadiene (**31**) was prepared simultaneously by Maier and Masamune and their respective groups by different routes. Masamune's route involved photoirradiation of the diazoalkane **29** and Maier's photoirradiation of the cyclopentadienone **30**, both at low temperatures (Fig. 2.32).

As can be seen from Figure 2.32, **31** behaves as a nucleophile, electrophile, or radical, depending on the electronic requirement of the reagent. This is presumably a reflection of the small energetic difference between the HOMO and LUMO in **31**. Tetra-*t*-butylcyclobutadiene (**32**) can be converted by photoirradiation into tetra-*t*-butyltetrahedrane (**33**), the first stable derivative of the tetrahedrane structure. Tetra-*t*-butyltetrahedrane is a fairly high melting solid that reverts at the melting point to the cyclobutadiene **32** (Fig. 2.33), the barrier to interconversion ΔE being 113 kJ mol^{-1}.

As we have already mentioned, the original synthesis of cyclobutadiene by Petit involved the oxidation of the iron tricarbonyl complex **23**. The suggestion that cyclobutadiene could be stabilized by complexing to a

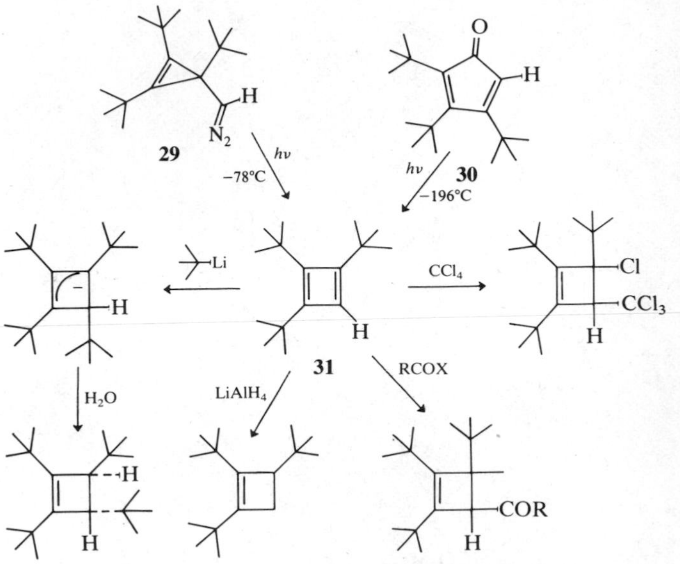

Figure 2.32

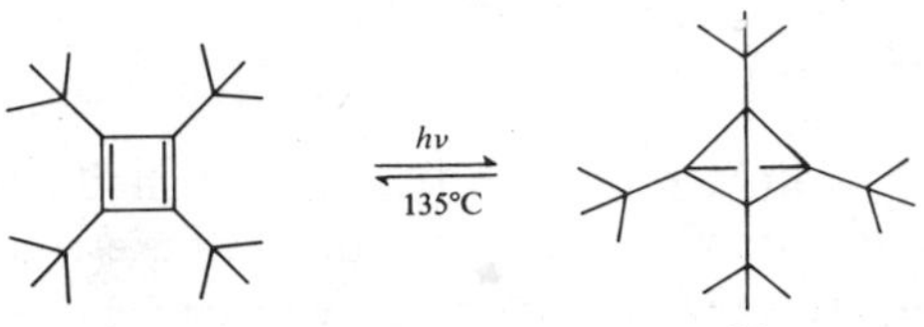

Figure 2.33

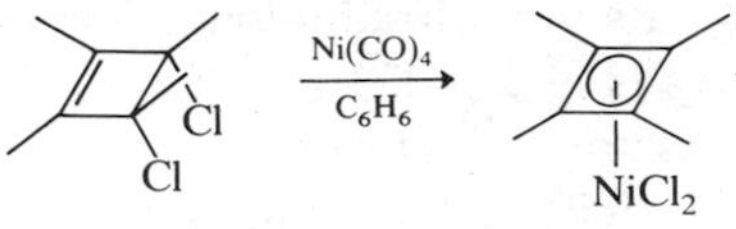

Figure 2.34

suitable transition metal was made by Longuet-Higgins and Orgel, and the prediction was verified by Criegee and Schröder who prepared the nickel chloride complex **35** by treatment of 1,2,3,4-tetramethyl-3,4-dichlorocyclobutene (**34**) with nickel tetracarbonyl in benzene (Fig. 2.34). The stabilization of **35** and of the many related systems that have subsequently been prepared is assumed to arise from the interaction of the molecular orbital of the ligand with those of the metal. A simple MO calculation indicates that this interaction produces 8 bonding orbitals and 1 nonbonding orbital. The total electron count should therefore be 16 or 18, depending on whether the NBMO is filled. In **35**, the cyclobutadiene supplies 4, the nickel 10, and the chlorine atoms 2 electrons, giving a total of 16 electrons. In the complex **23**, the iron supplies 8 and the carbonyl groups 6, which, together with the 4 electrons from cyclobutadiene, gives a total of 18 electrons. Filling the NBMO does not seem particularly important. Similar complexes of the cyclopentadienyl anion and related systems will be discussed in Chapter 5.

2.5. VALIDITY OF THE HMO METHOD

The HMO method, as we have seen, predicts a difference between $4n$ and $(4n + 2)$ π-electron systems in terms of their electronic configurations; but using ethene or cyclohexene as model systems, it does not satisfactorily account for the difference in resonance stabilization of the two types. In 1965, Dewar and Gleicher derived delocalization energies using a more sophisticated semiempirical molecular orbital method—the Pople-Pariser-Parr (PPP) approximation—and found that the energies alternated, with planar cyclobutadiene and cyclooctatetraene having negative delocalization energies (Fig. 2.35). However, as was subsequently pointed out by Schaad and Hess, besides changing the method of calculation, Dewar and Gleicher had also changed the reference system, adopting instead an acyclic polyene with the same number of double and single bonds as the cyclic polyene. They had calculated the energy of this polyene, and it was the change in model rather than the adoption of a more sophisticated MO method that had improved the results. Schaad and Hess were thus able to mirror the PPP results using the HMO method but with the new polyene reference structure (Fig. 2.36).

The HMO method, using the new model system, thus gives reasonable values of the delocalization energies and heats of formation for a variety of aromatic compounds. In order to be able to compare systems with different numbers of π electrons, it was suggested that the resonance energy per π electron (REPE) would be a more suitable parameter than the total resonance energy. Table 2.5 collects the REPEs for a number of systems as calculated by Dewar using the PPP method and by Hess and Schaad using the Hückel method; both groups use the equivalent acyclic polyene as reference structure. It has been suggested that resonance energies which

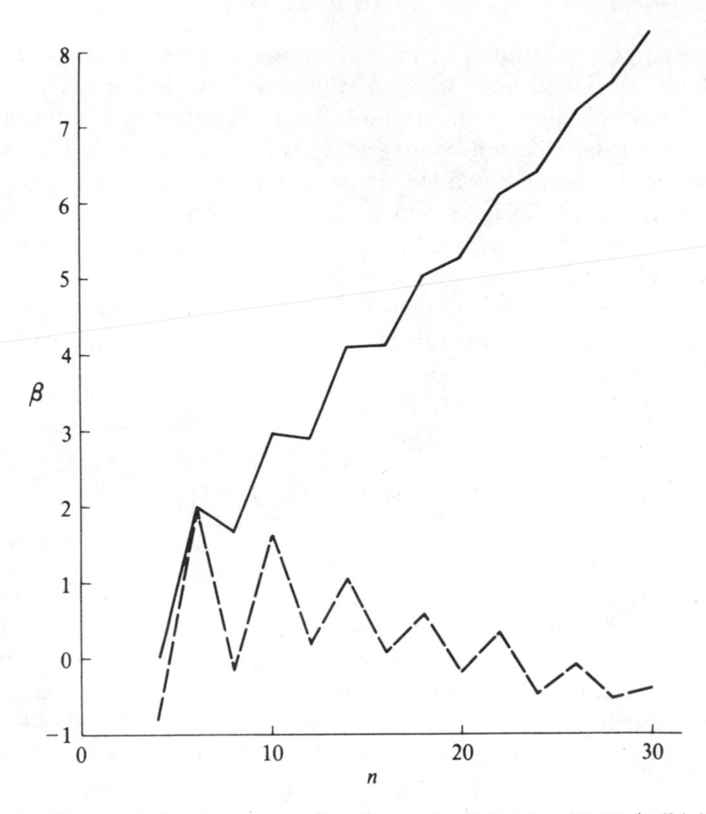

Figure 2.35. The delocalization energy (in β) calculated by the HMO (solid line) and Pople–Pariser–Parr (PPP) (broken line) approximations for monocyclic systems. Values for the PPP method taken from Dewar and Gleicher, *J. Am. Chem. Soc.*, 1965, **87**, 685.

TABLE 2.5

	REPE/β	
Compound	Dewar	Hess–Schaad
Benzene	0.120	0.065
Naphthalene	0.930	0.055
Azulene	0.017	0.023
Fulvene	0.006	−0.002
Heptalene	0.006	−0.004
Pentalene	0.001	−0.018
Cyclobutadiene	−0.136	−0.268

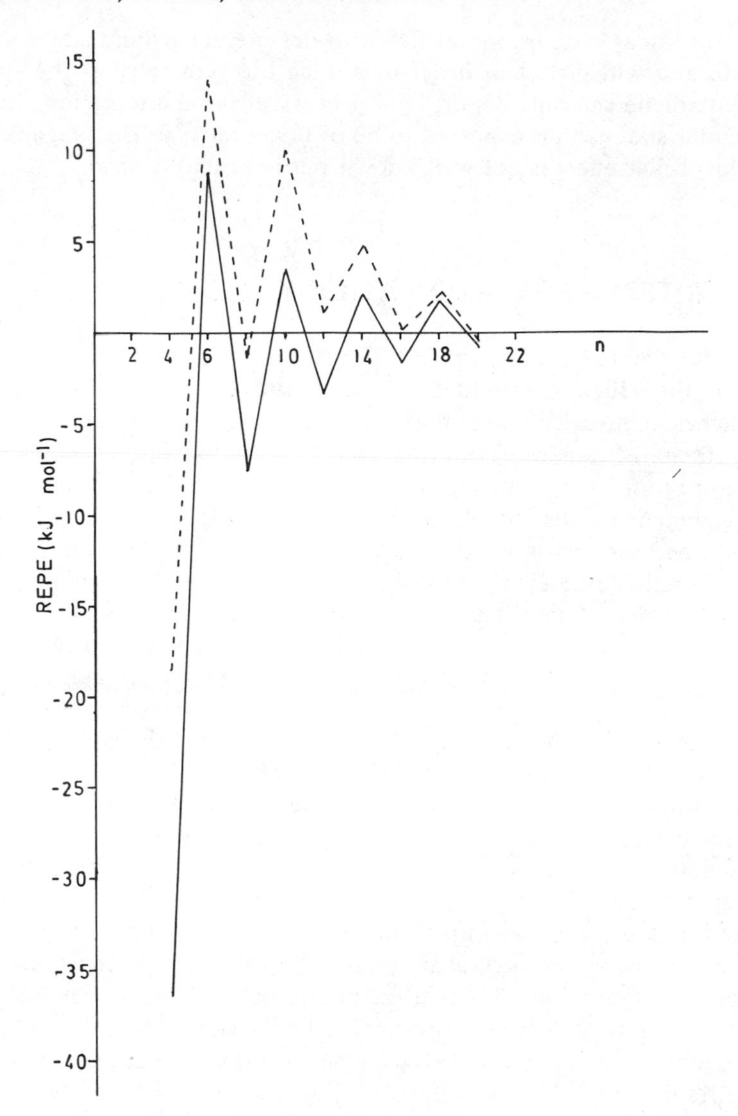

Figure 2.36. The resonance energy per electron (in kilojoules per mole) calculated by the PPP approximation (broken line) and the HMO approximation (solid line) using an acyclic polyene reference system. Data adapted from Schaad and Hess, *J. Chem. Educ.*, 1974, **51**, 640.

have the equivalent acyclic polyene as the reference system, calculated by any method, should be termed Dewar resonance energies.

In the simple HMO theory, cyclobutadiene and cyclooctatetraene both have degenerate ground states. The Jahn–Teller theorem postulates that, except for linear systems, molecules with degenerate ground states will be unstable and will distort in order to reduce the symmetry of the system. Cyclobutadiene can only distort by a process of bond alternation, and the rectangular state is thus expected to be of lower energy. The magnitude of the Jahn–Teller effect is not well known but is probably small.

2.6. CRITERIA FOR AROMATICITY

Continuing from Dewar's original PPP calculations and the reparameterization of the Hückel method by Schaad and Hess, a variety of other approaches to the calculation of resonance energies have been made. Aihara and Gutman, Milum and Trinajstic used an infinitely large cyclic polyene as reference structure, and Herndon introduced a valence bond method. The agreement on the calculated resonance energy (RE) for the various methods and models is good, and the calculated energies do appear to provide a reliable index of the stability and reactivity for the compounds for which calculations have been made. The occurrence of a positive calculated resonance energy, presumably above a prescribed limit, could then be used to define whether or not a compound was aromatic.

The most widely used method for determining aromaticity has been the observation of diatropicity in the ^{1}H NMR spectrum. This technique is easy to apply, is nondestructive, and requires only small quantities of compound—three attractive features to the experimentalist. Does the diatropicity, however, relate qualitatively or quantitatively to the degree of aromaticity of the compound and can it be correlated with the calculated REs?

Haddon has applied a simple model to the calculation of ring currents in annulenes using the previously reported application of the Biot–Savart law for the calculation of spacial magnetic fields. Model chemical shifts were assigned to different environments while geometries were obtained from x-ray crystallographic analyses wherever possible and a ring current geometric factor (RCGF) was derived. Two parameterizations were made, the most satisfactory being that in which the magnetic resonance integral β_M was estimated from the ring current of *trans*-15,16-dimethyldihydropyrene. A measure of the aromatic character (AC) of the system was taken from the value of k necessary to reproduce the ring current. The constant k relates to the resonance integral and measures the degree of bond alternation in the system. A value of $k = 1$ occurs if all the bonds are of equal length and k tends toward zero as bond alternation increases. For the model system 15,16-dimethyldihydropyrene $k = 1.000$; this and the values for other compounds are shown in Table 2.6.

TABLE 2.6. Calculated Ring Currents and Aromatic Character by Haddon's Method

Compound	Ring Current RC ($cm^2\ t^{-1}$ ppt)	Aromatic Character k
Benzene	−1.1861	1 +
Naphthalene	−1.1247	1 +
1,6-Methano[10]annulene	−0.7622	0.768
[12]Annulene	0.2137	0.582
Tridehydro[12]annulene	1.1811	0.749
[14]Annulene	−0.7451	0.779
trans-15,16-Dimethyl-dihydropyrene	−1.5495	1.000
[16]Annulene	0.6288	0.729
[16]Annulenyl dianion	−1.7861	0.937
[18]Annulene	−1.2043	0.837
[24]Annulene	0.8862	0.805

SOURCE: R. C. Haddon, *Tetrahedron*, 1972, **28**, 3613, 3635. For an explanation of k see text.

From the table it can be seen that aromatic ($4n + 2$) annulenes have values of $k > 0.75$, whereas antiaromatic $4n$ annulenes have values of $k < 0.75$. After 15,16-dimethyldihydropyrene, the [16]annulenyl dianion is the next most aromatic compound; it is considerably more aromatic than the isoelectronic [18]annulene. This is in accord with the observed physical and chemical properties (see Chapters 4 and 5). [12]Annulene is considerably more antiaromatic than [16]annulene, which again is in keeping with the known properties of the two compounds (see Chapter 4). The higher k value for tridehydro[12]annulene compared to [12]annulene probably reflects the steric nonbonded destabilization in the latter compound.

Haddon has subsequently presented a unified theory linking RE to reduced ring current (RC) determined in the manner described above. The relationship of RE to RC for the ($4n + 2$) annulenes is given by

$$\mathrm{RE} = \frac{\pi^2 \mathrm{RC}}{3S} \tag{2.5}$$

where S is the area of the ring and RE is the Dewar resonance energy. The REs derived in this way are in good accord with those observed by the PPP and modified HMO methods. A similar theoretical treatment has been presented by Aihara.

Other correlations between RCs and REs have been reported, and it now appears that a satisfactory theoretical connection between them can be made. How well k measures the degree of aromaticity quantitatively is still open to question, but it clearly does give an indication of the relative aromaticities of molecules. Presumably, the other properties that depend on the magnetic behavior of the molecule, such as magnetic anisotropy and diamagnetic exaltation, must also correlate with the resonance energies derived above.

Another ground-state property that could be used as a criterion to determine aromaticity is the C—C bond lengths. Applied to the annulenes, aromatic systems should have equal bond lengths, which should approach 139 pm, the length of the benzene C—C bond, whereas nonaromatic or antiaromatic annulenes should have bonds of alternating length. The C—C bonds in the higher annulenes are not all equivalent, however, and some variation in the bond lengths of the aromatic systems (but not alternation) is to be expected. This criterion would be difficult to apply to heterocyclic or polycyclic systems because of their lower symmetry. The method also requires an x-ray crystallographic determination, which is difficult to carry out on simple hydrocarbons.

Binsch has suggested as a criterion for aromaticity that the compound should not show strong first- or second-order double bond fixation. To determine these parameters, the planar sp^2 σ framework is set up with equal bond lengths (e.g., 150 pm), and the geometry is then examined after the π electrons are introduced. The total energy is developed as a Taylor series, which can be interrupted to give first-order terms (dependent upon ∂E) or second-order terms (dependent upon $\partial^2 E$). The bond orders can be derived from the first-order terms, and the difference in bond order between adjacent bonds measures the first-order double bond fixation. For second-order terms, a critical value was determined above which second-order bond fixation occurred. Benzene, as the quotation on the flyleaf indicates, exhibits neither first- nor second-order bond fixation.

Classical criteria such as lack of reactivity and stability can be discounted since these depend on differences in energy between ground and excited states and not on the ground-state energy itself. Similar objections apply to the electronic spectrum. Although this is useful for comparing molecules that are closely related in structure, such as the benzenoid polycyclic hydrocarbons, it is not a suitable means of distinguishing the presence or absence of aromaticity.

2.7. DEFINING AROMATICITY

Of the three compounds discussed in this chapter, benzene is the archetypal aromatic molecule, and cyclobutadiene and cyclooctatetraene may be taken as typical of antiaromatic and nonaromatic molecules, respectively. Benzene is the prototype of the Hückel ($4n + 2$) π-electron type and cyclobutadiene and cyclooctatetraene of the $4n$ π-electron type. These compounds support a classification scheme based on the Hückel Rule. A suitable definition on this basis would be that *aromatic systems are monocarbocyclic, conjugated molecules containing* ($4n + 2$) *out-of-plane* π *electrons.* Such a definition would have the merit that all of the terms can be precisely defined. However, it would have the disadvantage that many of the compounds normally considered aromatic, such as thiophene, pyridine, and naphthalene, would be excluded. As will be discussed in Chapter 4, problems also arise in monocyclic systems, both with regard to nonbonded interactions and the upper limit on the size of the ring, and therefore restrictions would have to be placed on the value of n. It is unlikely that a satisfactory definition of "aromatic compound" based on a structural concept will be forthcoming.

The older definitions that were based on the types of reactions undergone by benzene and its derivatives depend on the difference in free energy of the ground and transition states and are of little use in correlating ground-state properties, since a molecule might be resonance stabilized but extremely reactive. Dewar has defined aromatic molecules as *cyclic systems having a large resonance energy in which all the atoms in the ring take part in a single conjugated system.* This definition, which is very satisfactory in principle, is difficult to apply in practice. Although it is clear from the preceding discussion that benzene has a large RE and cyclooctatetraene has not, when more complex systems are investigated the interpretation of the term "large" often depends on the preconceived idea of the investigator. The establishment of the reliability of calculated REs, the concept of the REPE, and the finding that the magnetic properties can be related to the RE suggest a definition that incorporates these phenomena. Aromatic compounds could then be defined as *cyclic diatropic systems with a positive calculated Dewar RE in which all the ring atoms are involved in a single conjugated system.* Although this description is open to criticism it allows a formal definition of an aromatic system and provides physical properties that the experimentalist can determine.

FURTHER READING

Several of the references given at the end of Chapter 1 are also applicable to this chapter. For detailed descriptions of the molecular orbital theory, see the texts by Borden, Dewar, Salem, Streitwieser, and Yates. The application of the molecular orbital theory to electronic spectra is discussed in the following texts: C. Sandorfy, *Electronic Spectra and Quantum Chemistry*, Prentice-Hall, Englewood Cliffs, N.J., 1964; G. W. King, *Spectroscopy and Molecular Structure*, Holt, Reinhart, and Winston, New York, 1964; J. N. Murrell, *The Theory of the Electronic Spectra of Organic Molecules*, Methuen, New York, 1963. The text by Sandorfy has a very clear description of the application of group theory to the classification of electronic states.

For a detailed discussion of the 255-nm band of benzene see J. H. Callomon, T. M. Dunn, and I. M. Mills, *Phil. Trans. A.*, 1966, **259**, 60.

For a clear discussion of the problems involved in the comparison of theoretical and experimental IR spectra with particular reference to cyclobutadiene and other matrix isolated species, see B. A. Hess, L. J. Schaad, and P. Čársky, *Pure Appl. Chem.*, 1983, **55**, 253.

For a review of cyclobutadiene and its derivatives, see T. Bally and S. Masamune, *Tetrahedron*, 1980, **36**, 343. For stabilized cyclobutadienes, see R. Gompper and G. Seybold, in T. Nozoe, R. Breslow, K. Hafner, Shô Itô, and I. Murata (Eds.), *Topics in Nonbenzenoid Aromatic Chemistry*, Vol. 2, Hirokawa, Tokyo, 1977, p. 29.

The chemistry of cyclooctatetraene has been extensively reviewed. See G. Schröder, *Cyclooctatetraene*, Verlag Chemie, 1965; H. Röttelle in E. Müller (Ed.), *Houben-Weyl Methoden der Organischen Chemie*, Band V/1d, 423, ed. 4, Georg Thieme Verlag, Stuttgart, 1972; L. A. Paquette, *Tetrahedron*, 1975, **31**, 2855; G. I. Fray and R. G. Saxton, *The Chemistry of Cyclooctatetraene and its derivatives*, C.U.P., Cambridge, 1978.

For a comparison of PPP and HMO resonance energies, see L. J. Schaad and B. A. Hess, *J. Chem. Educ.*, 1974, **51**, 640.

For the calculation of ring currents from proton chemical shifts, see R. C. Haddon, *Tetrahedron*, 1972, **28**, 3613, 3635.

For theoretical studies on aromaticity, RE, and ring currents, see the text by Lewis and Peters (Further Reading, Chapter 1); R. C. Haddon, *J. Am. Chem. Soc.*, 1979, **101**, 1722; J. Aihara, *ibid.*, 1979, **101**, 558, 5913; *Pure Appl. Chem.*, 1982, **54**, 1115; P. Ilíc, B. Sinkovic, and N. Trinajstic, *Israel J. Chem.*, 1980, **20**, 258; W. C. Herndon, *ibid.*, 1980, **20**, 294; R. C. Haddon and T. Fukunaga, *Tetrahedron Lett.*, 1980, **21**, 1191; L. J. Schaad and B. A. Hess, *Pure Appl. Chem.*, 1982, **54**, 1097.

For a general review of the chemistry to be found in this and succeeding chapters see P. J. Garratt, Chapters 2.4 and 2.6 in J. Frazer Stoddard (Ed.), *Comprehensive Organic Chemistry*, Vol. 1, Sir Derek Barton and W. D. Ollis (series Eds.), Pergamon, Elmsford, N.Y., 1979. See also D. Lloyd, *Non-benzenoid Conjugated Cyclic Compounds*, Elsevier, Amsterdam, 1984.

3

BENZENE: ITS ISOMERS AND DERIVATIVES

3.1. INTRODUCTION

Benzene differs from the majority of the other compounds we will discuss by virtue of its high symmetry, D_{6h}. As was outlined in Chapter 2, this symmetry and its consequences could provide a reasonable, but highly exclusive, definition of aromaticity. That such a definition is too exclusive is readily demonstrated by reference to the derivatives of benzene, in which the high symmetry has been lost but most of the classical physical and chemical properties that separate benzene from other unsaturated compounds have been retained. In this chapter, we will examine in more detail the chemistry of benzene and its derivatives as a prelude to our discussion of less classically aromatic systems.

3.2. CHEMISTRY OF THE MONOCYCLIC BENZENOID HYDROCARBONS

3.2.1. Electrophilic Substitution

The classic reaction of benzene and its derivatives is electrophilic substitution, which was briefly mentioned in Chapter 2. The conventional description of this reaction is of the benzene ring providing electrons to the electrophilic reagent so that electron withdrawing substituents on the benzene ring decrease the rate of reaction and electron releasing substituents increase the rate. These substituent effects are not equally distributed to each carbon of the ring because the introduction of the substituent has reduced the symmetry of the system (e.g., toluene, C_{2v}). Long experience with electrophilic substitution of benzene derivatives showed that the positions 3 and 5 to the substituent [*meta*(*m*)] behaved differently from the 2, 4, and 6 positions [*ortho* (*o*) and *para* (*p*)]. Electron withdrawing groups deactivate the ring generally, but the 3, 5 positions less than the 2, 4, 6 positions, whereas electron donating groups activate the ring generally, but the 2, 4, 6 positions more than the 3, 5 positions. These experimental conclusions find a satisfying explanation in the mechanism outlined in Figure 3.1.

The tetramethylammonium group is electron withdrawing. Nitration with the nitronium ion could give the three possible σ intermediates (Wheland intermediates) shown. Of these, only that resulting from *meta*-attack does not have a partial positive charge on the carbon bearing the NMe_3^+ group, and the transition state to *meta*-substitution is therefore favored. The differences in reactivity are, however, small compared with the general level of deactivation of the ring, and the observed *meta*:*para* ratio, corrected for the different number of sites, shows that negligible discrimination occurs. With the less electronegative group V elements phosphorus, arsenic, and antimony as substituent, the ring is not as deactivated and the *meta*:*para* discrimination is greater. Toluene, in which the methyl group is electron donating so that the ring is generally activated, has a partial positive charge on the carbon bearing the methyl in the σ intermediate for *ortho*- and *para*-substitution but not for *meta* and thus the transition states to *ortho*, *para* are favored.

The course of events delineated in Figure 3.1 excludes attack at the carbon bearing the substituent, *ipso* attack, but this mode of reaction has recently been found to be more common than was previously supposed. Subsequent rearrangement is required to restore the aromatic system and, if ipso attack has occurred, the product ratios need not necessarily reflect the intrinsic reactivities of the *o*, *p*, and *m* positions. Nitration of *o*-xylene, investigated by Fischer and Wright, provided evidence for ipso attack under their conditions since the cyclohexadienyl adduct **1** was obtained (Fig. 3.2). It appears that ipso attack is important only when the electrophile is directed to this position by a second substituent, as in *o*-xylene.

The selectivity of an electrophile will depend on its reactivity; highly reactive electrophiles will show less regioselectivity than sluggish electrophiles. An anomaly between intermolecular and intramolecular selectivity was noted by Olah and co-workers, who found that nitration with nitronium

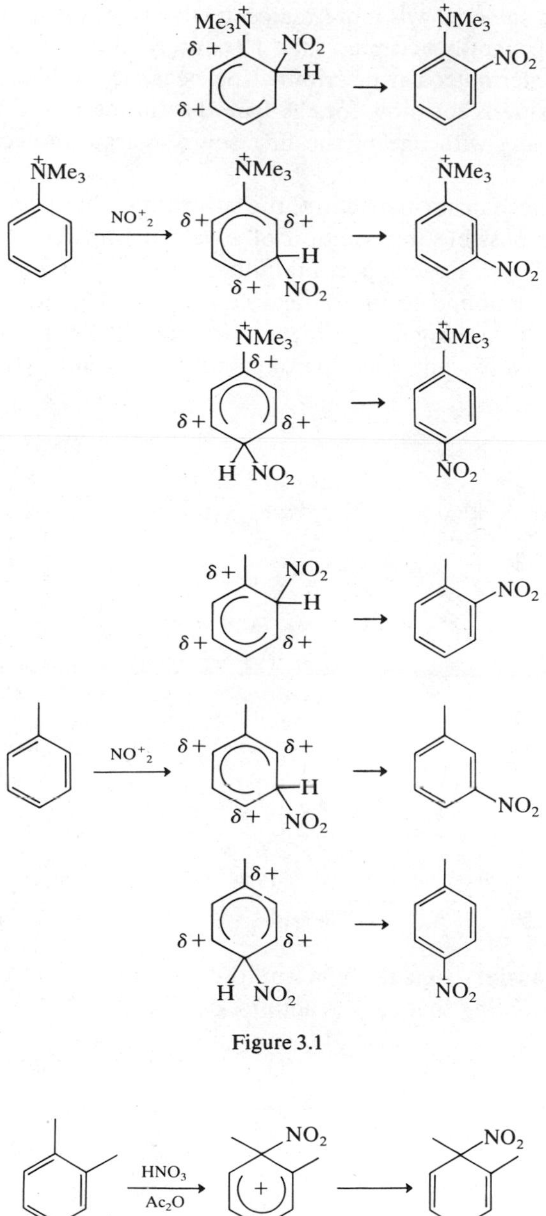

Figure 3.1

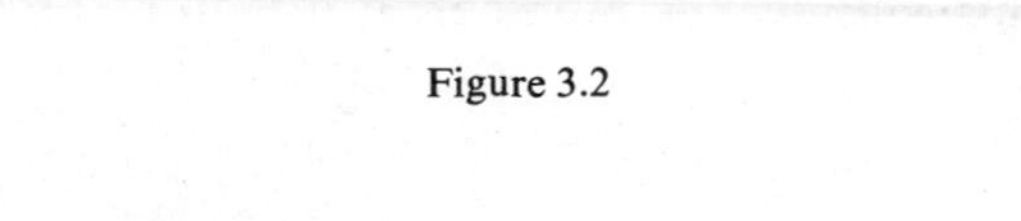

Figure 3.2

salts in organic solvents was more regioselective than substrate selective. It has now been generally accepted that this observation is a consequence of apparent low intermolecular discrimination because, in this rapid reaction, the rate of mixing is too slow for the true discrimination to be expressed, reaction occurring with one of the first few substrate molecules encountered.

The actual mechanism of electrophilic substitution has been rescrutinized of late for the possible intervention of electron transfer processes as an alternative pathway. Reactants containing a powerful oxidant can oxidize the aromatic compound to the radical cation, which then reacts with the nucleophile and undergoes further oxidation. This is illustrated for hydroxylation with aqueous peroxydisulphate ions [$(SO_3)_2O_2^{2-}$] in Figure 3.3.

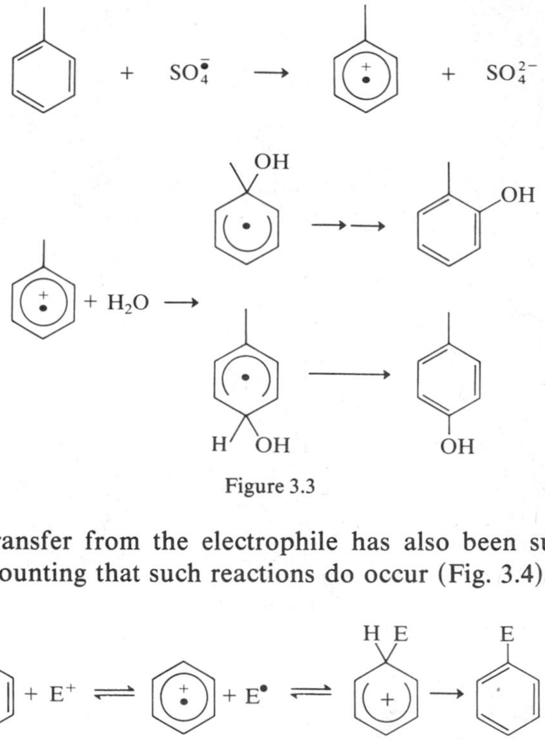

$SO_4^{\bullet-}$ SO_4^{2-} H_2O OH OH H OH OH

Figure 3.3

Electron transfer from the electrophile has also been suggested, and evidence is mounting that such reactions do occur (Fig. 3.4).

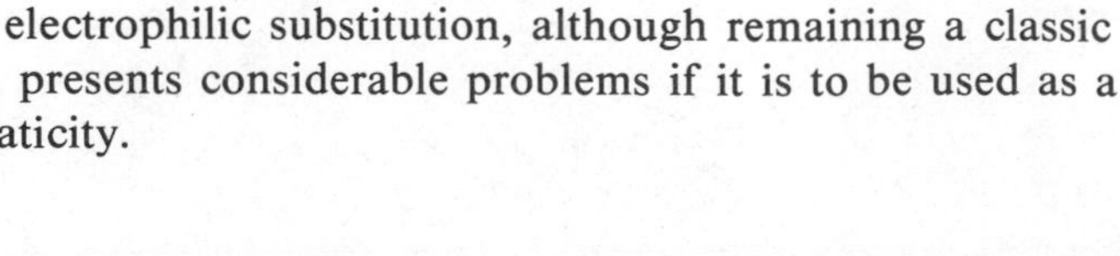

$+ E^+ \rightleftharpoons$ $+ E^\bullet \rightleftharpoons$ H E $\rightarrow$ E $+ H^+$

Figure 3.4

Thus, electrophilic substitution, although remaining a classic aromatic property, presents considerable problems if it is to be used as a criterion for aromaticity.

3.2.2. Nucleophilic Substitution

Benzene itself does not appear to undergo nucleophilic substitution, but benzene derivatives with electron withdrawing groups do. Usually, a leaving group other than hydrogen is preferred (e.g., F^-), but displacement of hydrogen can occur. The classic reaction of 2,4-dinitrofluorobenzene with amines is illustrated in Figure 3.5.

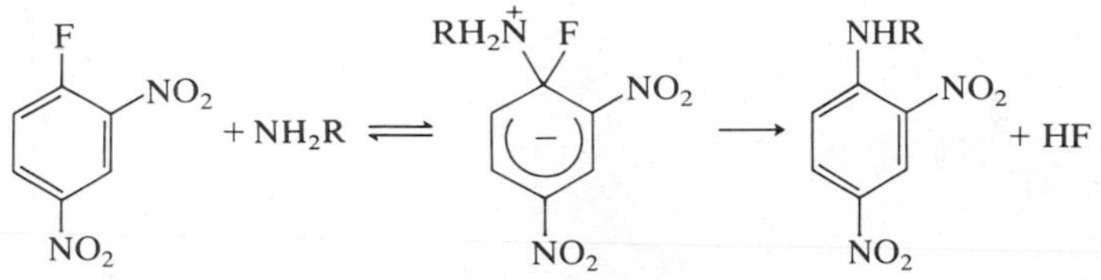

Figure 3.5

Reaction of halobenzenes with a strong base may involve an elimination–addition mechanism with the intervention of an aryne; these reactions are described in Section 3.4.

Again, recent work has uncovered a third, radical mechanism, in this case involving a chain reaction. Kim and Bunnett observed that 5- and 6-iodo-1,2,4-trimethylbenzene did not give the same ratio of 5- and 6-amino-1,2,4-trimethylbenzenes on treatment with KNH_2 in liquid ammonia and therefore could not completely proceed via an aryne intermediate. They suggested, by analogy with known processes in aliphatic substitution, a mechanism whereby an electron is transferred to the substrate to form a radical anion that can then exchange one nucleophile for another (Fig. 3.6). They called this an $S_{RN}1$ (substitution, radical, nucleophilic, monomolecular) reaction. The sequence is initiated by electron transfer to form the radical anion, which then undergoes a chain propagation process involving loss of the iodide ion in the rate-controlling step to form an [R—H] radical. This radical reacts with a nucleophile to form a second anion radical, which undergoes electron exchange with a substrate molecule to give the product and a new substrate radical anion, thus completing the chain. Several chain termination steps are available.

A variety of substrates, nucleophiles, and leaving groups has been investigated, and the initial radical anion has been generated photochemically. With a suitable choice of substrate, the photochemically generated radical has been shown to be a useful synthetic intermediate since intramolecular reactions can occur.

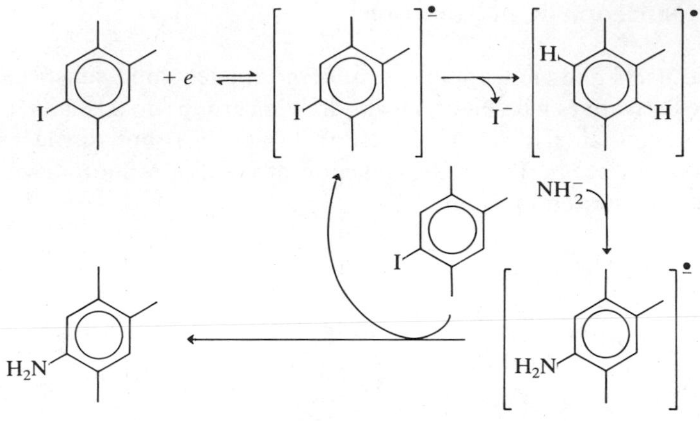

Figure 3.6

3.2.3. Homolytic Substitution

Aromatic substitution can, in certain circumstances, proceed by initial addition of a radical rather than the heterolytic or electron transfer processes described in the preceding sections. Benzene can be alkylated by alkyl radicals, conventionally generated by photolysis of alkylmercury(II) iodides (Fig. 3.7). The reaction again goes via a σ intermediate, in this case a radical, and side products arising from the dimerization of the intermediates have been observed.

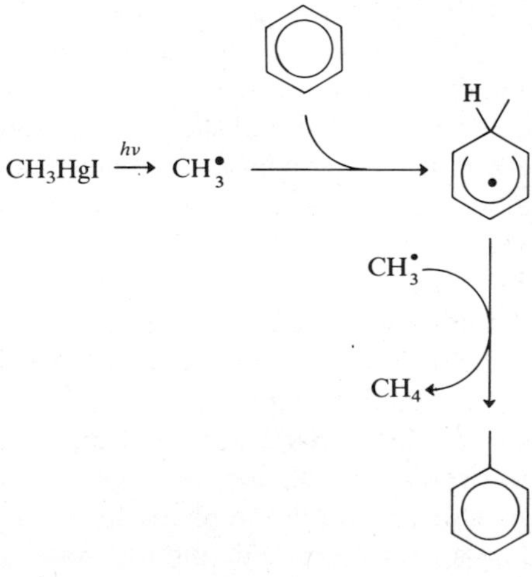

Figure 3.7

Amination of benzene can also occur homolytically in a process involving the amine radical cation as the reactive species, the reaction proceeding via a chain mechanism (Fig. 3.8).

$$R_2\overset{+}{N}HCl + Fe^{2+} \longrightarrow R_2\overset{+\bullet}{N}H + Fe^{3+} + Cl^-$$

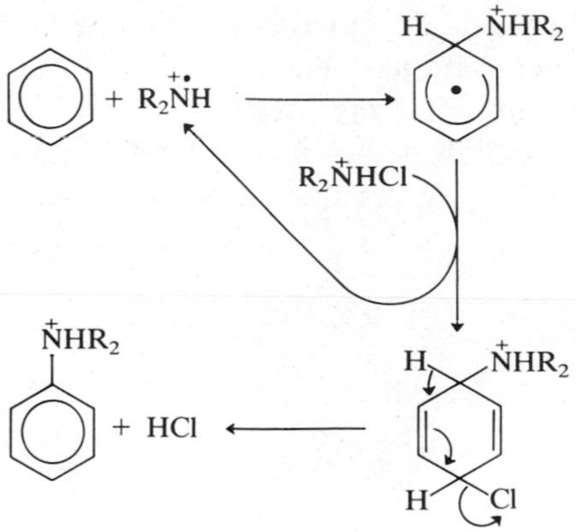

$$C_6H_6 + R_2\overset{+\bullet}{N}H \longrightarrow [C_6H_6(H)(\overset{+}{N}HR_2)]^{\bullet} \xrightarrow{R_2\overset{+}{N}HCl} C_6H_6(H)(Cl)(\overset{+}{N}HR_2) \longrightarrow C_6H_5\overset{+}{N}HR_2 + HCl$$

Figure 3.8

Arylation can also occur by a homolytic process, the aryl radicals frequently being generated from the corresponding diazonium salt. Again, the reaction takes place by way of a σ complex, which is then dehydrogenated by reaction with another diazonium ion in a radical chain process (Fig. 3.9).

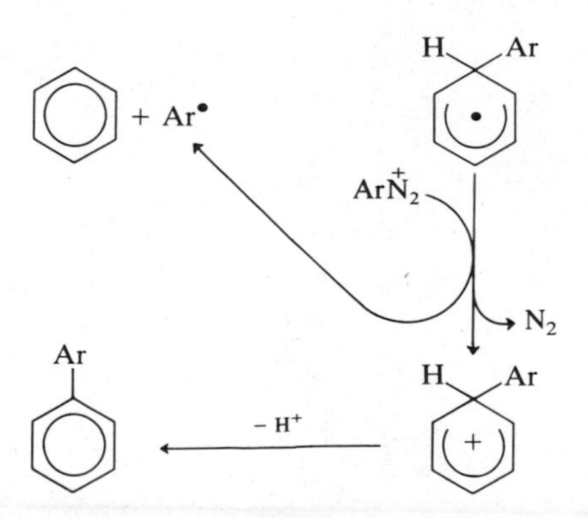

Figure 3.9

The generation of the initial aryl radical from the diazonium ion is complex.

3.2.4. Thermal Addition

Benzene reacts in a Diels–Alder fashion only under extremely forcing circumstances. A low yield of cycloadduct is obtained with hexafluorobut-2-yne (**2**), whereas toluene and *p*-xylene, with electron-donating substituents, react in reasonable yield. Benzyne (Section 3.4) will react with benzene to give benzobarrelene in low yield, whereas tetrachlorobenzyne, which is more electron demanding, gives tetrachlorobenzobarrelene in good yield (Fig. 3.10).

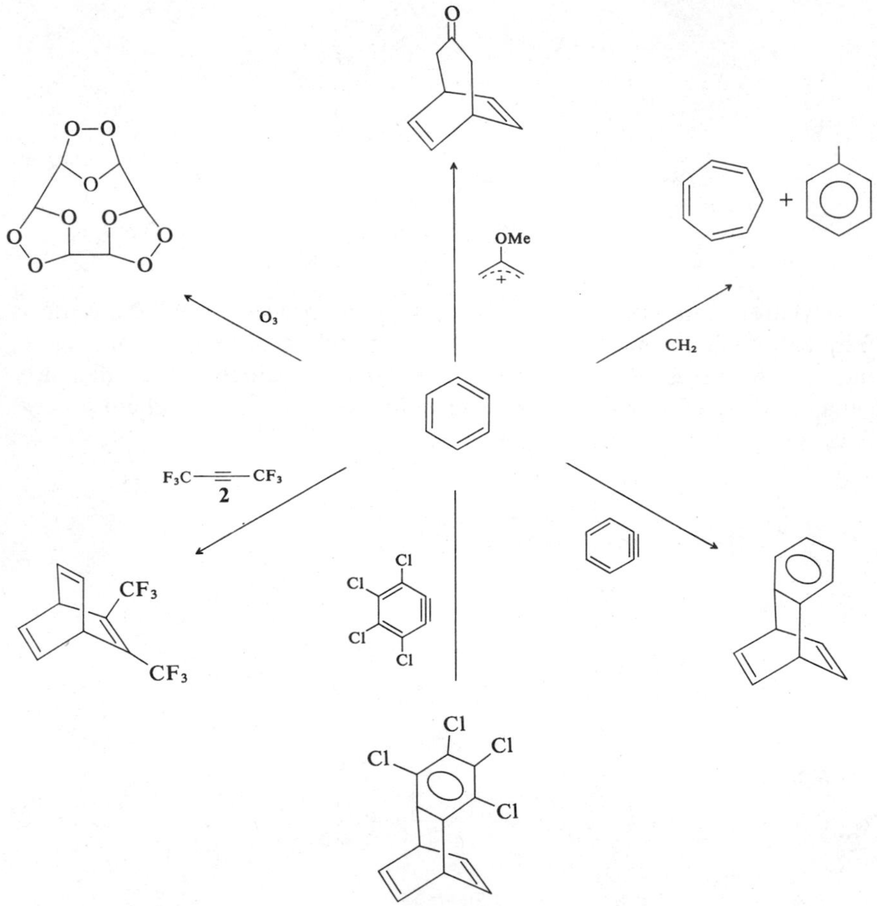

Figure 3.10

Thermal cyclization of benzene with the 2-methoxyallene cation occurs in low yield in a $4\pi + 3C\text{-}2\pi$ process. Reaction with ozone at 5–10°C gives a triozonide, which decomposes in water to give glyoxal; presumably this occurs by an initial 1,3-dipolar addition. Methylene, generated from a variety of sources, reacts with benzene to give cycloheptatriene and toluene.

In all of these cases, benzene shows a considerable resistance to the disruption of its structure, and in most, if not all, of these reactions benzene behaves as a nucleophile.

3.2.5. Photochemical Addition

Benzene is not, as was for a long time thought, photostable, but it can be converted into a number of valence tautomers (Section 3.3). It undergoes photochemical addition with a variety of alkenes and with molecules that would conventionally be classed as dienophiles (Fig. 3.11).

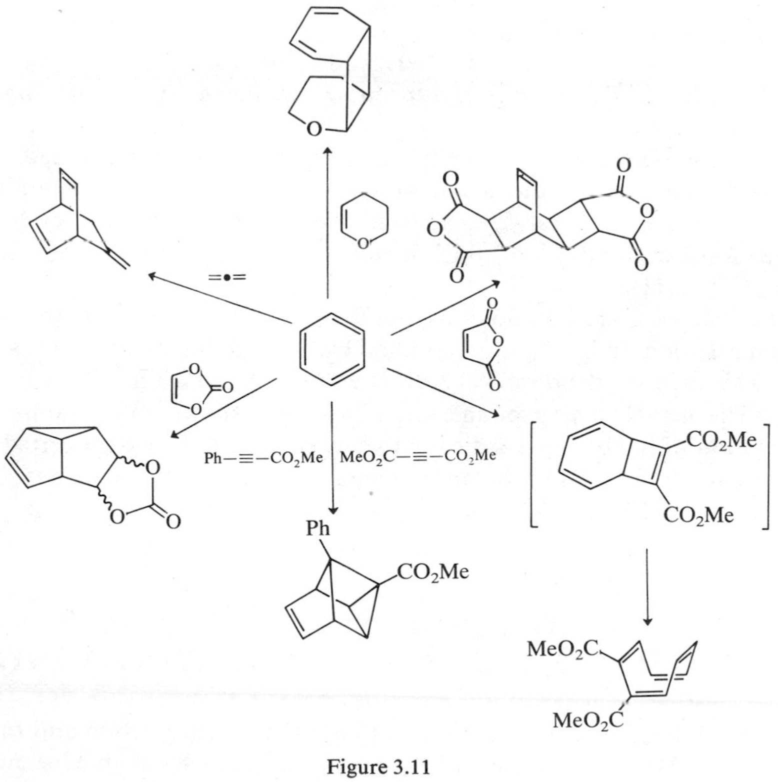

Ph—≡—CO_2Me | MeO_2C—≡—CO_2Me Ph CO_2Me CO_2Me CO_2Me MeO_2C MeO_2C

Figure 3.11

These photochemical reactions mainly occur by excitation of benzene through the 255-nm absorption band, which is the forbidden ${}^1B_{2u} + {}^1A_{1g}$ transition (Section 2.3.3). Photoaddition of maleic anhydride involves an excited state complex, which then collapses to the zwitterion, followed by a ring closure to the monoadduct. Subsequent thermal Diels–Alder addition to the cyclohexadiene gives the 2:1 adduct observed (Fig. 3.12).

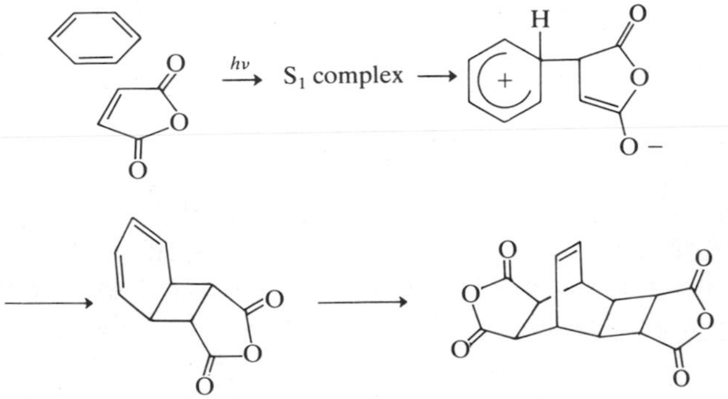

Figure 3.12

Competition in the thermal Diels–Alder reaction by other dienophiles can provide mixed adducts [e.g. with tetracyanoethene (TCNE) or N-phenylmaleïmide].

The photoaddition of dimethyl acetylenedicarboxylate proceeds by a similar route to give the monoadduct, which, in the absence of other dienophiles, undergoes valence isomerism to give dimethyl cyclooctatetraene-1,8-dicarboxylate (which is *not* in equilibrium with the 1,2-derivative) (Fig. 3.11).

Both of these are 1,2-additions, but the predominant pathway for photochemical addition is 1,3, as exemplified by the addition of vinylene carbonate. This involves the diradical **3** or its equivalent, which has the symmetry C_{2v}. The actual timing of the steps has been shown to be immaterial, formation of the benzene radical either preceding or being concerted with the photoaddition. The correlation diagram for the former process is shown in Figure 3.13. Bonds *b*, *c* appear to be formed in a concerted step.

3.3. VALENCE ISOMERS OF BENZENE: BENZVALENE, DEWAR BENZENE, PRISMANE AND BICYCLOPROPENYL

There are numerous ways of formulating a C_6H_6 hydrocarbon and four of formula $(CH)_6$. All of them will have a lower symmetry than benzene and

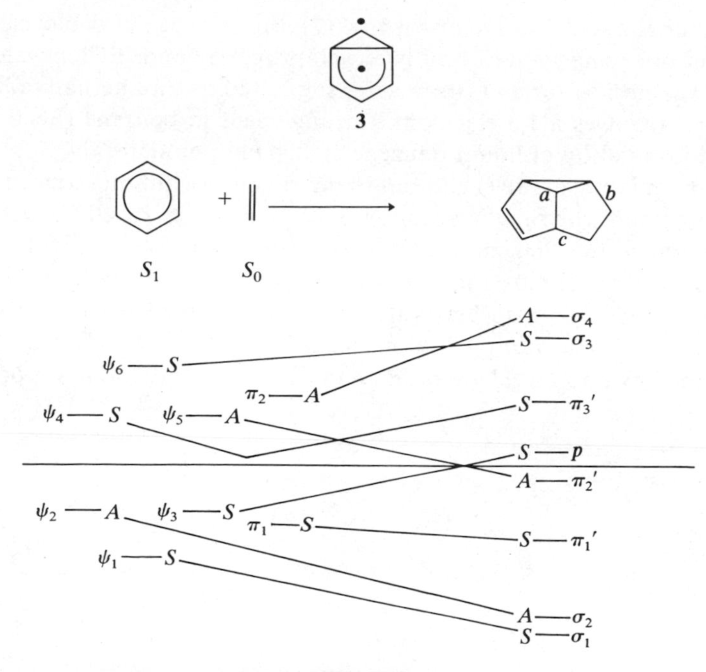

Figure 3.13

can thus be discounted as alternative benzene structures from the IR and Raman spectra. The four possible $(CH)_6$ isomers and a number of the more important C_6H_6 isomers are shown in Figure 3.14.

Prismane (**4**) and bicyclo[2.2.0]hexa-2,5-diene (Dewar benzene) (**5**) were both suggested as alternative benzene structures and are closely related to

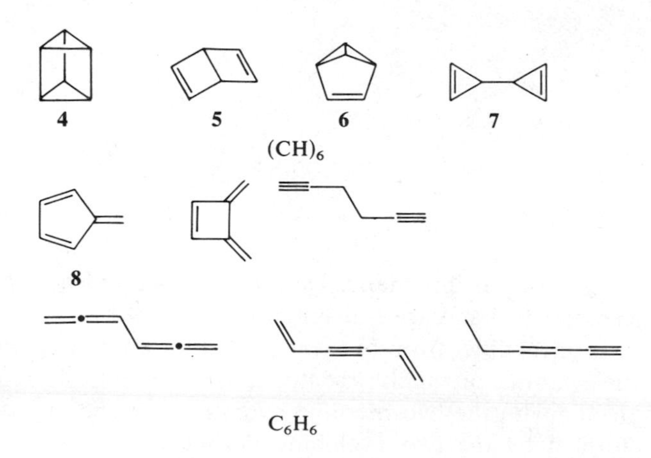

Figure 3.14

each other and to 3,3′-dicyclopropenyl (**7**), the cleavage of two cyclobutane bonds of prismane giving **7** and two cyclopropane bonds **5**. Benzvalene (**6**), which was not, as far as I am aware, suggested as an alternative benzene structure, involves a 1,3-electron rearrangement in benzene (Section 3.2).

Photoirradiation of liquid benzene at 254 nm populates the S_1 state and results in a mixture of fulvene (**8**) and benzvalene (**6**), limiting concentrations being obtained when the irradiation is carried out at 50–60°C. Irradiation over the range 165–200 nm gave a mixture of Dewar benzene, fulvene, and benzvalene (1:2:5) plus a little biphenyl. Similar wavelength irradiation in the vapor phase gave no benzvalene or Dewar benzene but a mixture of fulvene and *cis* and *trans*-1,3-hexadiene-5-yne (Fig. 3.15). Dewar benzene has been shown to arise by irradiation of the 204-nm band and population of the S_2 state.

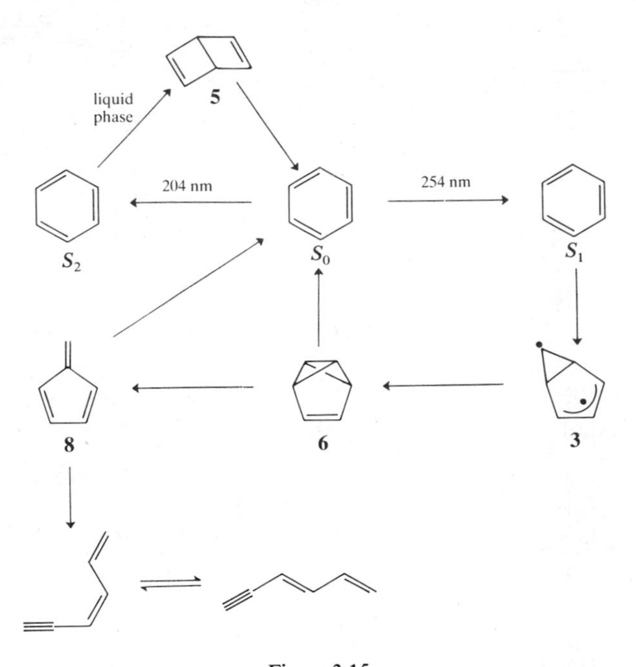

Figure 3.15

Rational syntheses of prismane, Dewar benzene, and benzvalene have all been accomplished and are illustrated in Figure 3.16.

All three isomers differ from benzene in that they are not planar. Dewar benzene, the synthesis of which predated its photochemical preparation, is a bent molecule with the two methine hydrogens on the opposite side of the apex formed by the two cyclobutene rings. Dewar benzene is best

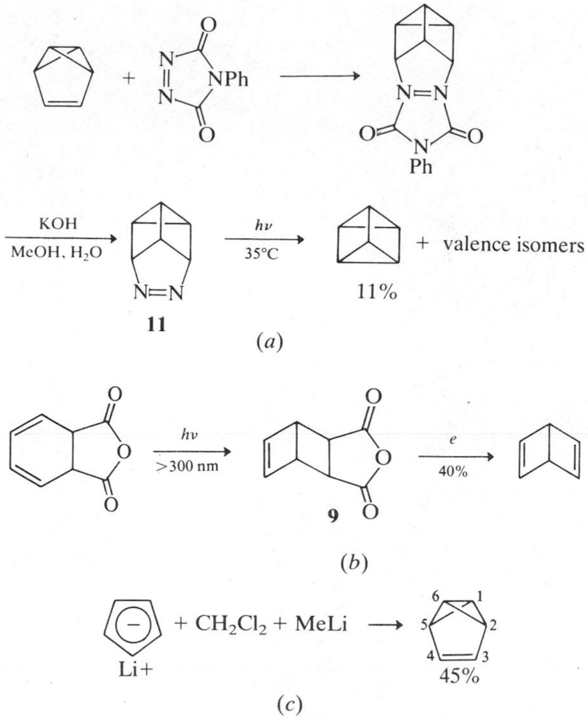

Figure 3.16. (*a*) Prismane: Katz and Acton, 1973; (*b*) Dewar benzene: van Tamelen and Pappas, 1963; (*c*) benzvalene: Katz, Wang, and Acton, 1971.

generated from the photoanhydride **9** by electrolytic oxidative decarboxylation. The ^1H NMR spectrum shows two bands at δ 3.84 (q, 2H) and 6.45 (t, 4H), with a coupling constant of 0.7 Hz. Reduction with di-imide gave bicyclo[2.2.0]hexane, which thermally rearranged to biallyl, and thermolysis of **5** gave benzene. The thermal rearrangement to benzene is not orbital-symmetry allowed, involving an antiaromatic transition state (Chapter 11), the enthalpy of activation being 154 kJ mol^{-1}. Nonequivalent substitution at the bridgehead positions facilitate the rearrangement to benzene, and 1-chlorobicyclo[2.2.0]hexadiene (**10**, Fig. 3.21) has an enthalpy of activation of 80 kJ mol^{-1}, a striking change for a single substitution. The chemistry of Dewar benzene is outlined in Figure 3.17.

The Katz synthesis (Fig. 3.16) has rendered benzvalene (**6**) the most readily accessible of these three $(CH)_6$ isomers. Benzvalene can be isolated pure as an explosive compound, even small samples detonating violently, but it can be readily handled in solution. The ^1H NMR spectrum shows

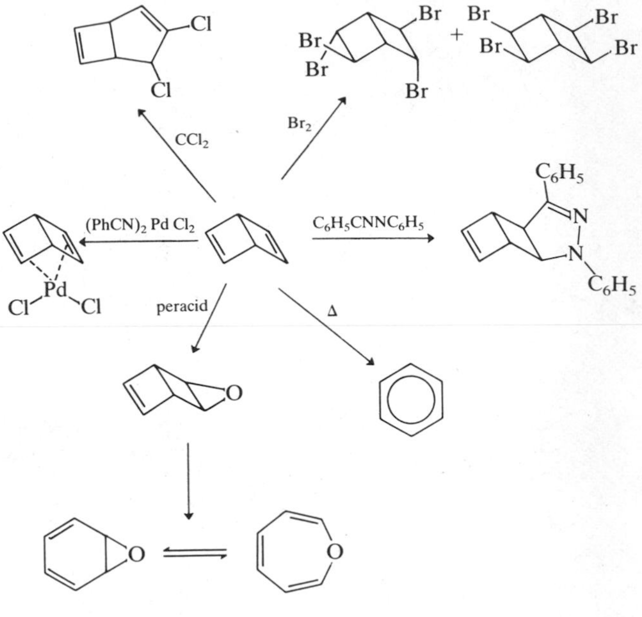

Figure 3.17

triplets at δ 5.95 and 3.53 and a quartet at δ 1.84, which are assigned to the olefin, 1,6- and 2,5-hydrogens, respectively. The ^{13}C NMR spectrum has the C-1,6 signals at δ 48.3, which is 45.9 ppm lower field than the equivalent carbons in bicyclo[2.2.0]hexane. This low field shift of the C-1,6 and H-1,6 atoms arises from an interaction of the alkene π^* orbital and the a_2 orbital of the bicyclobutane system. The electronic spectrum shows a low intensity maximum at 217 nm, again derived from interaction of the alkene and bicyclobutane systems; this is further reflected in the low value (8.55 eV) of the first ionization potential. The central single bond is short (about 145 pm) and is believed to have 90% *p*-character. The dipole moment, 0.88 D, is somewhat higher than bicyclobutane, and the low frequency of the C=C stretch, 1556 cm^{-1}, also reflects the interaction of the alkene and bicyclobutane systems. A considerable chemistry of **6** has been revealed, which is outlined in Figure 3.18.

The half-life of benzvalene at 30°C is 48 h. The rearrangement to benzene has an enthalpy of activation of 108 kJ mol^{-1}, and the reaction enthalpy, as measured for the silver catalyzed rearrangement, is -282 kJ mol^{-1}. Clearly, benzvalene is thermodynamically much less stable than benzene,

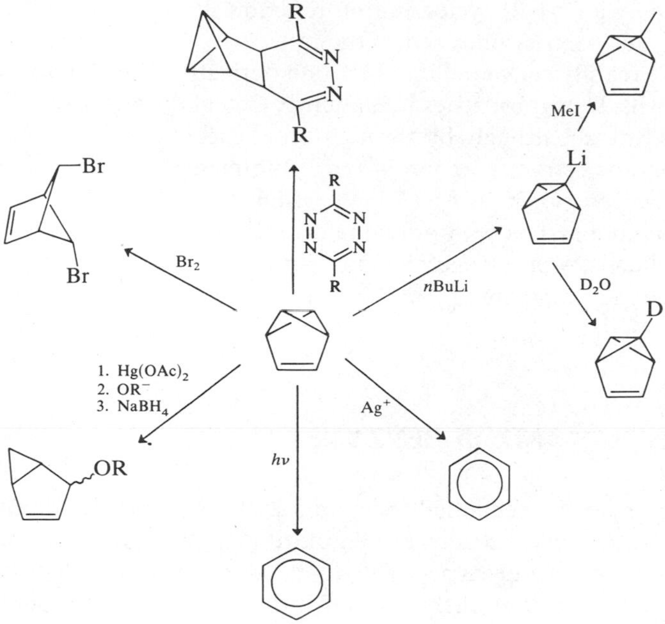

Figure 3.18

and this, combined with the low enthalpy of activation, cxplains its detonation. The reaction is Woodward–Hoffmann forbidden, but calculation and the experimental findings suggest it is a one-step concerted process. It cannot occur via Dewar benzene since the latter decomposes to benzene with chemiluminescence, whereas benzvalene does not.

Prismane is obtained in only low yield from the photoirradiation of **11**. The photoirradiation must be carried out at ambient temperatures or above since the populated S_1 state requires thermal activation for cleavage of the C—N bond. This process is inhibited at low temperatures: Intersystem S^1 to T^1 crossing occurs and 1,2-diazacyclooctatetraene results. Under strictly controlled conditions, photoirradiation of **11** leads to Dewar benzene (45%) and prismane (11%), together with some benzvalene. Little chemistry is known because of its inaccessibility. It is an explosive, colorless liquid, the ^{1}H NMR spectrum showing a single resonance at δ 2.28 and the ^{13}C NMR at δ 30.6 ($J_{^{13}C-H}$ = 179.7 Hz). It has little or no electronic absorption above 233 nm.

3,3′-Bicyclopropenyl (**7**) differs from the other isomers in that it is not a fused bicyclic compound, although it is clearly conceptually related to prismane by a (2 + 2) cycloaddition reaction. It has also, unlike the other isomers, not been synthesized, although derivatives have been prepared. These are readily converted into benzene derivatives in a highly exothermic process with Dewar benzenes intervening. Calculations suggest that the two π bonds interact strongly by through-bond mechanism. This is supported by the photoelectron spectrum of the 3,3′-dimethyl derivative in which the first ionization potential is 8.76 eV, which is considerably less than the corresponding system with only one double bond.

The chemistry of fulvene, the most important C_6H_6 hydrocarbon related to benzene and prepared from it by photoirradiation, will be described in Chapter 7.

3.4. *o*-, *p*-, AND *m*-BENZYNE

As was previously mentioned (Section 3.2.2), nucleophilic addition to benzene derivatives may occur by an elimination–addition mechanism involving dehydrobenzene (*o*-benzyne). This is the mechanism by which chlorobenzene is converted into phenol on an industrial scale. The intervention of *o*-benzyne **12** in this type of reaction was originally suggested by Wittig and was substantiated by the labeling studies of Roberts and co-workers, who showed that chlorobenzene labeled with ^{14}C at the carbon bearing the chlorine gave, on treatment with sodamide in liquid ammonia, a 1:1 mixture of aniline labeled at the carbon bearing the amino group and at the adjacent carbon (Fig. 3.19).

Figure 3.19

Much later, *o*-benzyne was identified by its IR spectrum in an argon matrix. A considerable chemistry has developed in which *o*-benzyne is used as a reactive intermediate in synthesis.

Two other forms of benzyne are theoretically possible by dehydrogenation at the 1,4 or 1,3 positions. Clearly, unlike *o*-benzyne, *p*-benzyne (**13**) and *m*-benzyne (**14**) cannot be formulated as simple triple bond derivatives. Both compounds can be envisaged as diradicals or as bicyclic structures with long 1,4 or 1,3 bonds. Alternatively, the systems may lose their benzene origins and adopt a new σ framework as valence tautomers of the "aromatic" species. Butalene (**15**) is the structure related to *p*-benzyne, and bicyclo[3.1.0]hexa-1,3,5-triene (**16**) is the structure related to *m*-benzyne (Fig. 3.20).

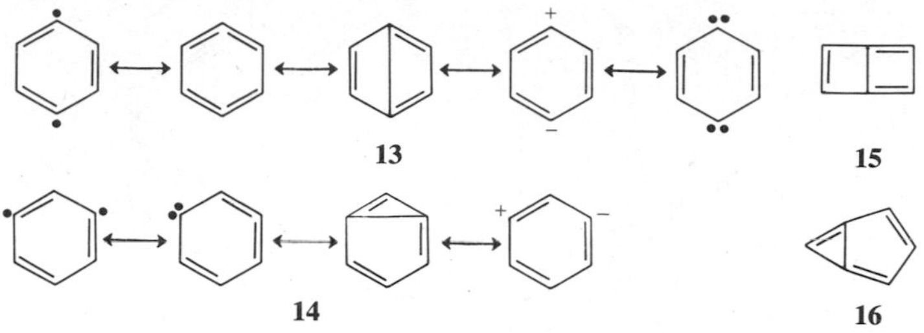

13 15

14 16

Figure 3.20

There have been a number of attempts to prepare *p*- and *m*-benzyne or their tautomeric equivalents. Initial studies paralleled those on *o*-benzyne, with benzenediazonium-4-carboxylate and benzenediazonium-3-carboxylate being decomposed photochemically and the products directly investigated by mass spectrometry and flash-absorption optical spectroscopy. In the case of the 4-carboxylate, a relatively long-lived species was observed at m/e 76, which was assigned to *p*-benzyne, whereas the 3-carboxylate showed only a short-lived species at m/e 76.

Chemical routes to both species or their tautomeric equivalents have been investigated. Treatment of 1-chlorobicyclo[2.2.0]hexa-2,5-diene (**10**) with lithium diethylamide in diethylamine gave diethylaniline, presumably via butalene (**15**). When the reaction was carried out in the presence of diphenylisobenzofuran (**17**), the adduct **18** was obtained (Fig. 3.21). The

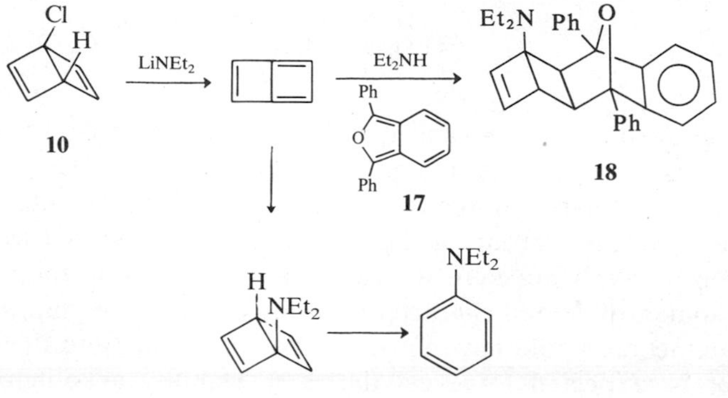

Cl H 10 $LiNEt_2$ Et_2NH Ph O Ph 17 Et_2N Ph O Ph 18 H NEt_2 NEt_2

Figure 3.21

degenerate rearrangement of 3-hexene-1,5-diyne goes through a transition state in which C-1,3,4, and 6 are chemically equivalent and which could be *p*-benzyne (Fig. 3.22).

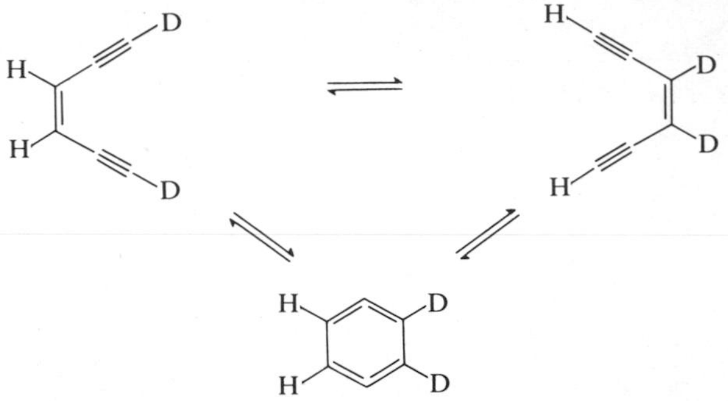

Figure 3.22

Some attempts have also been made to generate *m*-benzyne. Reaction of the dibromide **19**, readily available from benzvalene, with potassium *t*-butoxide and dimethylamine at −75°C gave mainly 6-dimethylaminofulvene (**20**) (Fig. 3.23). Examination of alternative pathways and deuteration

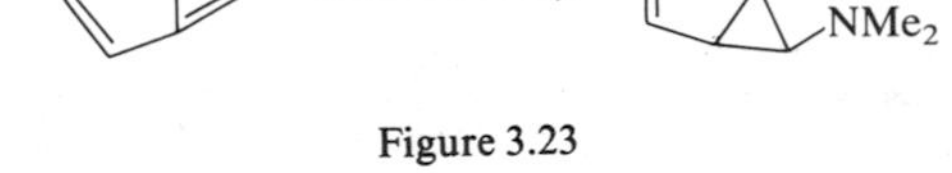

Figure 3.23

studies suggested that a mechanism via bicyclo[3.1.0]hexa-1,3,5-triene was the most satisfactory in explaining these results.

The bonding situation in the three benzynes is very different. *o*-Benzyne may have decreased overlap of the *p*-orbitals from a normal triple bond, and there is probably not equivalent electron density inside and outside the ring, but some triple bond character certainly occurs. In *p*- and *m*-benzyne orbital interaction would have to occur inside the ring, and these systems might not be expected to reveal the same stability as *o*-benzyne, and alternative nonbenzenoid arrangements could be expected to become relatively more favored (Fig. 3.24).

Figure 3.24

3.5. CYCLOPHANES

Benzene is most readily distorted by out-of-plane deformations. The serendipitous discovery of paracyclophane (**21**), soon followed by its intentional synthesis by Cram and Steinberg, has exposed an enormous wealth of related chemical compounds with a multitude of novel properties. x-Ray crystallographic analysis of **21** showed that the benzene rings are bent into boat forms, the bond angle between the linked out-of-plane atoms and the plane of the remaining four carbons being 12.6° (Fig. 3.26).

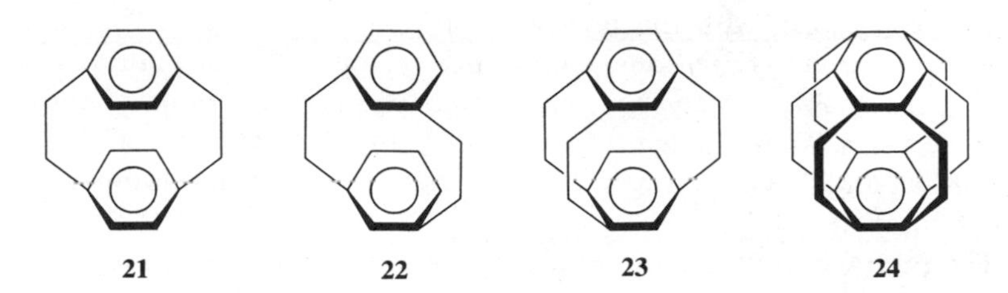

Figure 3.25

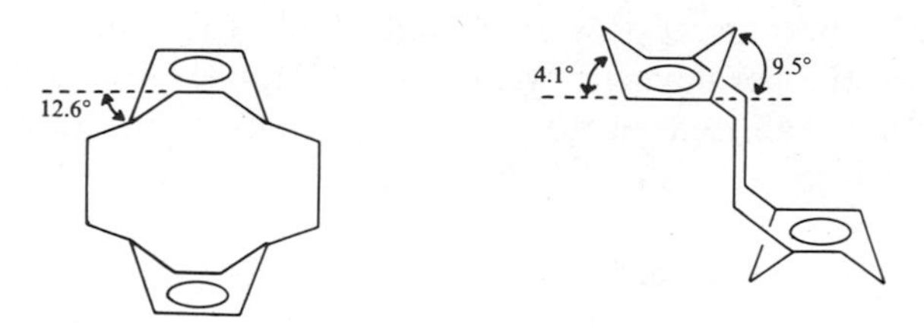

Figure 3.26

Paracyclophane has chemical properties dependent on the proximity of the two benzene rings, but these are nevertheless benzenoid properties and, despite the distortion, π delocalization still occurs. Metacyclophane (**22**) is less distorted than paracyclophane, the benzene rings adopting a staggered conformation, but the rings are still in the boat form, the atom between the bridges being further out of the four carbon plane than the carbon *para* to it.

More stringent strain can be introduced into the system by the addition of more bridges that limit the distortions available to the benzene rings. In [2.2.2](1,2,4)cyclophane* (**23**) the rings are distorted into twist boats and are inclined at 13.3° to one another. The ultimate cyclophane, superphane ([2_6](1,2,3,4,5,6)cyclophane (**24**) has recently been prepared by Boekelheide and co-workers, and now the benzene rings have been forced to return to the plane and this molecule, like benzene, has D_{6h} symmetry.

The compounds we have so far discussed have only two connected rings, but the possibility of forming stacks of such rings is obviously available. A variety of such systems has been prepared. The rings may be bonded to one another in either the *para* or *meta* sense, so that numerous isomers of any layered structure are possible. One of the first such compounds to be prepared was **25**, and x-ray crystallographic analysis showed that the outer benzene rings were "bent-in" boats and the central rings were twisted boats. The nonbonded distances between benzene rings are in the range 274–313 pm, which is slightly longer than the interbenzene distances in superphane (262.4 pm), with the nonlinked atoms further apart. These multilayered compounds are considerably more strained than paracyclophane itself, presumably largely because of the further distortions of the inner benzene rings. Thus, the triple layer paracyclophane **26** has a strain energy of 246 kJ mol^{-1}, almost twice the value of paracyclophane itself (131 kJ mol^{-1}).

The interaction of the benzene rings is clearly shown in the electronic spectra, which become progressively red shifted and less structured with the increasing number of benzene rings in the stack. The ^1H NMR spectra also show the effect of this interaction and the distortion of the inner rings, the inner ring protons appearing at high field. Thus, the protons on the inner rings of the hexalayered cyclophane **27** appear at δ 4.80.

Strain can be inflicted on the benzene rings by connection of the *para*- or *meta*-positions with simple alkyl chains to form the [*n*]cyclophanes (Fig. 3.28).

In the case of the [*n*]-cyclophanes, it might be expected that when *n* is small and the consequent strain large, valence tautomerism to the bicyclohexadiene might occur. In the case of the longer chain [*n*]paracyclophanes ($n = 8, 9, 10$, and 12), decreasing the chain length leads to an increase in

* The numbers in square brackets [ ] are the number of atoms in the bridges, and the numbers in curved brackets ( ) are the positions of attachment of the bridges.

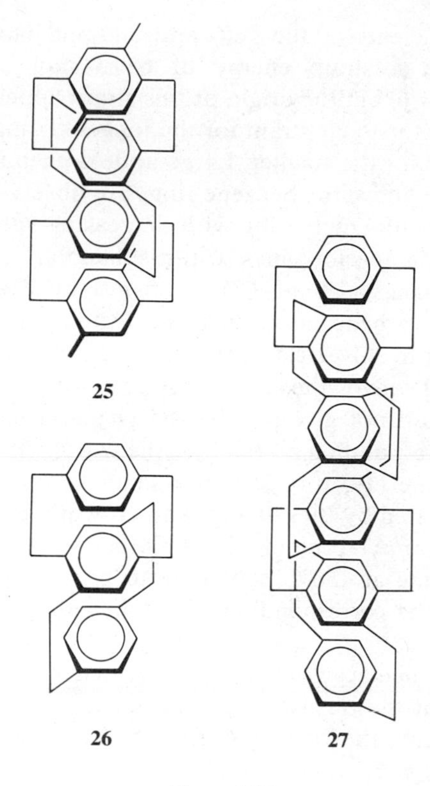

Figure 3.27

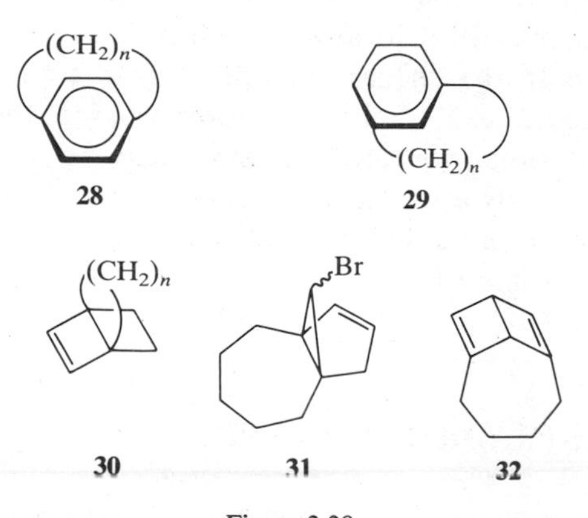

Figure 3.28

the extinction coefficients of the 260- and 230-nm bands. Strain energy calculations suggest a strain energy of 65 kJ mol^{-1} for $n = 10$ and 120 kJ mol^{-1} for $n = 6$, but the origin of this strain appears to be different, being largely bridge torsional strain for the longer chains and deformation of the benzene rings for the smaller. Large upfield chemical shifts of chain protons forced to lie above the benzene rings are observed in the ^1H NMR spectra, the upfield shifts increasing with decreasing value of n.

The smallest [n]paracyclophanes so far isolated are the [6] (**28**, $n = 6$) and [7]paracyclophanes (**28**, $n = 7$), prepared by Maitland Jones and co-workers using a carbene insertion route. Attempts to prepare smaller [n]paracyclophanes by silver catalyzed valence isomerism of 1,1′-bridged 1,1′-bicyclopropenyl gave the corresponding bicyclohexadienes (**30**). Pyrolysis of these did not give the desired [n]paracyclophanes, whereas pyrolysis of the corresponding bicyclohexadiene of [6]paracyclophane did.

[5]Metacyclophane (**29**, $n = 5$) is the smallest known member of the series and was prepared by Bickelhaupt and co-workers by base rearrangement of the cyclopropyl bromide **31**. [5]Metacyclophane rearranges to benzocycloheptatriene at 150°C. Attempts to prepare [4]metacyclophane by this route gave the corresponding bicyclohexadiene **32**, itself a highly strained compound.

The bridge in [5]metacyclophane is indicated by NMR studies to be fixed on one side of the phenyl ring, whereas that in [6]metacyclophane flips above and below the ring at 77°C. The benzene rings in all of the compounds with $n \leqslant 8$ are severely strained and undergo Diels–Alder reactions fairly readily with strongly electron deficient dienophiles to give bicyclo[2.2.2]octatrienes. Other reactions reveal the interaction of the benzene ring and the bridge leading to transannular products and bridge rearrangement and cleavage. The physical properties of the various cyclophanes are nevertheless consistent with the view that the benzene rings, however distorted, are still delocalized, diatropic and show no tendency to bond localization. In the [n]paracyclophanes, it appears that when the strain becomes too acute, 1,4-bond formation provides the means of escape from the strained situation, and this may also be true for the smaller [n]metacyclophanes. No multilayered or multibridged cyclophane yet appears to have been prepared in which a benzene ring has had to resort to such valence tautomerism.

3.6. OTHER STRAINED BENZENES

Benzene is resistant to bond angle and bond length deformation, and the introduction of such strain might be expected to raise the ground-state energy of the delocalized system and offer the possibility that alternative arrangements of electrons and nuclei may become of lower energy. In the cyclophanes we have seen that such effects do occur and systems with additional σ bonds become preferred to π systems with diminished delocalization. In-plane distortions should more easily provide such opportunities. Two main methods of introducing in-plane strain can be envisaged: (1) the substitution of large, bulky groups on the benzene ring, which will sterically interact forcing changes in bond length and possibly bond angle, and (2) *o*-annelation by small rings introducing bond angle distortion and possible changes in bond length.

van Tamelen and Pappas realized the synthetic possibilities of steric substitution and exploited it in the first preparation of a derivative of Dewar benzene. Photoirradiation of 1,2,4-tri-*t*-butylbenzene (**33**) gave 1,2,5-tri-*t*-butylbicyclo[2.2.0]hexa-2,5-diene (**34**), and the reaction was thermally reversible at 200°C. The ^{1}H NMR spectrum of **34** showed the two olefinic resonances at δ 6.13 and 6.05, the latter being coupled to the methine proton signal at δ 3.20, and there were signals for two different types of *t*-butyl groups. The 1,2- (**35**) and 1,2,4,5-tetra-*t*-butylbenzenes (**36**) have been prepared. Their strain energies relative to the *para*- or *meta*-isomers are approximately 92 kJ mol^{-1} for both molecules. Apparently, the introduction of the second pair of *t*-butyl groups has not greatly increased the strain, possibly because the bond deformations made to accommodate the first pair serve to accommodate the second. The tetra-*t*-butyl isomer **36** remains as a planar system, and calculations indicate that the twisted structure has

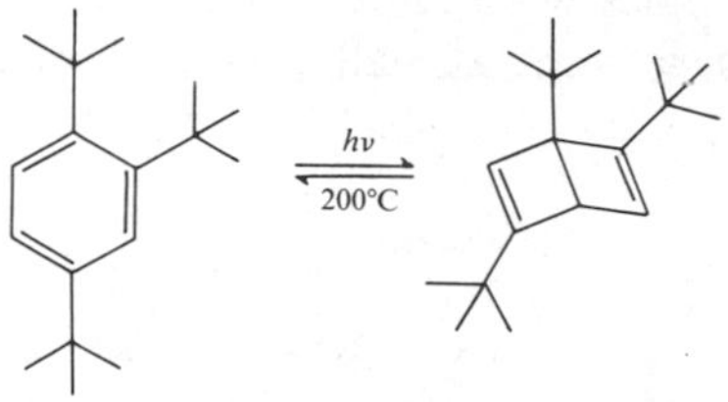

hν
200°C

33 **34**

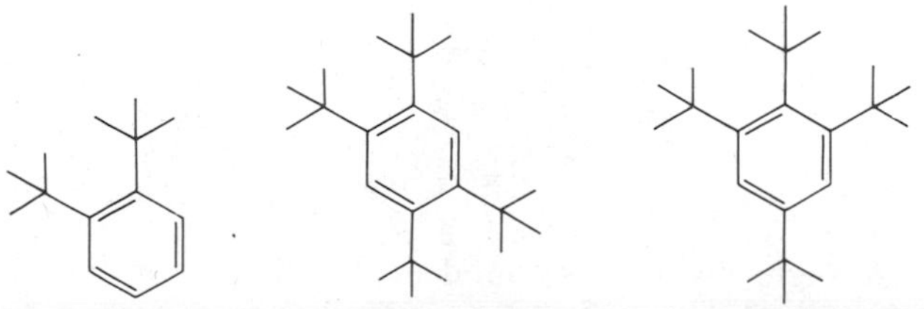

35 **36** **37**

Figure 3.29

a higher energy. The preliminary results of an x-ray crystallographic determination gave the structure shown in Figure 3.30. The benzene ring is flattened and the bond between the *t*-butyl groups has lengthened, moving the groups apart. However, the data were accumulated at room temperature and the *R* factor is 9.4%. 1,2,3,5-Tetra-*t*-butylbenzene (**37**) shows a large bathochromic shift in the electronic spectrum compared to other tetrasubstituted benzenes, but its ring geometry is not known.

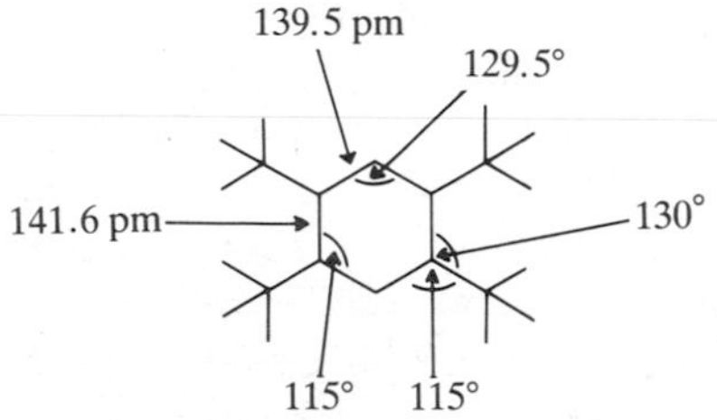

Figure 3.30

Annelation with small rings causes considerable distortion to the bond angles and bond lengths of the benzene nucleus. Several cyclopropabenzenes are now known following the original synthesis of a 7,7-disubstituted derivative by Anet and Anet and the subsequent preparation of the parent system **38** by Vogel and co-workers. An x-ray crystallographic analysis of the diester **39** showed that the bond common to both rings was short (133.3 ppm) and that the *ortho*-bond angles were more acute (109°) (Fig. 3.31). The system is not quite planar, there being a small angle between the planes of the two rings.

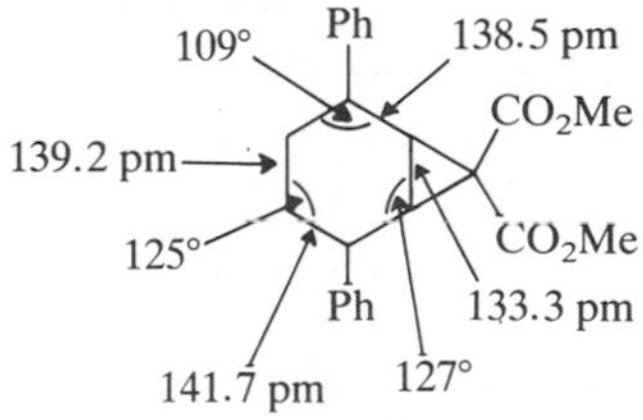

Figure 3.31

The bond angle distortions found in this derivative had already been suggested by Günther and co-workers from NMR studies on the parent compound and derivatives. The one bond $^{13}C—^{13}C$ coupling constant between C-1,7 in **38** is small (20.8 Hz), whereas that between C-1,6 is large (87.1 Hz) when compared to other annelated benzenes. Based on the Walsh model, the hybridization changes indicated in Figure 3.32 would predict both the $^{13}C-^{13}C$ couplings and the ^{1}J ($^{13}C-^{1}H$) coupling constant. An increase in the *p*-character of the ring bonded atoms and a decrease in the *p*-character of the *ortho* C—H bonds would result in a decrease of the C—C—C and an increase in the C—C—H bond angles, in accord with

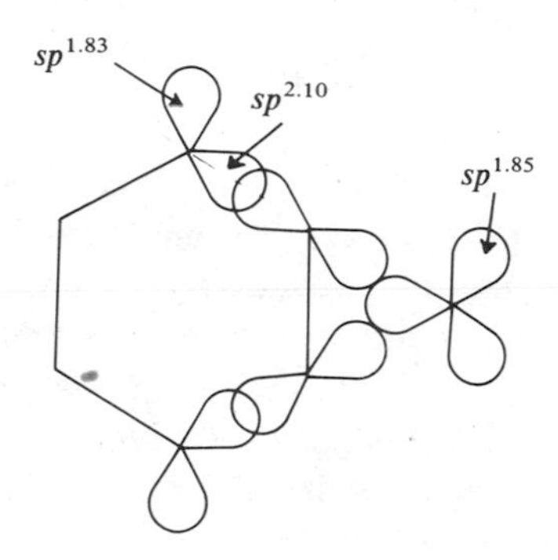

Figure 3.32

the x-ray study. These results also fit the Finnegan-Streitwieser model explaining the acidity of benzenoid protons *ortho* to small fused rings.

The fusion to benzene of more than one small ring would further increase the strain and might provide evidence for the elusive Mills-Nixon effect. This was a suggestion made in 1933 to account for some anomalous results between tetralins and indans in which it was supposed that the two Kekulé structures would lose their equivalence and one be favored.* Although the original results found another explanation, the concept has remained, finding some support through theoretical calculations. Doubly annelating benzene in a 1,2:4,5 fashion produces a compound with identical Kekulé structures, whereas annelating in a 1,2:3,4 fashion does not (Fig. 3.33). Comparison of isomers annelated in this way might provide evidence for the Mills-Nixon effect.

A number of highly strained, small ring annelated derivatives of benzene has been prepared, but no evidence supporting the Mills-Nixon effect has been adduced. The strain present in these compounds can be demonstrated by the ^{13}C NMR spectra; thus the chemical shift of the nonannelated

* The concept of the structure of benzene held by Mills and Nixon was closer to the "mechanical motion" of Kekulé rather than modern theory, and they treat the two Kekulé structures as tautomeric forms.

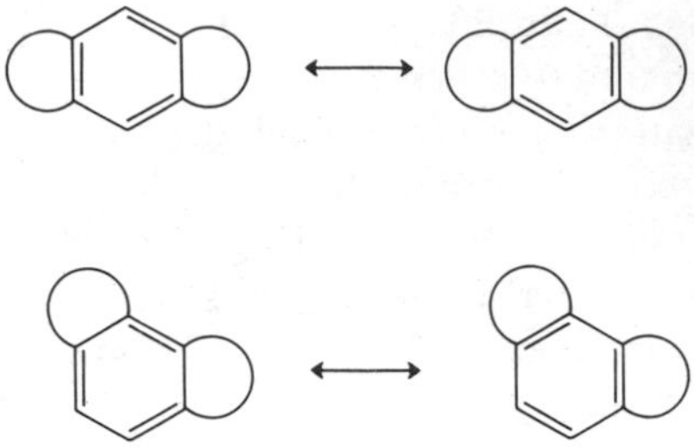

Figure 3.33

benzene atoms in **40** is at δ 110.0, which is a higher field than even the parent benzocyclopropene, δ 114.7. Presuming the model in Figure 3.32, a further increase in the *s*-character of the C—H bond can be expected with the introduction of the cyclobutane ring with its preference for *p*-character in the ring bonds. The positional isomer, **41**, has a ^{13}C NMR spectrum, which is fully in accord with predictions derived from the spectra of **40** and **43**, and the Mills–Nixon effect is absent. The electronic spectrum of **40** shows a red shift in comparison to that of **42**, whereas that of **41** differs little from that of **43**. The three-membered rings in all of these systems are readily cleaved by electrophiles, and a study of the substitution patterns suggests that reaction occurs via the σ orbitals of the cyclopropene ring rather than the π system.

Tricyclobutabenzene **44** is a thermally stable compound, and its position in the $C_6(CH_2)_6$ manifold is of considerable current interest. Dicyclobutabenzocyclopropene (**45**) has very recently been prepared, and this is the currently known benzene system most highly strained by annelation.

Evidence for the intervention of both cyclobuta- (**46**) and cyclopropa-*o*-benzynes (**47**) has been obtained, and these compounds must have very

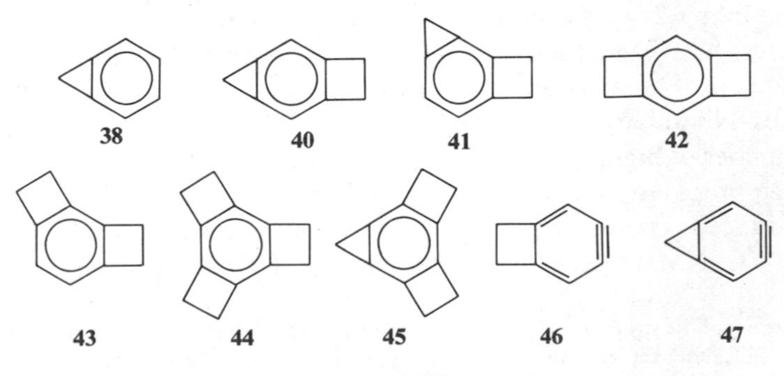

Figure 3.34

large strain energies. Their generation and the preparation of **45** may encourage further examination of the possibility of isolating dicyclopropabenzenes.

FURTHER READING

The chemistry of benzene and its derivatives is enormous. For general accounts see the following: S. Coffey and M. F. Ansell (Eds.), *Rodd's Chemistry of Carbon Compounds, Aromatic Compounds*, Vol. 3A and Supplement 3A, ed. 2, 1971, 1983, Elsevier, New York; J. Frazer Stoddart (Ed.), *Comprehensive Organic Chemistry*, Vol. 1, Sir Derek Barton and W. D. Ollis (series Eds.), Pergamon Press, Elmsford, N.Y., 1979; C. Grundmann (Ed.), *Houben-Weyl Methoden der Organischen Chemie*, Band 5/2b, ed. 4, Georg Theile Verlag, Stuttgart, 1981.

For an account of electrophilic substitution in deactivated systems, see J. H. Ridd, *Chem. Soc. Spec. Publ.*, 1967, **21**, 149.

For the $S_{RN}1$ reaction, see J. F. Bunnett, *Accounts Chem. Res.*, 1978, **11**, 413; *Aromatic Substitution by the $S_{RN}1$ Mechanism*, R. A. Rossi and R. H. de Rossi (Eds.), ACS, Washington D.C., 1983.

For the argument concerning diffusion control and regiospecificity, see G. A. Olah, *Accounts Chem. Res.*, 1971, **4**, 240, and J. H. Ridd, *ibid.*, 1971, **4**, 248.

For the photochemistry of benzene, see D. Bryce-Smith and A. Gilbert, *Tetrahedron*, 1976, **32**, 1309; *ibid.*, 1977, **33**, 2459.

For the chemistry of benzvalene, see M. Christl, *Angew. Chem. Int. Ed.*, 1981, **20**, 529.

For valence bond isomers stabilized by the CF_3 group, see Y. Kobayashi and I. Kumadaki, *Accounts Chem. Res.*, 1981, **14**, 76.

For 1,4-dehydroaromatics, see R. G. Bergman, *Accounts Chem. Res.*, 1973, **6**, 25; for recent *ab initio* MO studies, see L. Radom, R. H. Nobes, D. J. Underwood, and W-K. Li, *Pure Appl. Chem.*, 1986, **58**, 75.

For strained benzenes, see D. J. Cram and J. Cram, *Accounts Chem. Res.*, 1971, **4**, 204; A. Greenberg and J. F. Liebman, *Strained Organic Molecules*, Academic Press, New York, 1978; V. Boekelheide, *Accounts Chem. Res.*, 1980, **13**, 63; R. P. Thummel, *ibid.*, 1980, **13**, 70; P. M. Kheen and S. M. Rosenfeld (Eds.), *The Cyclophanes*, Vols. 1 and 2, Academic Press, New York, 1983; W. E. Billups, B. E. Arney, and L-J. Lin, *J. Org. Chem.*, 1984, **49**, 3436; for the recent synthesis of [5]paracyclophane, see L. W. Jenneskens *et al.*, *J. Am. Chem. Soc.*, 1985, **107**, 3716.

For a recent paper and references on the Mills–Nixon effect, see R. H. Mitchell, P. D. Slowey, T. Kamada, R. V. Williams, and P. J. Garratt, *J. Am. Chem. Soc.*, 1984, **106**, 2431.

4

THE ANNULENES

4.1. INTRODUCTION

Cyclobutadiene, benzene, and cyclooctatetraene are members of an homologous series of fully conjugated monocyclic hydrocarbons of general formula $(C_2H_2)_n$. Such systems have been termed *annulenes* and a prefixed number, [*n*], has been added to indicate the ring size. In this nomenclature, benzene becomes [6]annulene and cyclooctatetraene [8]annulene.

The annulenes have been the subject of a number of theoretical discussions, and one of the earliest successes of the MO theory was to explain the change in chemical properties on going from benzene to cyclooctatetraene (see Chapter 2). Hückel's Rule predicts that those monocyclic hydrocarbons with $(4n + 2)$ π electrons will have properties similar to those of benzene, whereas those with $4n$ π electrons will exhibit properties similar to cyclooctatetraene. However, this rule does not consider geometric factors, and Mislow pointed out that steric interactions would be particularly important in planar medium ring compounds in which the bond angle deformations have been minimized. Thus, in the planar, di-*trans* form of [10]annulene (**1**) acute interaction occurs between the inner hydrogens on the *trans* double bonds. Mislow predicted that [30]annulene (**3**) would be the first annulene after benzene that could attain a sufficiently planar conformation for the π electrons to be delocalized. Subsequently, other authors suggested that Mislow had overestimated the steric requirements, and they proposed that the internal hydrogen interactions in [18]annulene

(**2**) might be sufficiently small for this system to exhibit aromatic properties (Fig. 4.1).

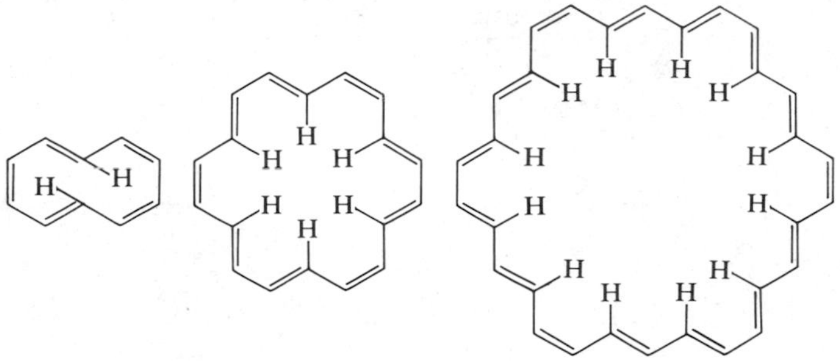

1 **2** **3**

Figure 4.1

Beside the question of steric interaction, a later theoretical treatment by Longuet-Higgins and Salem indicated that as the ring size increased, the difference in energy between the delocalized and localized bond configuration models decreases, until in the region of rings containing 30 carbon atoms the system with localized bonds becomes energetically favored. Thus, in the larger rings the difference between the $4n$ and $(4n + 2)$ π-electron annulenes disappears, and the larger annulenes would be expected to have properties similar to acyclic polyenes rather than to those of the smaller annulenes. Dewar and Gleicher later predicted that configurations with localized bonds would be favored for the annulenes with more than 22 carbon atoms (see Chapter 2).

4.2. ANNULENES

In order to provide an experimental test for these predictions, Sondheimer and collaborators began, in 1956, to investigate the preparation of a number of annulenes. The general procedure of the Sondheimer group is illustrated by their synthesis of [18]annulene (Fig. 4.2). The first step in the synthesis involves the oxidative coupling of α,ω-diacetylenes, which, besides linear products, gives cyclic "monomers," "dimers," "trimers," and higher cyclic "oligomers."* The total amount of cyclic products and relative proportions

* All of these cyclic compounds are actually *not* "oligomers," having for each acetylene coupling two hydrogens *less* than the uncoupled starting materials.

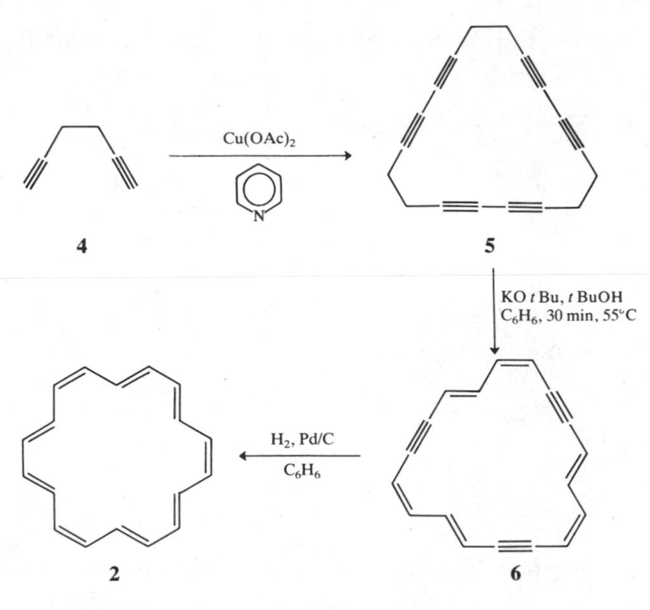

Figure 4.2

of the rings of various sizes depend upon the nature of the monomer and the reaction conditions used. The coupling reaction may be carried out either under Glaser conditions (oxygen in the presence of copper(I) chloride and ammonium chloride) or Eglinton conditions [copper(II) acetate in pyridine].

With the correctly chosen α,ω-diynes, the resulting cyclic products on treatment with strong base (usually potassium *t*-butoxide in *t*-butanol) are rearranged to the fully conjugated dehydroannulenes. In the case of the rearrangement of the $C_{18}H_{12}$ "trimer" **5** (Fig. 4.2), three dehydro[18]annulenes were obtained; two of these were isomeric with the trimer and one was an oxidation product containing two fewer hydrogens. Such dehydrogenations have been found to occur frequently during this type of rearrangement. In Figure 4.2 only the major monocyclic product, 1,7,13-tridehydro[18]annulene (**6**), is shown. Partial hydrogenation of the dehydro[18]annulene over palladium on charcoal gave crystalline [18]annulene as dark reddish-brown needles, having a main electronic maximum at 369 nm (ε 303,000).

[18]Annulene is a $(4n + 2)$ π-electron system and should, by Hückel's Rule, be aromatic. The ^1H NMR spectrum (Fig. 4.3) in perdeuterotetrahydrofuran at 20°C shows two broad bands at δ 8.94 and -2.0, which integrate for 12 and 6 protons, respectively. This is the spectrum expected for an

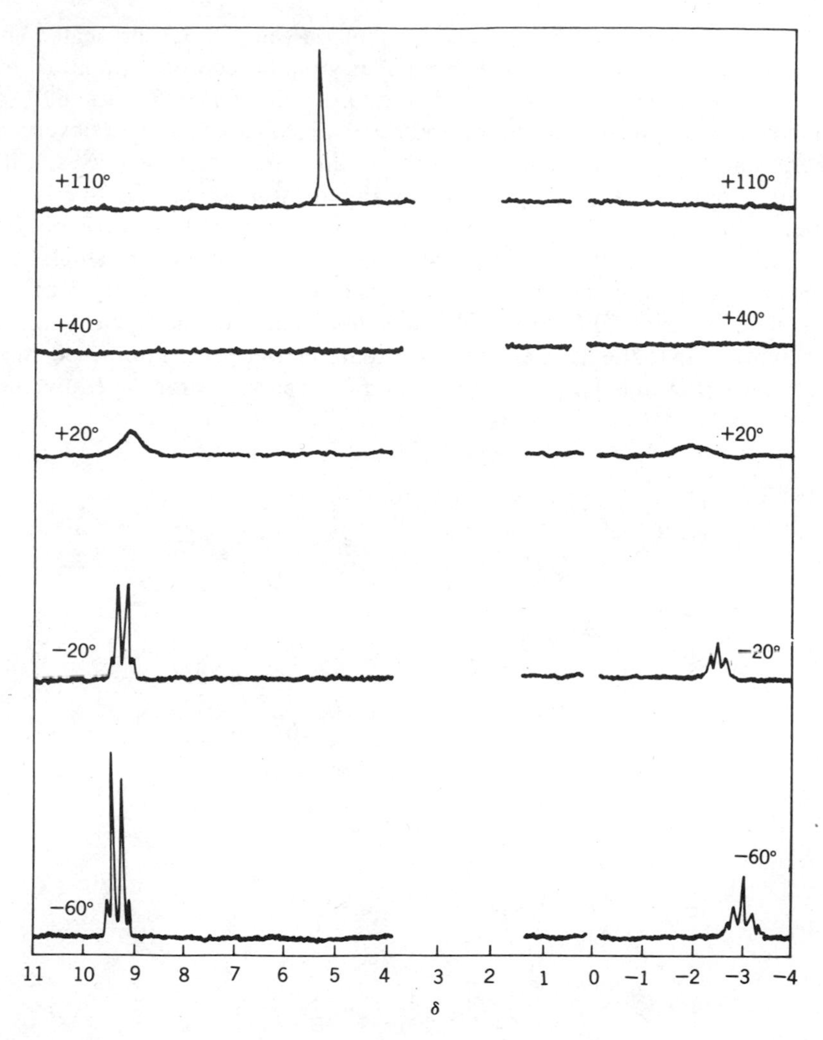

Figure 4.3. ^{1}H NMR spectra of [18]annulene taken at 60 MHz at various temperatures.

aromatic, delocalized π-electron system with 12 outer and 6 inner protons, which is diamagnetically anisotropic. The spectrum was unexpectedly poorly resolved, the broad bands showing no fine structure. This feature was explained when it was realized that the spectrum is temperature dependent. At −60°C, the spectrum (Fig. 4.3) exhibits multiplets at δ 9.28 and −2.99, which is consistent with an $A_{12}X_6$ system. At +110°C, however, only a single resonance was observed at 5.45, indicating that at this temperature the inner and outer protons have become equivalent on the NMR time scale. The molecule is thought to consist of three nearly planar conformational forms that are interconverting (Fig. 4.4). The rate of interconversion at −60°C is sufficiently slow for discrete inner and outer proton signals to be observed, whereas at higher temperatures the rate of interconversion increases, until at 110°C only an average position of the proton is seen in the NMR experiment (Fig. 4.3). Such an interpretation is supported by NMR double irradiation experiments; thus, whereas at −60°C irradiation of the high field band causes the low field band to collapse to a singlet, irradiation of the high field band at 20°C causes the low field band to disappear. This is apparently due to the transfer of the irradiated inner proton to the position of an outer proton before spin relaxation has occurred, which is equivalent to irradiating the outer protons.

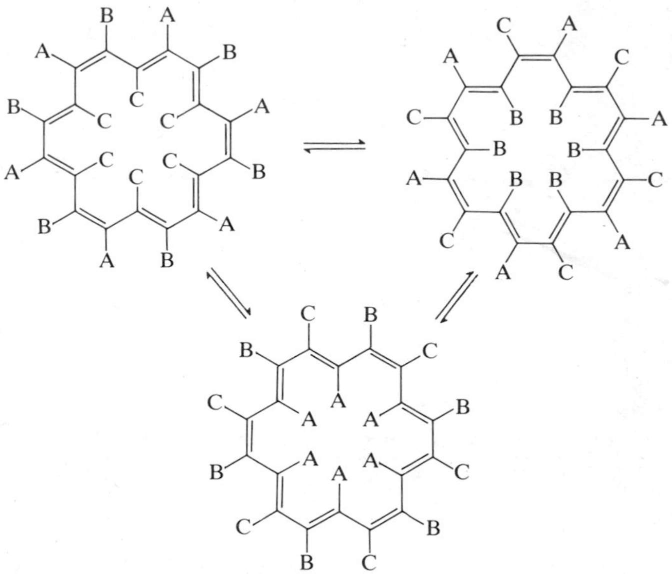

Figure 4.4

The interchange of the internal and external protons in [18]annulene, which requires the equivalent of rotation around a C=C, is a process of low activation energy (about 60 kJ mol^{-1}). The process probably involves bond rotation and bond shift, similar to that described for cyclooctatetraene in Chapter 2, but such a mechanism involves subtle arguments regarding the nature of delocalization. This mobility of [18]annulene may be reflected in the low ratio of ring current to calculated maximum ring current (0.58) (see Chapter 2, Section 2.6).

It is of interest to point out that if the rotational barrier in [18]annulene had been only a little lower, a single resonance signal would have been observed at room temperature at a position (δ 5.45) more typical of a cyclooctatetraene type olefin than benzene. Such a misleading observation was made with [14]annulene, which was at first thought to be a nonaromatic compound on the basis of a single resonance in the ^{1}H NMR spectrum.

An x-ray crystallographic analysis of [18]annulene was interpreted to show it was a nearly planar molecular that did not have alternate single and double bonds. Two types of bonds were discerned, however, the "cisoid" type (142 pm) and the "transoid" type (138 pm), but these are not alternate. On this basis, the structure of [18]annulene is best represented by Figure 4.5, but considerable doubt has more recently been put on this interpretation, since theoretical calculations suggested that [18]annulene should have a bond alternated structure. Baumann and Oth have now, however, examined the low temperature electronic spectrum of [18]annulene and concluded that this could only be accounted for on the basis of a delocalized structure. The 450 nm band was attributed to a B_{1u} and the 790 nm band to a B_{2u} transition.

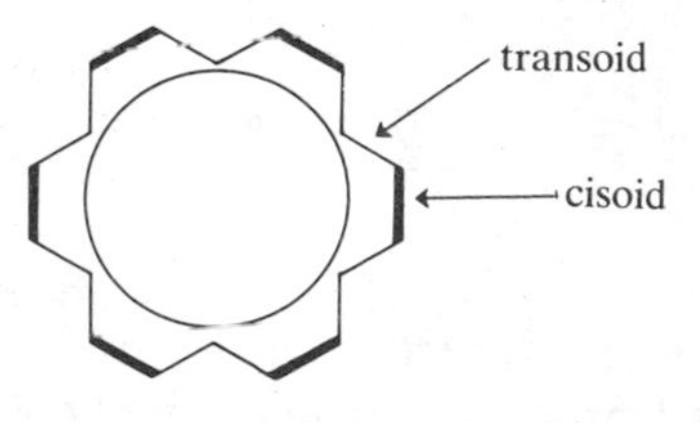

Figure 4.5

As regards chemical behavior, [18]annulene is not a particularly aromatic compound in the classical sense. Under special conditions it can be nitrated or acylated, but these reactions may not have the mechanism of electrophilic substitutions. The monosubstituted products have been well characterized and are of interest in that the conformation with the substituent inside the ring is of higher energy. This is clearly shown by the temperature dependence of the ^{1}H NMR spectra of these compounds. The ^{1}H NMR spectrum of nitro[18]annulene (**7**) at −70°C (Fig. 4.6) shows a low field band at δ 10.5-9.3 (11H) and a high field band at δ −3.0 to −4.0 (6H). On warming to +30°C, the high field band and *part* of the low field band coalesce; at +100°C a new band at about δ 3.3 (12H) has appeared, and a low field band remains at δ 8.8 to 8.2 (5H).

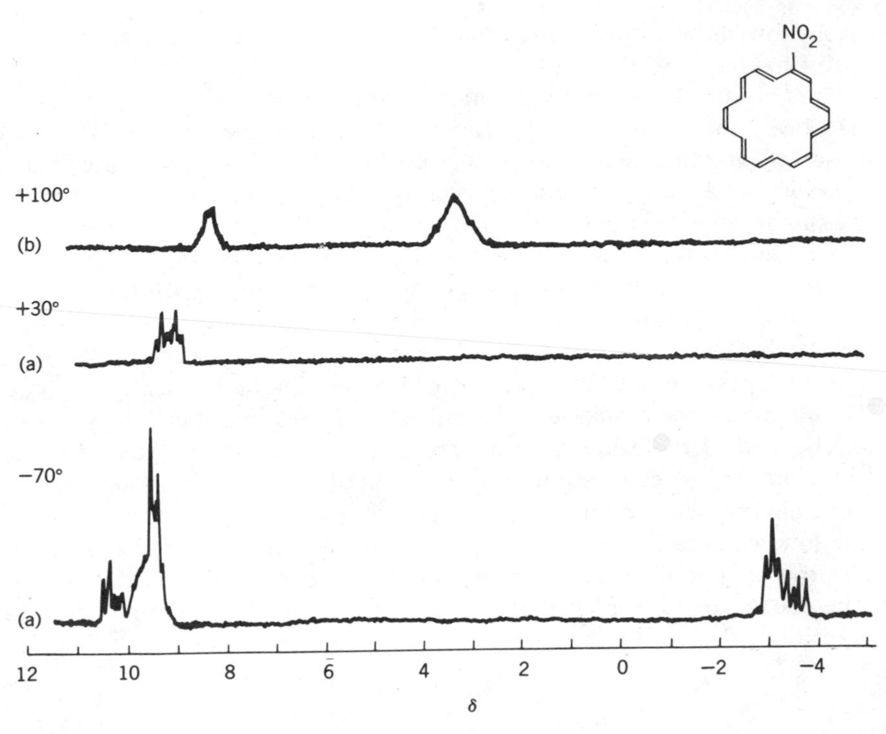

Figure 4.6 ¹H NMR spectra of nitro[18]annulene taken at various temperatures: (*a*) in perdeuterotetrahydrofuran and (*b*) in perdeuterotoluene.

It appears that due to the higher energy of the conformation with an inner nitro group, the 5 protons associated with the nitro group *retain their distinct external character*, and only 12 of the protons show a temperature-dependent spectrum. The two interconverting conformations, together with the third noncontributing conformation, are shown in Figure 4.7. In this figure, it is the 5 protons of type A that are external and temperature independent.

More recently, Neuberger, Schröder, and Oth have prepared 1,2-difluoro (**8a**) and 1-chloro-2-fluoro[18]annulene (**8b**) by the route outlined in Figure 4.8. The two substituents have a cisoid relationship to each other and occupy outer positions on the ring. Consequently, the ¹H NMR spectrum is temperature invariant, with the conformation in which both groups are external being of lower energy than the other two.

[18]Annulene decomposes thermally at 130°C in dimethylformamide to give **11** and benzene, probably via the pathway outlined in Figure 4.9. The initial step is a transannular ring closure to a valence tautomer such as **9**

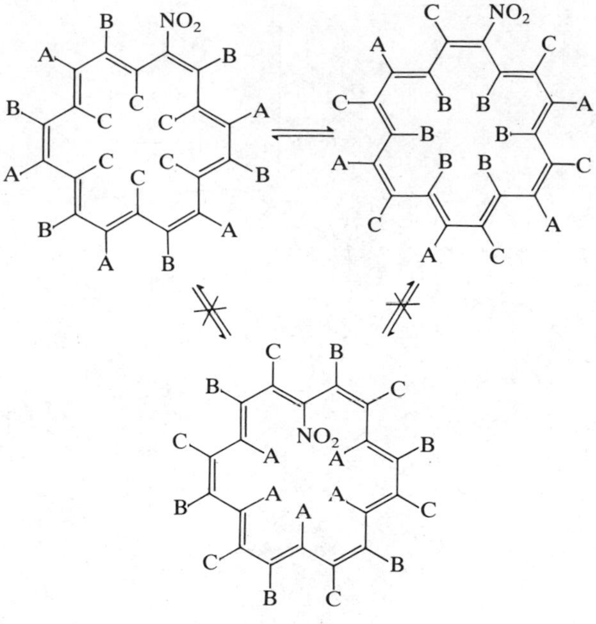

B NO_2
A B
C
B C C A
C C B
A C
B A
A B

C NO_2
A C
C B B A
B B
A B B C
C A
A C

C B
B C
NO_2
C A A B
A A
B C
A
C B
B C

Figure 4.7

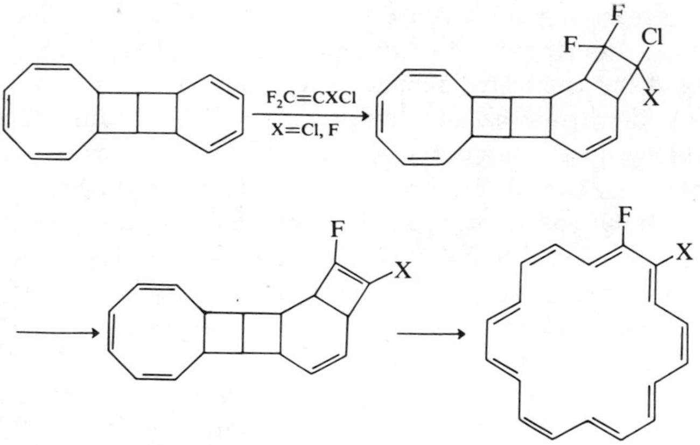

F
F Cl
X
$F_2C{=}CXCl$
$X{=}Cl, F$

F
X

F
X

8a X = F
8b X = Cl

Figure 4.8

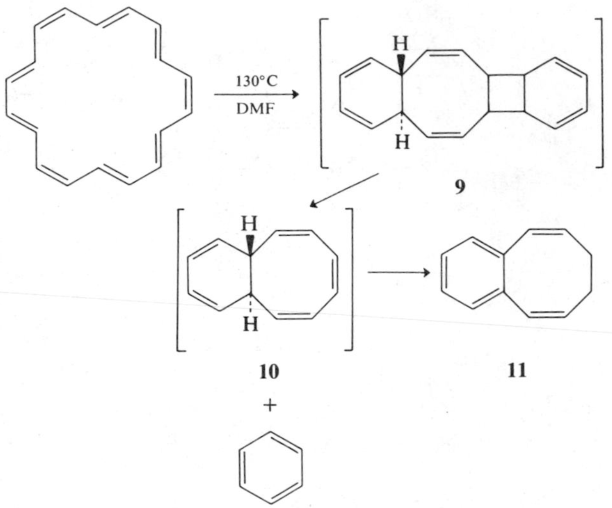

Figure 4.9

followed by ring cleavage to benzene and **10**, the latter compound then undergoing two 1,5-hydrogen shifts to give **11**. Amusingly, this sequence represents the conversion of one C_6H_6 isomer, 1,5-hexadiyne, into another, benzene.

[18]Annulene reacts readily with bromine and maleic anhydride, reactions not expected for an aromatic system of the benzene type. It is, however, clear that the reaction pathways available to [18]annulene via rearrangement and valence tautomerism are considerably greater than those available to benzene, and the chemistry of macrocyclic aromatic annulenes might be expected to differ from that of benzene in the same way that the chemistry of the macrocyclic alkenes differs from that of cyclohexene.

[16]Annulene (**12**) was originally synthesized by Sondheimer and Gaoni by a route similar to that used for [18]annulene, but a superior synthesis was subsequently described by Schröder and Oth starting from the dimer (**13**) of cyclooctatetraene. Both of these synthetic schemes are shown in Figure 4.10. The same configurational isomer of [16]annulene (**12a**), a brown, crystalline compound with a main UV maximum at 284 nm, was obtained by either route. This is not unexpected, since the barrier to interconversion of internal and external protons in [16]annulene is low (about 38 kJ mol^{-1}), and the adoption of the thermodynamically most stable configuration from any initial configuration will thus occur.

[16]Annulene is a $4n$ π-electron system and thus should not, according to Hückel's Rule, be aromatic. The ^1H NMR spectrum taken at 35°C (Fig.

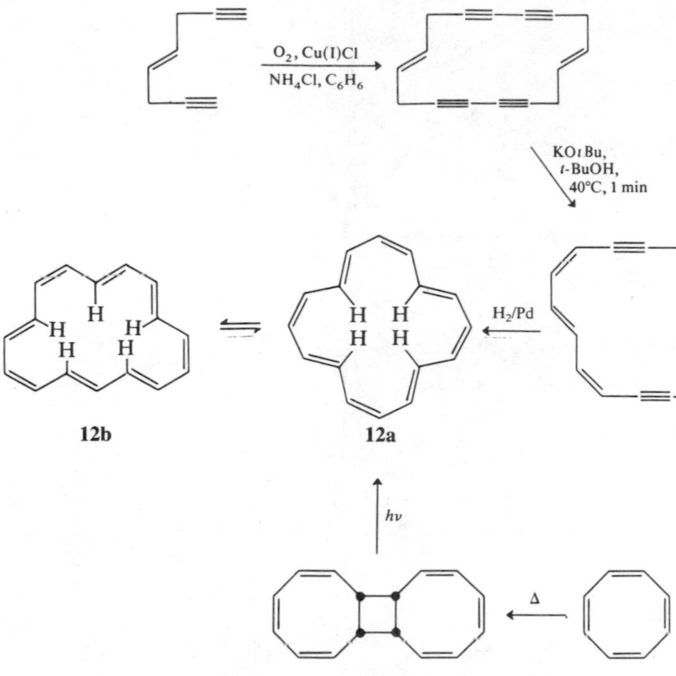

Figure 4.10

4.11) shows a single peak at δ 6.73 at a reasonable position for a cyclic olefin, although at a lower field than the averaging single peak of [18]annulene. As soon as the temperature dependence of the NMR spectrum of [18]annulene was observed, it became of urgent interest to investigate the thermal behavior of [16]annulene. The low temperature spectrum of [16]annulene, taken by Schröder and Oth, showed that at −120°C the spectrum consisted of two bands at δ 10.43 and 5.40, which integrate for 4 and 12 protons, respectively. Thus, although this behavior is superficially similar to that of [18] annulene, the integration shows that in [16]annulene the *inner* protons are at *low* field and the *outer* protons are at *high* field.

The complete reversal of the position of the protons in the $4n$ compared to the $(4n + 2)$ annulenes has been explained as being due to the superposition of a *paramagnetic* ring current upon the diamagnetic moment of the $4n$ π-electron systems. This paramagnetic current arises from the mixing of electronically excited states with the ground state of $4n$ π-electron system. The interaction is magnetic-dipole allowed, and the resulting paramagnetic ring current should be large because the difference in energy between the

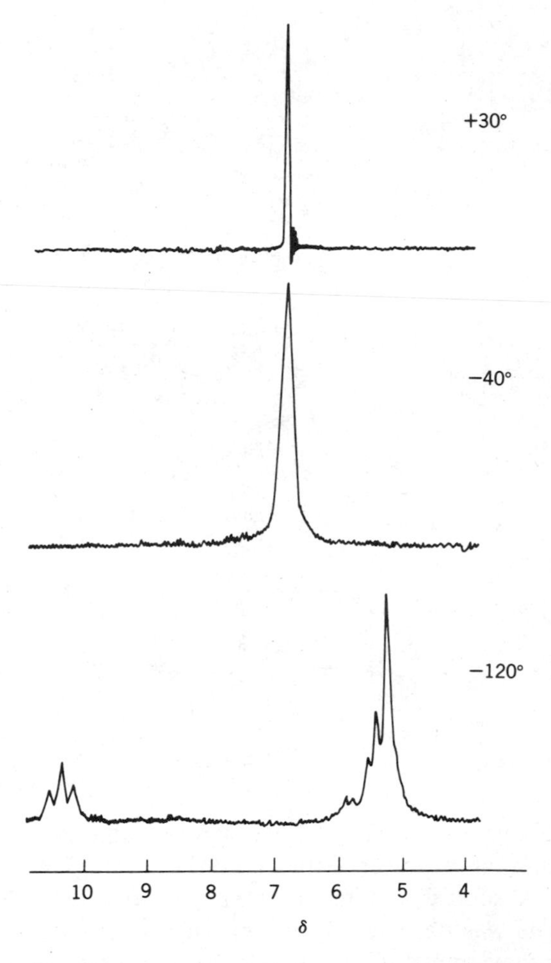

Figure 4.11 ^1H NMR spectra of [16]annulene in 50% CS_2, 50% CD_2Cl_2 taken at 60 MHz at various temperatures.

highest occupied and lowest unoccupied MO is small. In the $(4n + 2)$ π-electron annulenes, the paramagnetic ring current effect is small since the states concerned have considerably different energies and the transition between them is magnetic-dipole forbidden. The occurrence of a paramagnetic ring current was not observed in cyclooctatetraene, presumably because the molecule is far from planar, but the high field position of the ring proton in the ^1H NMR spectrum of tri-*t*-butylcyclobutadiene was attributed to this cause. As we shall see, the higher $4n$ annulenes up to and including [24]annulene exhibit this effect. I suggested in the earlier version of this book that it would be appropriate to classify conjugated systems into three types, depending on the nature of the shielding effect observed in the ^{1}H NMR spectrum. *Paratropic* systems will exhibit a paramagnetic ring current shielding in the ^{1}H NMR spectrum, *diatropic* systems a diamagnetic ring current shielding, and *atropic* systems no shielding effects arising from ring currents. This is a phenomenal description and removes it from the connotations evoked by the expression aromaticity. However, it will usually be the case that paratropic systems are antiaromatic, diatropic systems aromatic, and atropic systems nonaromatic (see Chapter 2, Sections 2.3 and 2.6). These terms, particularly the first two, have been widely adopted by workers in the field and will be used throughout the subsequent chapters of this book.

An x-ray crystallographic analysis of [16]annulene showed that the molecule has alternate double and single bonds and is significantly nonplanar, although the deviation from planarity is not large. The four inner protons are arranged alternately on opposite sides of the mean molecular plane, thus minimizing the hydrogen-hydrogen interactions.

In solution, [16]annulene has been shown to exist mainly as isomer **12a**, but this is in equilibrium with a small amount of the configurational isomer **12b**. The interconversion of **12a**$\leftrightharpoons$**12b** occurs by a combination of *cis-trans* isomerism similar to that in [18]annulene, together with a bond alternation process similar to that which occurs in cyclooctatetraene. A similar interconversion of isomers has been observed in [12]annulene.

At the present time, little is known about the chemistry of [16]annulene and no examples of electrophilic substitution have been reported. [16]Annulene valence isomerizes thermally via the conformational isomer **12c** to the tricyclic hydrocarbon **14** and on photoirradiation gives the stereomer **15**, both reactions being orbital-symmetry allowed (Fig. 4.12).

[16]Annulene and [18]annulene are systems in which the steric interactions are not excessively large and in which the distinction between the $4n$ and $(4n + 2)$ types should still be clear. Although the properties of these systems are more complex and interesting than may have been thought initially, nevertheless the Hückel Rule does appear to be valid in predicting a difference in properties of the two classes of annulenes of this ring size. The problems may now be posed of the role of steric interaction in the smaller annulenes and the limiting size for delocalization in the larger annulenes.

In planar [10]annulene, the problem of steric strain is acute. Three planar configurational isomers of this system may be considered: the all-*cis* **16**, the mono-*trans* **17**, and the di-*trans* **1** configurations (Fig. 4.13). In the all-*cis* configuration **16**, the bond angle strain is high (internal angle 144°), but there are no internal hydrogen interactions, whereas in the di-*trans* config-

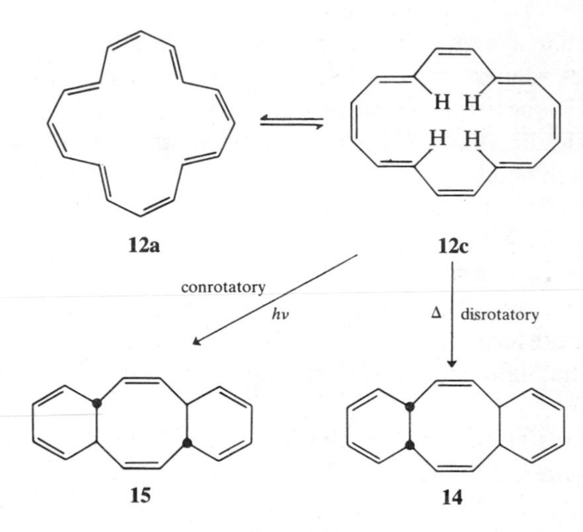

Figure 4.12

uration **1** there is no bond angle strain but the 1,6-hydrogen interactions are acute. The mono-*trans* form **17** has some of the advantages (and disadvantages!) of both of the other configurations.

Numerous attempts to prepare [10]annulene showed it to be an elusive quarry, but the all-*cis* **16** and mono-*trans* **17** isomers were eventually captured by Masamune and co-workers. van Tamelen and Burkoth had irradiated *trans*-9,10-dihydronaphthalene (**18**), which, according to the Woodward–Hoffmann rules, should give **1** or **16**. Some evidence for the formation of [10]annulene was provided, particularly in the isolation of cyclodecane by di-imide reduction. The photoirradiation of the dihydronaphthalenes was then extensively investigated by Masamune, who eventually succeeded in preparing crystalline samples of **16** and **17** at −80°C. Irradiation of *cis*-9,10-dihydronaphthalene (**19**) at −60°C gave mono-*trans*-[10]annulene (**17**), which thermally rearranged at −25°C to *trans*-9,10-dihydronaphthalene (**18**) (Fig. 4.14). Irradiation of **18** at −70°C gave all-*cis*-[10]annulene (**16**), which rearranged thermally at −10°C to **19**. All of these

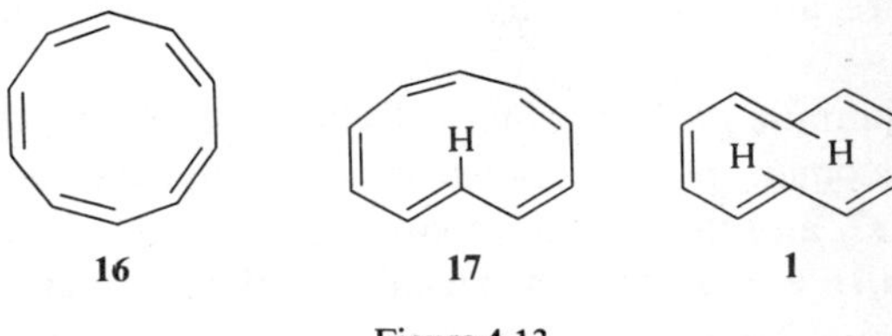

Figure 4.13

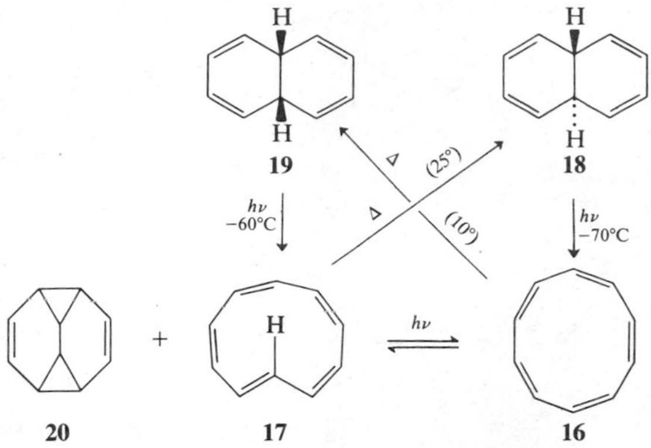

H H H H 19 18 20 17 16 $h\nu$ −60°C $h\nu$ −70°C Δ (25°) Δ (10°) $h\nu$

Figure 4.14

transformations are in accord with the Woodward–Hoffmann rules. Preparatively, it was found that **16** and **17** could be most readily isolated by irradiation of **19** at −60°C, removal of the tetracyclic diene **20** which is also formed by crystallization at −80°C, and chromatography of the enriched [10]annulene solution on alumina at −80°C. The all-*cis* isomer **16** shows a single resonance in the ^{1}H NMR spectrum at δ 5.67 and a single ^{13}C NMR proton-decoupled signal at δ 130.4; both spectra were temperature independent over the range −40–−160°C. By contrast, the ^{1}H NMR and ^{13}C NMR spectra of **17** were both temperature dependent between −40 and −100°C. At −40°C, the ^{1}H NMR spectrum of **17** showed a singlet at δ 5.86 and the ^{13}C NMR spectrum a singlet at δ 131.

The *cis* isomer **16** is a nonplanar molecule, probably in equilibrium between two nonequivalent conformations, whereas the mono-*trans* isomer **17** has the *trans* bond migrating around the ring at a rate that is not observable on the ^{1}H NMR time scale at temperatures above −100°C. Clearly, neither **16** nor **17** is an aromatic compound, both exhibiting NMR spectra more characteristic of polyenes. The collective effects of bond angle and nonbonded strain outweighs any gain in energy attained by delocalization, and both isomers adopt nonplanar conformations.

The electronic spectra of **16** and **17** are surprisingly different, the mono-*trans* isomer showing a much more intense band at about 260 nm.

[12]Annulene (**24**) has been prepared by photoirradiation of any of its valence tautomers **21**, **22**, or **23** at −100°C (Fig. 4.15). The molecule is only stable at low temperature, and the ^{1}H NMR spectrum shows two signals of equal intensity at δ 6.88 and 5.97, which are attributed to the protons on the *cis* and *trans* double bonds, respectively. The equivalence of these protons is due to the rapid conformational interconversion of **24a** and **24b**

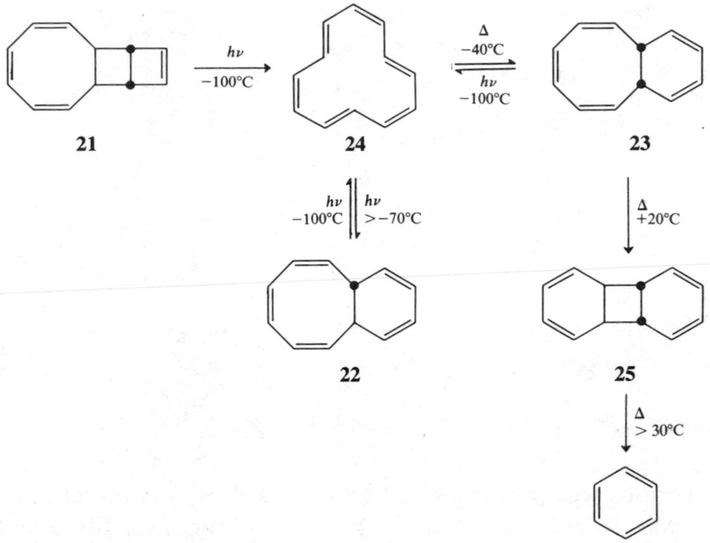

Figure 4.15

(Fig. 4.16) with a low activation energy (ΔG = 23 kJ mol^{-1}), presumably because of overcrowding of the hydrogens in the center of the ring. The bond shift process does not appear to occur at this temperature. The chemical shift of the protons suggests that [12]annulene is only slightly paratropic.

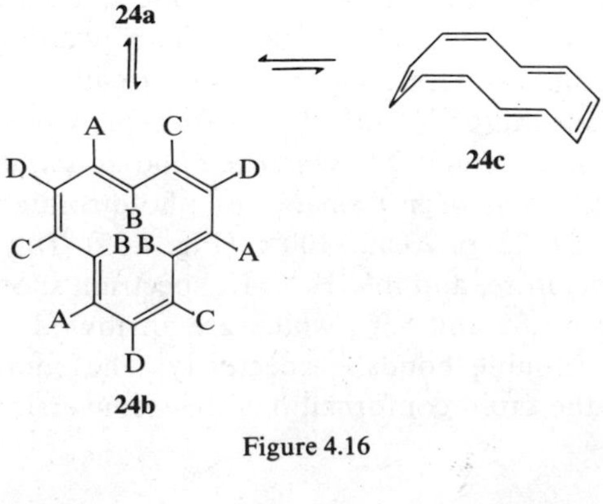

Figure 4.16

At −40°C, [12]annulene rearranges to the bicyclic tautomer **23**, which at +20°C undergoes further rearrangement to **25**. The compound **25** on photoirradiation or heating above 30°C gives benzene. These rearrangements of **24** to **22** and **23** and the reverse reactions are not orbital symmetry allowed, and it is presumed that [12]annulene has the configuration **24** in equilibrium with a small amount of the configuration **24c** (Fig. 4.16). This equilibrium is a process similar to that discussed for [16]annulene, and the rearrangements of **24c** would be orbital-symmetry allowed.

Similar interactions occur in [14]annulene as occur in [10]annulene and [12]annulene, but now these interactions are not so severe. The synthesis of [14]annulene has been described by Sondheimer and Gaoni (Fig. 4.17). This synthesis involves the cyclization of the $C_{14}H_{14}$ hydrocarbon tetradeca-4,10-diene-1,7,13-triyne (**26**) under the Eglinton conditions, followed by the usual base rearrangement and catalytic hydrogenation procedure.

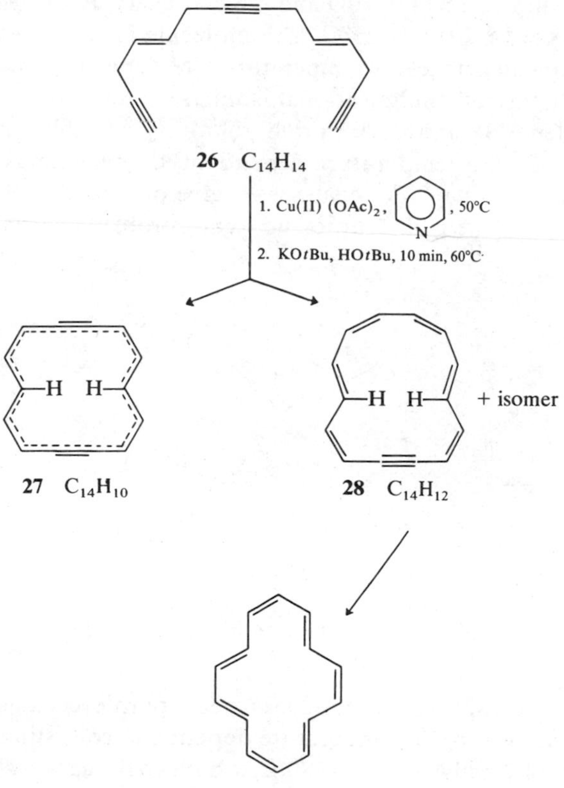

26 $C_{14}H_{14}$

1. Cu(II) (OAc)$_2$, pyridine, 50°C
2. KO*t*Bu, HO*t*Bu, 10 min, 60°C

27 $C_{14}H_{10}$

28 $C_{14}H_{12}$ + isomer

29

Figure 4.17

In solution, [14]annulene appears to be a mixture of two configurational isomers, which can be chromatographically separated from each other but each of which rapidly equilibrates at room temperature to a mixture of both isomers. Crystalline [14]annulene is the pure, major component of the isomeric mixture, and a preliminary x-ray crystallographic analysis indicated that the molecule is near planar, centrosymmetric, and of the configuration indicated by **29**. The ^{1}H NMR spectrum of the major isomer of [14]annulene at room temperature showed only a single band at δ 5.58; but on cooling, this band disappeared and on further cooling to −60°C two new bands appeared at δ 7.6 (10H) and δ 0.0 (4H) due to the outer and inner protons, respectively. [14]Annulene is diatropic and it shows the same behavior as [18]annulene, except that in this case the barrier to proton exchange is lower (about 46 kJ mol^{-1}), and a time-averaged spectrum is observed at room temperature.

[20]Annulene (**30**) was obtained as brown-red needles by Sondheimer's method. The ^{1}H NMR spectrum is temperature dependent, consisting of two broad multiplets (δ 13.9–10.9 and δ 6.6–4.1) at −105°C, which coalesce to give a singlet (δ 7.18) at 25°C. The molecule is clearly paratropic and from the complexity of the low temperature NMR spectrum, almost certainly exists as a mixture of configurational isomers.

[22]Annulene (**31**) is a dark purple, relatively unstable compound. The ^{1}H NMR spectrum is temperature dependent, the two bands at δ 9.65–9.3 and δ −0.4 to −1.2 at −90°C coalesce to give one band at δ 5.65 at 65°C. Clearly, [22]annulene is diatropic and again probably exists as a mixture of isomers.

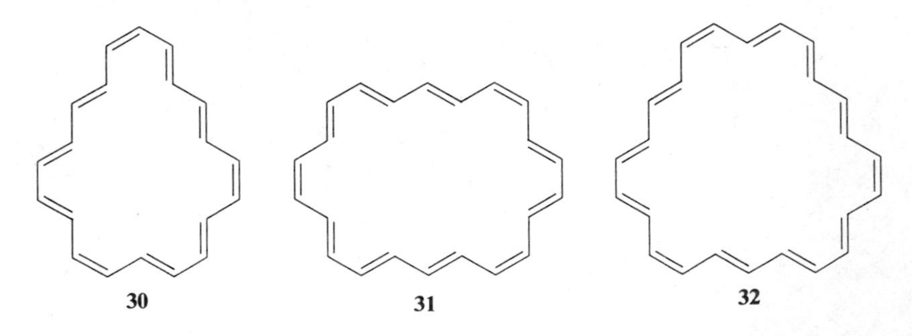

Figure 4.18

[24]Annulene (**32**) was obtained as deep purple crystals, and the ^{1}H NMR spectrum was again temperature dependent, consisting of two multiplets (δ 12.9–11.2 and δ 4.73) at −80°C, which coalesce to a singlet (δ 7.25) at 30°C. [24]Annulene is paratropic and a mixture of isomers.

[26]Annulenes and [28]annulenes have not been synthesized, and although [30]annulene has been prepared the ^{1}H NMR spectrum was not studied. In these larger annulenes, the question of bond localization becomes important, but up to and including [24]annulene the alternation of properties predicted by Hückel's Rule is observed. Dehydro[26] and [30]annulenes have been prepared, and from their ^{1}H NMR spectra (see Section 4.3) it appears that structures with localized bonds become favored at about this ring size, in accord with theoretical predictions. The properties of the larger annulenes would thus be expected to be independent of the number of π electrons.

Table 4.1 lists the known annulenes and indicates their ^{1}H NMR spectral type. As can be seen, the alternation of diatropic and paratropic behavior is only disturbed by [8] and [10]annulene, which are atropic because of their preference for a nonplanar structure.

TABLE 4.1. The ^{1}H NMR Spectral Characteristics of the Annulenes

Compound	Type
Cyclobutadiene	Paratropic (?)
Benzene	Diatropic
Cyclooctatetraene	Atropic
[10]Annulene	Atropic
[12]Annulene	Paratropic
[14]Annulene	Diatropic
[16]Annulene	Paratropic
[18]Annulene	Diatropic
[20]Annulene	Paratropic
[22]Annulene	Diatropic
[24]Annulene	Paratropic

Figure 4.19 shows the plot of the position of maximum absorption in the electronic spectrum against ring size (n), and it can be seen that between $n = 14$ and $n = 24$ the $(4n + 2)$ π-electron annulenes absorb at longer wavelength than their $4n$ neighbors. The extinction coefficients also increase with ring size, but the $(4n + 2)$ π-electron compounds have more intense bands than the neighboring $4n$ systems.

The higher annulenes, unlike benzene, are fluxional molecules that undergo configurational change with low inversion barriers (Table 4.2). The barriers are higher in the $(4n + 2)$ than in the comparable $4n$ annulenes, and the $(4n + 2)$ systems exist in one configurational form.

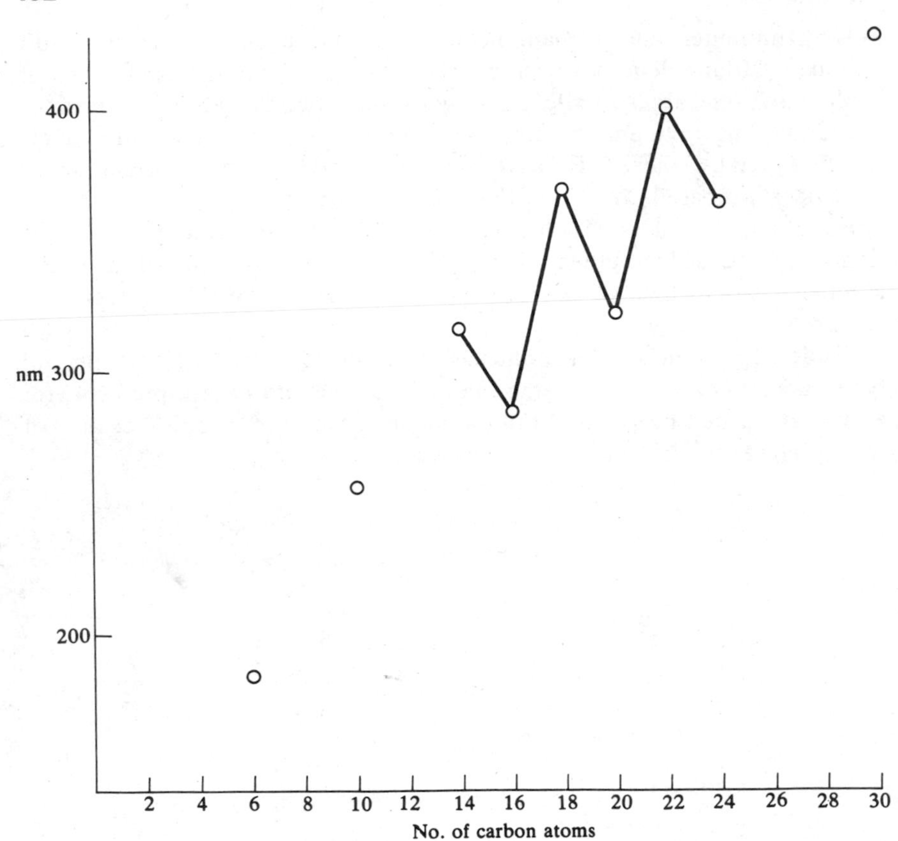

Figure 4.19. Maximum absorption in electronic spectrum with ring size for the annulenes. From Garralt, *Comprehensive Organic Chemistry*, Vol. 1, Pergamon Press, 1979.

As regards chemical behavior, the larger annulenes are much more reactive than benzene. [18]Annulene is the only system that undergoes apparent electrophilic substitution, even the other $(4n + 2)$ annulenes being too reactive for such products to be isolated. Nevertheless, it is clear that with regard to their physical properties, Hückel's Rule, predicting an alternation of behavior in the annulene series with the number of π electrons, has been amply confirmed.

4.3. DEHYDROANNULENES

In the course of the discussion of annulenes given in the preceding section, we have seen that the immediate precursors to the annulenes in the Sond-

TABLE 4.2. Inversion Barriers to Proton Exchange in the Annulenes

Annulenes	ΔG^* (kJ mol^{-1})	Ref
[12]	23.0	*a*
[14]	42.43, 45.2	*a, b*
[16]	36.2, 36.0	*a, b*
[18]	61.50, 56.1	*a, b*
[20]	41.0	*c*
[22]	53.6	*b*
[24]	46.0	*b*

[a] J. F. M. Oth, *Pure Appl. Chem.*, 1971, **25**, 573.
[b] I. C. Calder and P. J. Garratt, *J. Chem. Soc.* (*B*), 1967, 660.
[c] B. W. Metcalf and F. Sondheimer, *J. Am. Chem. Soc.*, 1971, **93**, 6675.

heimer syntheses were compounds with a greater degree of unsaturation. These compounds were termed dehydroannulenes, with the prefixes mono, bis, and tri indicating the number of triple bonds.* Thus, compound **6** is 1,7,13-tridehydro[18]annulene (Fig. 4.2), and 1,9-bisdehydro[16]annulene is shown in Figure 4.11.

The triple bonds contribute two electrons to the out-of-plane π system, the other two electrons being in an in-plane, orthogonal orbital. Consequently, the dehydroannulenes should be of the same Hückel type as the corresponding annulene. The dehydroannulenes thus exhibit the same type of magnetic behavior as the annulene of the same ring size but, because of the triple bonds, the σ framework is much more rigid. As a consequence, the fluxional behavior is damped, and nonaveraged ^{1}H NMR spectra are usually obtained at room temperature. Also as a probable result of the rigidity of the system, the $(4n + 2)$ systems are more diatropic and the $4n$ systems more paratropic than the corresponding annulenes. The chemical behavior of the $(4n + 2)$ dehydroannulenes is often more typical of a classic aromatic compound than is that of the equivalent annulene. For example, 1,8-bisdehydro[14]annulene (**27**) (Fig. 4.17) can be readily nitrated, acety-

* Neither the sequence of prefixes nor the method of designating unsaturations is, from a nomenclature standpoint, sound. Other, more consistent prefix terms have been used, for example, no prefix to indicate one unsaturation, then bis, tris, tetrakis, and more correct dehydrogenation nomenclature has been suggested, e.g., didehydro to indicate one unsaturation and tetradehydro to indicate two unsaturations. However, since the nomenclature is trivial and the majority of papers used the original system, I will conform to that usage.

lated, and sulphonated under conditions that decompose [14]annulene. The x-ray crystallographic analysis of **27** indicates that it is centrosymmetric and best represented by the formula illustrated, with a triple bond length of 120.8 pm and the remaining bonds in the range 137.8–140.3 pm.

Paratropicity is also strongly emphasized. Thus, 1,5-bisdehydro[12]annulene (**33**) and 1,5,9-tridehydro[12]annulene (**34a**) are both strongly paratropic, and this is dramatically illustrated by the ^{1}H NMR spectrum of 9-bromo-1,5-bisdehydro[12]annulene (**34b**), which shows the inner proton at δ 16.4!

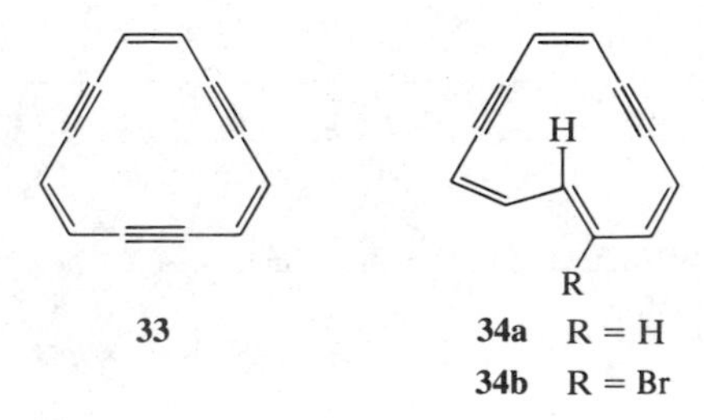

Figure 4.20

Krebs prepared cyclooctatrienyne as an unstable intermediate that could, however, be trapped in a variety of reactions. Whether cyclooctatrienyne is planar is unknown, although the electron spin resonance (ESR) spectrum of the radical anion derived from it is best explained by a planar structure. Benzannelated cyclooctatrienynes are planar, paratropic systems (see Chapter 9).

Nakagawa and co-workers devised a route toward the specific syntheses of dehydroannulenes. They were initially particularly interested in the preparation of compounds analogous to **27** in which the two Kekulé structures are equivalent with the transposition of the cumulene and acetylene linkages (Fig. 4.21).

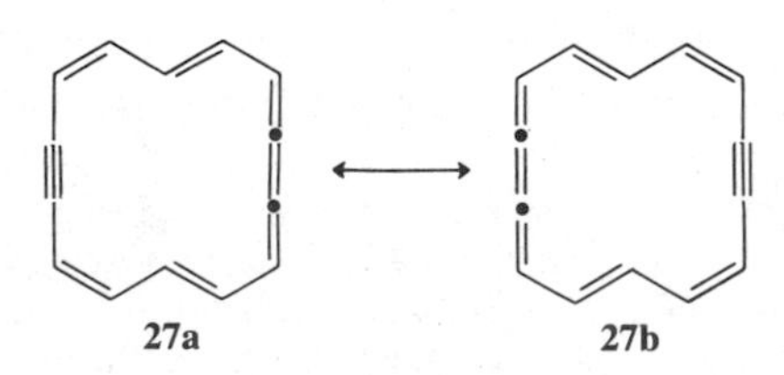

Figure 4.21

The Nakagawa synthesis of tetradehydro[18]annulenes is outlined in Figure 4.22. A 3-substituted-2-penten-4-ynal was condensed with the appropriate ketone to give the corresponding dienyne, which was then coupled under Eglinton conditions. Addition of lithium acetylide gave the diol, which was cyclized by oxidative coupling, again under the Eglinton conditions, to give **33a** as a mixture of diastereomers. Reaction of **33a** with tin(II) chloride in HCl gave the desired tetradehydro[18]annulene (**34c**). The room temperature ^{1}H NMR spectrum of **34c** ($R^1 = R^2 =$ Me) showed a triplet at δ −5.24 assigned to the two inner protons and a doublet at δ 9.66 assigned to the equivalent outer ring protons, with the methyl protons at δ 2.58. Clearly the molecule is diatropic.

Modification of the method outlined in Figure 4.22 gave a tetrasubstituted tetradehydro[22]annulene and tetrasubstituted bisdehydro[14], [18], [22],

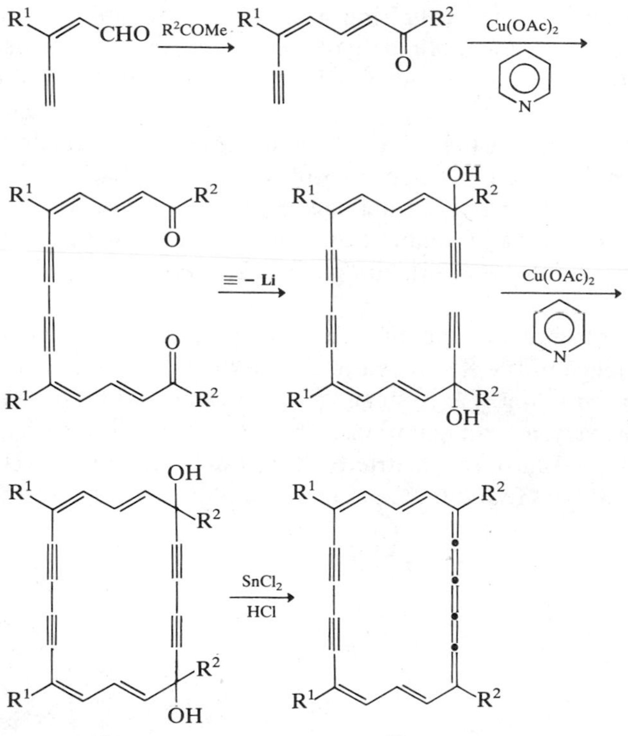

R^1 = Me, Ph, *t*-Bu; R^2 = Me, Ph, *t*-Bu

Figure 4.22

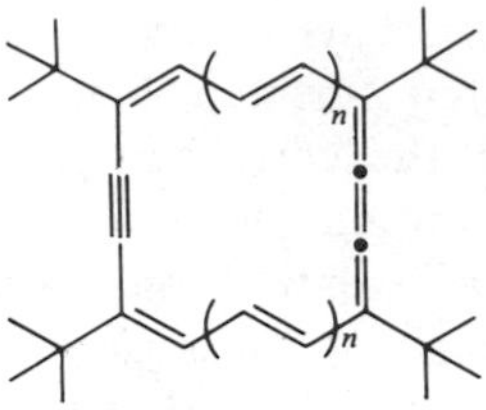

35a	$n = 1$	[14]
35b	$n = 2$	[18]
35c	$n = 3$	[22]
35d	$n = 4$	[26]
35e	$n = 5$	[30]

Figure 4.23

and [26]annulenes. All of these compounds are diatropic, with the diatropicity decreasing with increasing ring size. It is not possible to synthesize an exactly similar series of $4n$ dehydroannulenes since equivalent Kekulé structures require an odd number of atoms between each set of four carbons of the allene acetylene system (**35**; $2n + 1$ atoms in each bridge).

In order to compare the $4n + 2$ dehydroannulenes with their $4n$ counterparts, Nakagawa synthesized a series of tridehydro[$4n$]annulenes in which the two Kekulé structures are not equivalent (Fig. 4.24). The molecules in the [$4n$] series were all found to be paratropic, with low field inner protons and high field outer protons; the paratropicity decreased with increasing ring size.

In order to show that the difference in properties did not arise from the nonequivalence of the Kekulé structures in the second series, a number of related $4n$ and ($4n + 2$) systems was prepared, with the triple bonds arranged in various configurations. Thus, a symmetric bisdehydro[16]annulene (**36**) and an unsymmetric trisdehydro[18]annulene (**37**) were prepared, and **36** was found to be paratropic and **37** diatropic (see Table 4.1).

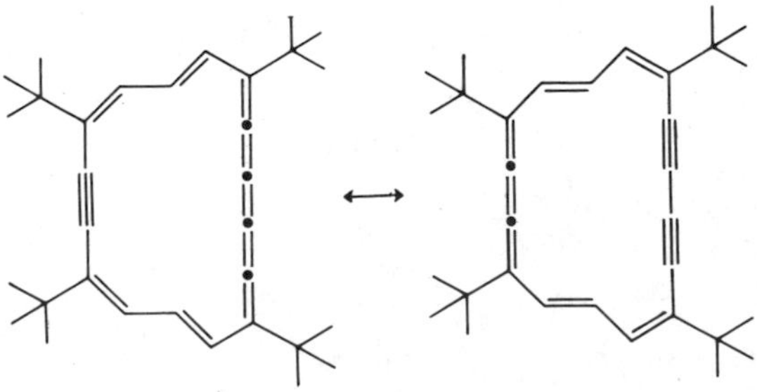

Figure 4.24

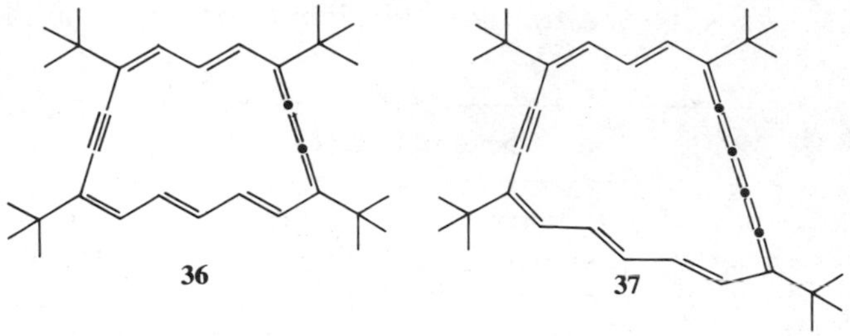

Figure 4.25

The trisdehydro[18]annulene (**37**) did show one effect of the longer sequence of double bonds, the ^{1}H NMR spectrum now being temperature dependent because of *cis, trans* isomerism of the central double bond. This compound can be best represented by a tautomerism between the two structures shown in Figure 4.26, in which the A, B protons are interchanged between inner and outer positions, which is a similar, but reduced, manifestation of the phenomena observed in [18]annulene.

The equivalence of Kekulé structures is not a requirement for electronic interaction in either the $4n$ or the $4n + 2$ annulenes, and the small differences in ring current between dehydroannulenes of the same ring size are insignificant and probably merely reflect different degrees of rigidity in the systems. The ring currents, as measured by the difference, $\Delta\delta$, between outer and inner protons is greater for the dehydroannulene than the equivalent annulene and again reflects the greater rigidity of the dehydrosystems (see Table 4.3 and Section 4.4).

The dehydroannulenes are highly colored substances, ranging from deep green to dark violet, and have a rich electronic spectrum extending beyond 1000 nm in the case of the bisdehydro[26] and [30]annulenes. The position of the main maximum absorption varies with both ring size and substitution.

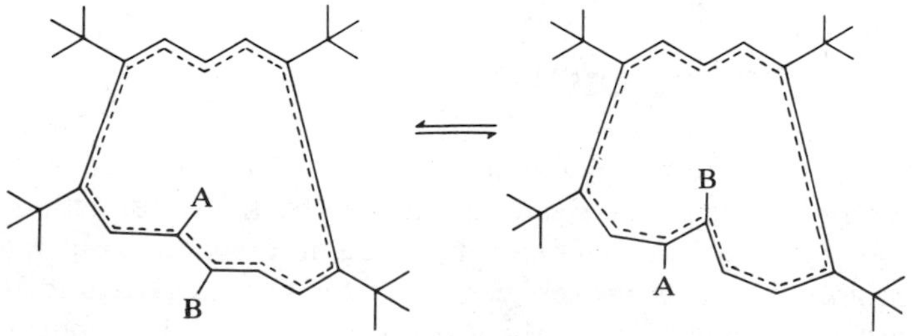

Figure 4.26

TABLE 4.3. ^1H NMR Spectral Data for Some Tetra-*t*-butylbisdehydroannulenes

	Chemical Shift (δ)		
Bisdehydroannulene	Inner Protons	Outer Protons	$\Delta\delta$ = (outer–inner)
[14]	−4.44	9.32	13.76
[18]	−3.42	9.87	13.29
[22]	−0.83	9.16	9.99
[26]	1.95	8.23	6.28
[30]	3.5	7.5	4.0

SOURCE: M. Nakagawa, *Pure Appl. Chem.*, 1975, **44**, 885.

Ascending a series leads to a regular bathochromic shift, and the short wavelength absorption bands become broader and exhibit less fine structure. In the set of bisdehydro[18]annulenes shown in Figure 4.27, successive substitution of phenyl for *t*-butyl causes a bathochromic shift of the main maximum.

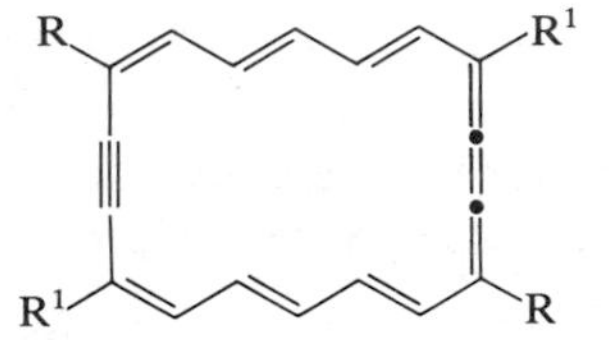

R, R^1, R^1, R

$R = R^1 = t$-Bu 372 nm (447,000)
R = *t*-Bu, R^1 = Ph 401 nm (330,000)
$R = R^1$ = Ph 431 nm (506,800)

Figure 4.27

4.4. BRIDGED ANNULENES

We have seen in the preceding section that the nonbonded destabilizing interaction of inner protons can be removed by the introduction of triple bonds. A second method of removing these interactions is to replace the inner protons by an atom (or group of atoms) that is bound to all the carbons bearing such protons, thereby producing a bridged annulene. Such systems may also, as did the dehydroannulenes, benefit from a greater rigidity of the σ framework. The bridged annulenes, unlike the annulenes and dehydroannulenes, are polycyclic molecules in which, for the terminology annulene to be satisfied, complete delocalization must be capable of occurring around the periphery.

A range of bonding situations can be envisaged in the bridged annulenes, and it is not surprising that such systems have proliferated. Further, because of the rigidity of the σ framework, changes in π-electron distribution free from change in molecular structure can be more readily investigated. Since the bridges need not be in the same plane as the peripheral ring system, geometric isomerism may occur and deviations of the π system out the plane be readily manipulated.

The difference between the monocyclic and bridged annulenes is most graphically illustrated by the 10 π-electron systems. Whereas [10]annulene in both the all-*cis* and mono-*trans* forms has the properties of a polyene (Section 4.2), 1,6-methano[10]annulene (**41**) is an aromatic system. This compound was synthesized by Vogel and Roth by the route outlined in Figure 4.28.

Isotetralin (**38**), prepared from naphthalene by the Birch reduction, was converted to the tricyclic chlorohydrocarbon **39** by reaction with

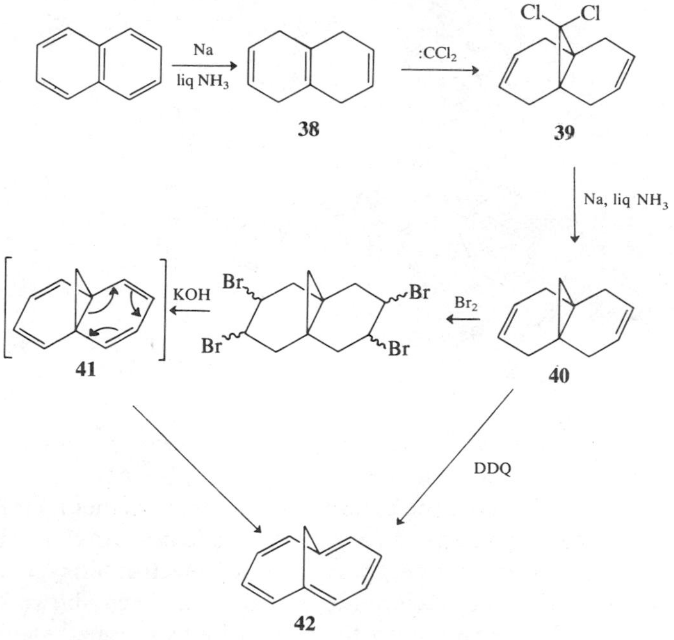

Na / liq NH_3 → **38** → $:CCl_2$ → **39** (Cl, Cl) → Na, liq NH_3 → **40** → Br_2 → (Br, Br, Br, Br) → KOH → [**41**] → **42**; **40** → DDQ → **42**

Figure 4.28

dichlorocarbene. The geminal chlorines were removed with sodium in liquid ammonia, and the resulting diene was brominated and dehydrobrominated with potassium hydroxide. The presumed resulting tricyclic norcaradiene spontaneously tautomerizes to the aromatic, bicyclic 1,6-methano[10]annulene (**42**). Subsequently, it was found that the diene **40** could be more efficiently converted to **42** with dichlorodicyanoquinone (DDQ).

1,6-Methano[10]annulene is a diatropic system, the ring protons in the ^{1}H NMR spectrum appearing as an AA′BB′ system at low field [δ 7.27 (4H) and 6.95 (4H)] with the internal hydrogens on the methylene bridge as a high field singlet (δ −0.52). The broad band decoupled ^{13}C NMR spectrum shows four lines, the three types of ring carbons being at low field [δ, 114.6, C-1; 128.7, C-2; 126.1, C-3] with the methylene bridge carbon at high field (δ 34.8). The electronic spectrum has a main maximum at 256 nm (ε 68,000). These data favor the delocalized structure for this molecule.

An x-ray crystallographic analysis of 1,6-methano[10]annulene-2-carboxylic acid (**43**) (Fig. 4.29) confirms the delocalized structure and shows the molecule to have a reasonably flat perimeter with nonalternating "benzenoid" type bonds (138–142 pm).

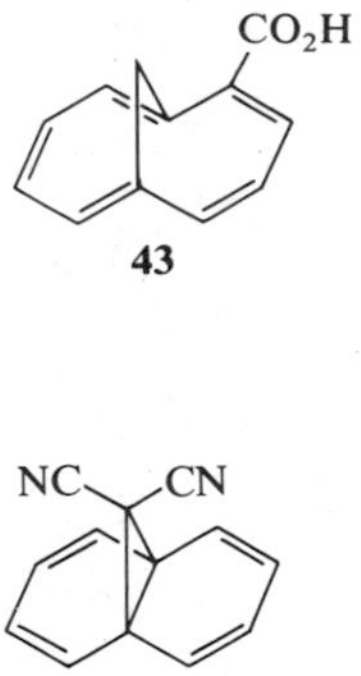

CO_2H

43

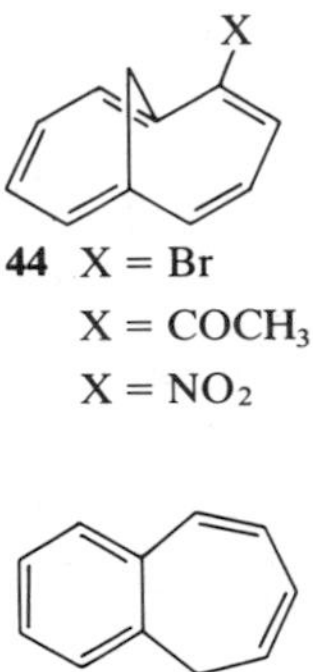

X

44 X = Br
X = $COCH_3$
X = NO_2

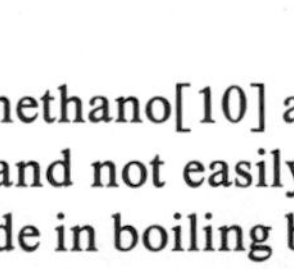

NC CN

45

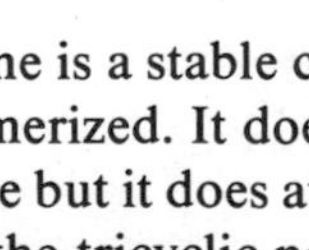

NC CN

46

Figure 4.29

Chemically, 1,6-methano[10] annulene is a stable compound, insensitive to oxygen and heat and not easily polymerized. It does not form an adduct with maleic anhydride in boiling benzene but it does at higher temperatures, the adduct formed being derived from the tricyclic norcaradiene structure. 1,6-Methano[10]annulene undergoes apparent electrophilic substitution with bromine or N-bromosuccinimide, but this has been shown to occur by syn addition of the bromine to the same side as the methylene bridge followed by dehydrobromination. Acylation occurs with acetic anhydride in the presence of tin(IV) chloride, and nitration occurs with copper(II) nitrate in acetic anhydride. In all these cases, reaction takes place to give the 2-substituted derivatives **44** (Fig. 4.29).

As was previously mentioned, the formation of **42** appears to involve the valence tautomerism of a tricyclic norcaradiene intermediate. Calculations suggest that the energy difference between **42** and **41** is similar to that between cycloheptatriene and norcaradiene and that substitution of electron-withdrawing groups on the methylene bridge should, as in the norcaradiene case, favor the tricyclic structure. The 11,11-dicyano derivative **45** has been prepared, which rapidly rearranges to the heptatriene **46** in solution, but it has been crystallized at low temperature. The crystals are less susceptible to rearrangement, and an x-ray crystallographic analysis has shown that the compound exists in the tricyclic norcaradiene form in the solid state. The ^{1}H and ^{13}C NMR spectral data at low temperature are best accommodated by the tricyclic structure also being the form present in solution.

Although the norcaradiene isomer cannot be observed in equilibrium with the bicyclic isomer for 1,6-methano[10]annulene itself, calculations suggest that it should be present to the extent of 0.05% at room temperature. When **42** is treated with the cobalt complex **47**, the resulting cobalt complexes **48**, **49** have the annulene ligand in the tricyclic form (Fig. 4.30).

42 + **47** → **48** major + **49** minor

Figure 4.30

The preference for the bicyclic over the tricyclic structure in the 1,6-bridged compounds appears to require a one-atom bridge. Thus, whereas the oxygen **50** and nitrogen **51** 1,6-bridged compounds exist in the bicyclic annulene form and are diatropic, 4a,8a-ethanonaphthalene (**52**) and 4a,8a-(methanoxymethano)naphthalene (**53**), with two-and three-atom bridges, exist in the tricyclic, nondelocalized atropic form (Fig. 4.31). In the compounds with one-atom bridges, the tricyclic structure is probably

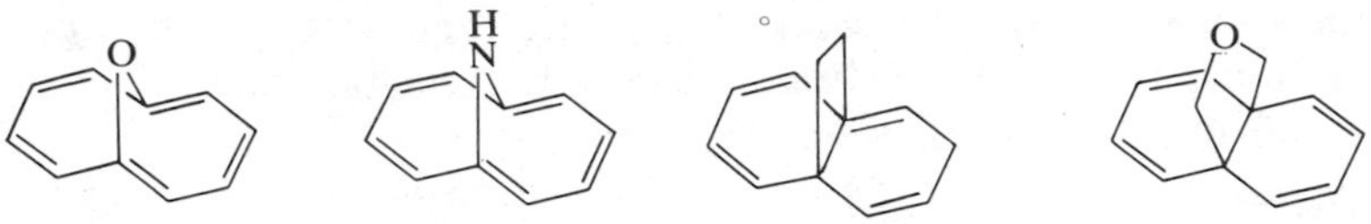

50 51 52 53

Figure 4.31

destabilized by the occurrence of two norcaradiene entities that raise the ground-state energy above that of the delocalized form.

A second bridged [10]annulene, 1,5-methano[10]annulene (**54**) has been prepared by Masamune and Scott and their respective co-workers. The two syntheses are outlined in Figure 4.32. Masamune's involves the addition of

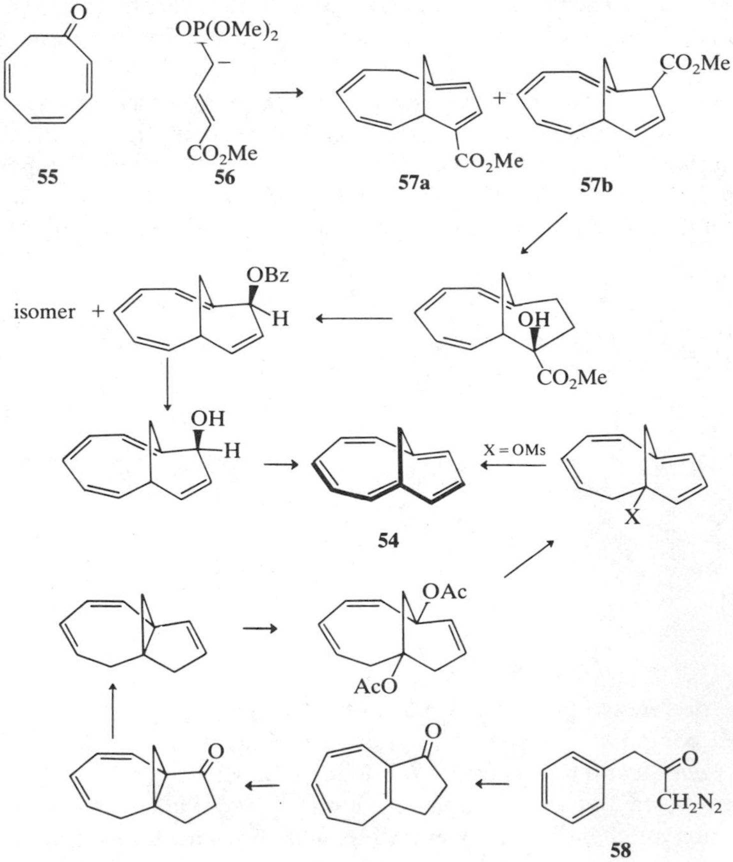

O OP(OMe)$_2$ CO_2Me + CO_2Me CO_2Me
55 56 57a 57b
isomer + OBz H OH CO_2Me
OH H X = OMs X
54
OAc AcO
O O O CH_2N_2
58

Figure 4.32

the phosphonate **56** to cyclooctatrienone (**55**) to give the strained dihydro[10]annulene **57**, whereas Scott's involves an intramolecular carbene addition of the diazoketone **58**, derived from dihydrocinnamic acid.

In the ^{1}H NMR spectrum, the ring protons resonate between δ 6.8 and 8.1 and the bridge protons at δ −0.7 to −1.2, and the compound is clearly diatropic. The ^{13}C NMR spectrum shows the ring protons at low field (δ 125–161), with the methylene protons at δ 34.7, and the spectrum is temperature invariant, so that if there was interconversion between two structures, the rate of interconversion would be very high. 1,5-Methano[10]annulene is an orange compound, and the electronic spectrum has a maximum at 482 nm. It undergoes cycloaddition with TCNE and dimethyl acetylenedicarboxylate (DMDA), in the latter case to form a derivative of 1,7-methano[12]annulene. The ^{1}H and ^{13}C NMR spectra are temperature invariant and the electronic spectrum [266 (ε 16,000), 295 (ε 25,000), 364 (ε 9000) and 480 nm (ε 600)] shows a bathochromic shift compared to the corresponding 1,6-methano[10]annulene derivative, which is reminiscent of the difference in electronic spectra of azulene and naphthalene. A number of derivatives of 1,5-methano[10]annulene have been prepared by variation of the reaction sequences shown in Figure 4.32.

The two systems **42** and **54** are clearly diatropic and aromatic. Are they perturbed [10]annulenes or homonaphthalene and homoazulene, respectively? The question is effectively an enquiry into the degree of transannular orbital interaction between the bridgehead carbons (Fig. 4.33); if there is a large interaction, the molecules resemble the bicycloaromatic system; if small, they resemble the hypothetic "aromatic" [10]annulene.

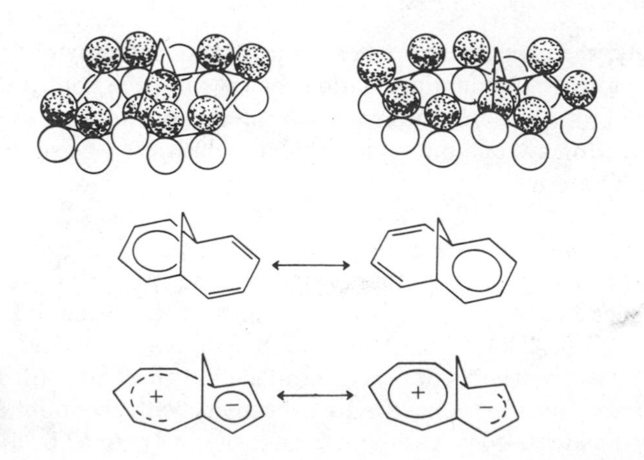

Figure 4.33. Atomic orbital arrangement in 1,6- and 1,5-methano[10]annulenes.

The weight of accumulated evidence now favors the view that these compounds are best described as homoazulene and homonaphthalene rather than [10]annulenes (see Chapter 10). Thus, the electronic and photoelectronic spectra of 1,6-methano[10]annulene are most readily interpreted by a model with a significant transannular interaction. These findings were supported by examination of the absorption, polarized fluorescence and magnetic circular dichroism of **42** and its higher homologues, and the transannular resonance integral was estimated to be about 40% of the standard value for aromatic carbons.

The delocalization energies (DE) of **42** and **54** have been determined from the enthalpies of hydrogenation. Both DEs are small, **42** being 47 kJ mol^{-1} and **54** being 27 kJ mol^{-1}, but it is noteworthy that the DE of homoazulene is much closer to that of azulene (67 kJ mol^{-1}) than is that of homonaphthalene to naphthalene (170 kJ mol^{-1}). This may reflect the greater peripheral delocalization of homoazulene compared with azulene, which would not be reflected in the comparison of homonaphthalene and naphthalene.

The structure of a third, so far unknown, bridged [10]annulene, 1,4-methano[10]annulene (**59**) poses intriguing questions (Fig. 4.34). Torsional strain would be greater in a planar, peripheral σ framework and, more

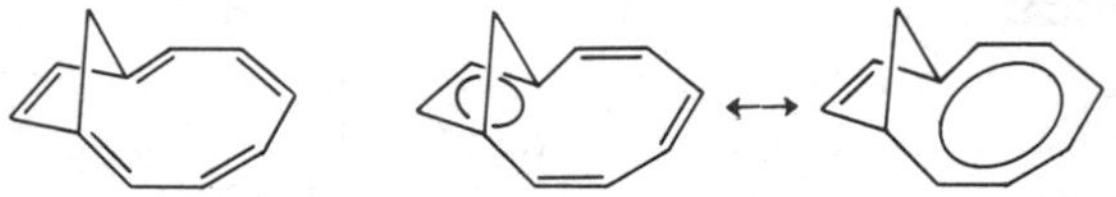

59

Figure 4.34

interestingly, the transannular interaction would not provide a homoaromatic but rather a homoantiaromatic species, since the contributing structures would be $4n$ systems. Transannular interaction would thus destabilize the 10 π-electron system, but would this destabilization be so great that **59** would not be aromatic?

Two other bridged [10]annulenes are known. Cyclo[3.2.2]azine (**60**) is diatropic, the protons resonating at low field (δ 7.9–7.2) in the ^1H NMR spectrum. 11-Methyl-1,4,7-methano[10]annulene (**61**), more recently prepared by Rees and co-workers, is a yellow oil that shows the ring hydrogens as an AB_2 (δ 7.53–7.83; H-8,9,10) and AB system (δ 7.89–7.92; H-2,3,5,6) in the ^1H NMR spectrum, with the central methyl group at δ −1.67. Clearly, **61** is a diatropic system, and the low reactivity with dienophiles suggests that it is aromatic. It does readily rearrange thermally to **62** by a 1,5-methyl sigmatropic shift (Fig. 4.35).

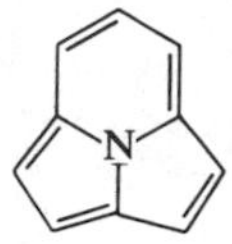

60 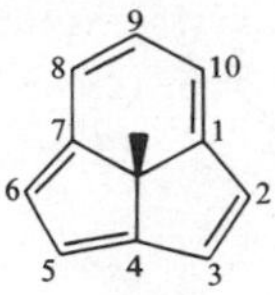 61 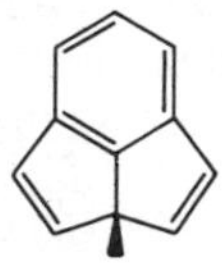 62

Figure 4.35

Both 1,7-methano[12]annulene (**63**) and 1,6-methano[12]annulene (**64**) have been synthesized by Vogel and co-workers. Both compounds were prepared from **40**, the precursor of 1,6-methano[10]annulene, the former by dibromocarbene addition and ring expansion and the latter by thermolysis of the polycyclic system **65**, itself the product of a putative ring expansion (Fig. 4.36).

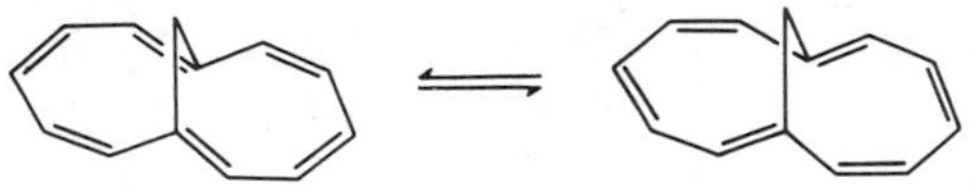

63a 63b

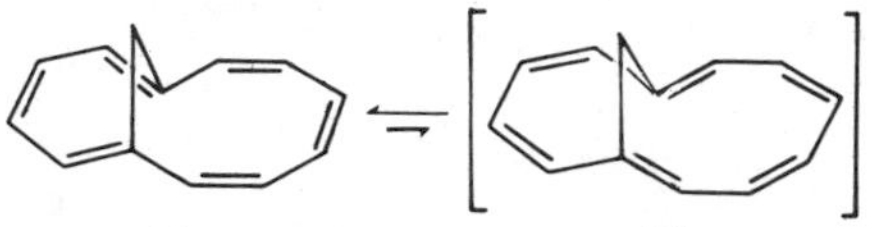

64a 64b  65

Figure 4.36

1,7-Methano[12]annulene (**63**), a deep purple oil, could also be considered a homoheptalene (see Chapter 10), and its chemical and physical properties support this latter view. The room temperature ^{1}H NMR spectrum shows a singlet for the methylene bridge at δ 6.16 and two multiplets for the ring protons at δ 5.5 and 5.2, and **63** is thus paratropic. The ^{1}H NMR spectrum is temperature dependent, the signals for the ring protons broadening at −60°C, whereas the ^{13}C NMR spectrum shows five lines at room temperature, which split into seven lines at −118°C. Homoheptalene is undergoing a bond alternation between the equivalent structures **63a** and **63b**, and this process is slow on the NMR time scale at low temperature.

1,6-Methano[12]annulene (**64**), a low melting orange solid, has a temperature-independent ^{1}H NMR spectrum with an AA′BB′ system at δ 6.17, 5.63, a multiplet at δ 5.73, with the methylene protons as an AB system at

δ 2.29 and 7.00. The compound exists exclusively as **64a** and does not show bond alternation to the tautomer **64b**. It exists in a nonplanar conformation, such as **64c** but, despite this lack of planarity, is quite strongly paratropic.

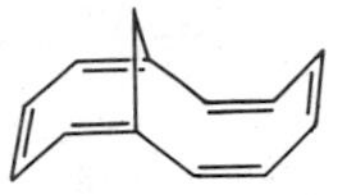

64c

Figure 4.37

Cycl[3.3.3]azine (**66**), a bridged [12]annulene and the higher homologue of **60**, is a brown crystalline solid. It shows in the ^{1}H NMR spectrum the expected A_2B pattern consisting of a triplet at δ 3.65 due to protons 2, 5, and 8, with a doublet at δ 2.07 for the remaining six protons and is clearly paratropic.

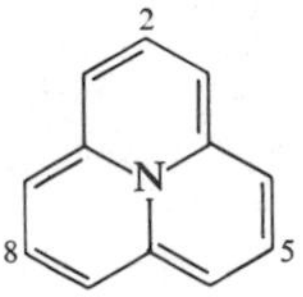

66

Figure 4.38

Three types of bridged [14]annulenes have been reported, one based on anthracene, one on pyrene, and the third on the dicyclopentaheptalene structure. Vogel and co-workers have made a variety of molecules based on the anthracene type, using a synthetic sequence based on that devised for the synthesis of 1,6-methano[10]annulene. Syn-1,6:8,13-bisoxido[14]annulene (**67**), a red crystalline compound, is a typical example of the substances prepared by this route. It is thermally stable and diatropic, and an x-ray crystallographic analysis has shown that the molecule has a

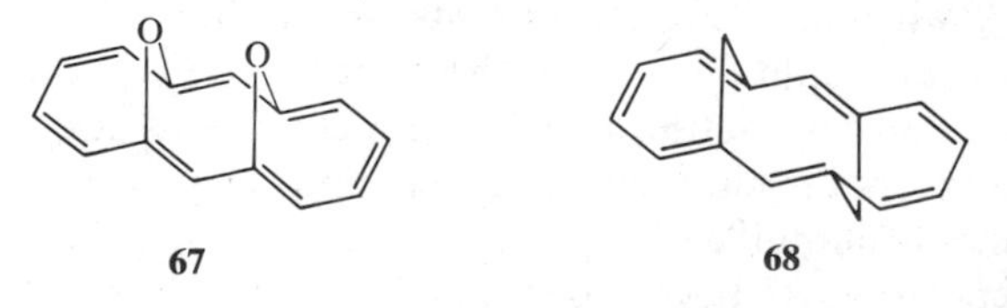

Figure 4.39

reasonably planar perimeter with bonds of similar length (about 139 pm).

1,6:8,13-Bismethano[14]annulene, like the dioxide, could exist in two forms, one with the bridges on the same side, syn-1,6:8,13-bismethano[14]annulene (**69**) and the other with bridges on opposite sides, anti-1,6:8,13-bismethano[14]annulene (**68**). The anti-isomer was prepared by the above route and is an atropic system, presumably because orbital overlap is prevented in this configuration. Bond shift does occur, possibly via a delocalized transition state, in a similar manner as observed for cyclooctatetraene (Chapter 2). The syn isomer could not be synthesized by this route but more recently has been prepared by a new general synthetic scheme, which is illustrated in Figure 4.40. The ^{1}H NMR spectrum of **69**

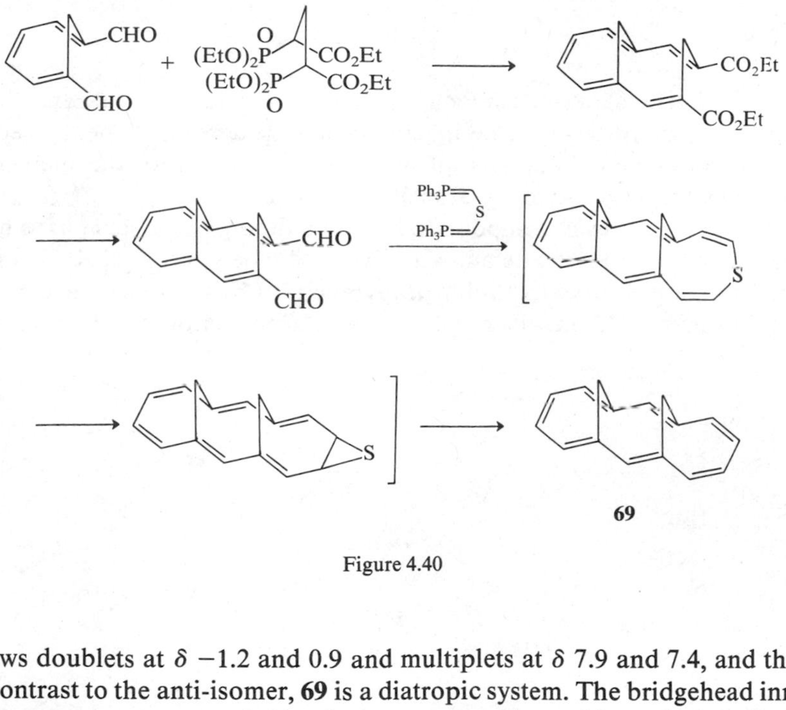

Figure 4.40

shows doublets at δ −1.2 and 0.9 and multiplets at δ 7.9 and 7.4, and thus, in contrast to the anti-isomer, **69** is a diatropic system. The bridgehead inner hydrogens are very crowded, and an x-ray crystallographic analysis indicates they are only 178 pm apart. The result of this steric strain is illustrated by the reaction with bromine to give the addition product **70** in which these nonbonded interactions have been relieved by a change in hybridization (Fig. 4.41).

By contrast, in **71** ($n = 1$), in which the nonbonded hydrogens have been replaced by a bridging atom, bromination gives the 2-bromo substitution

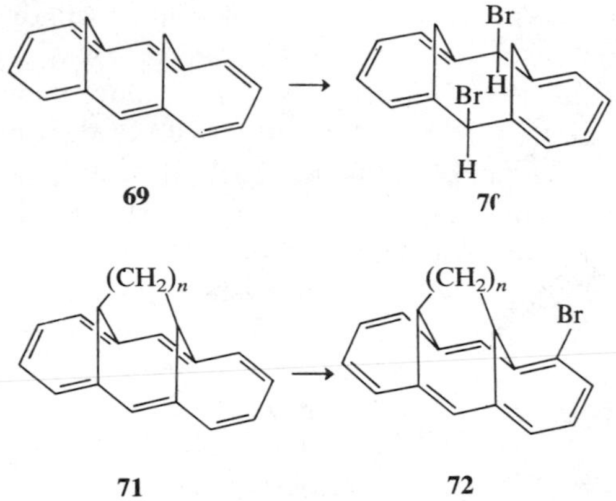

Br

Br H

H

69 → 70

$(CH_2)_n$ $(CH_2)_n$ Br

71 → 72

Figure 4.41

product **72**. The series of compounds **71** with $n = 0, 1, 2,$ and 3 has been prepared, and although the periphery becomes progressively bent, spectral and structural determinations all indicate that even with the increasing strain these remain diatropic, aromatic compounds.

The second series of compounds, based on the pyrene system, have been synthesized by Boekelheide and associates. Mitchell and Boekelheide prepared the parent *trans*-15,16-dihydropyrene (**75**) by the reaction sequence show in Figure 4.42, in which the key step was the extrusion of sulphur by

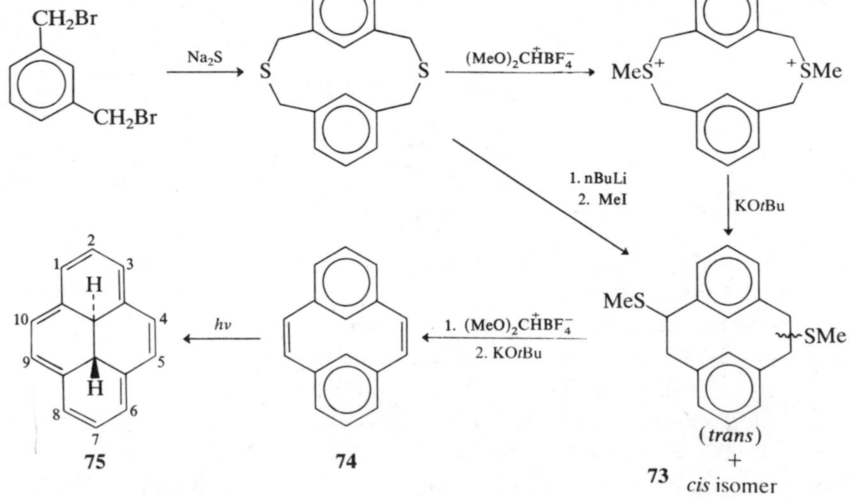

CH_2Br CH_2Br → Na_2S → S S → $(MeO)_2\overset{+}{C}HBF_4^-$ → MeS^+ ^+SMe

1. nBuLi 2. MeI

KO*t*Bu

MeS ~SMe (*trans*) + *cis* isomer 73

← 1. $(MeO)_2\overset{+}{C}HBF_4^-$ 2. KO*t*Bu ← 74 ← *hν* ← 75

1 2 3 4 5 6 7 8 9 10 H H

Figure 4.42

the Stevens rearrangement (i.e., **73** → **74**). The compound **75** can only be formed in the presence of pyrene, but its physical properties are readily observed. Solutions in cyclohexane are deep green, and the ^{1}H NMR spectrum shows signals at δ 8.58 (H-4,5,9,10), 8.50 (H-1,3,6,8), and 8.02–7.89 (H-2,7) for the ring protons, whereas the inner protons H-15,16 appear at δ −5.49. Compound **75** is thus strongly diatropic. Boekelheide and Philips had earlier prepared *trans*-15,16-dimethyldihydropyrene (**76**) by a long synthetic route, but this compound is now more easily accessible via the Stevens rearrangement sequence. The compound **76** is diatropic, showing the ring proton absorptions at δ 8.87–7.98 and the internal methyl groups at δ −4.25. The molecule exhibits a large diamagnetic anisotropy, and an x-ray crystallographic analysis of the 2,7-diacetoxy derivative shows that the bonds do not alternate in length, varying between 138.6 and 140.1 pm.

Application of Haddon's method (see Chapter 2, Section 2.6) for calculating ring currents to **76** gave chemical shifts very close to those observed. Compound **76** was consequently used as a model both in Haddon's original and subsequent calculations. In these terms, **76** is the most diatropic nonbenzenoid compound known. *trans*-15,16-Dimethyldihydropyrene undergoes a range of electrophilic substitution reactions, substitution occurring at position 2, and thus in both chemical and physical properties it acts as an aromatic compound. An interesting valence tautomerism occurs between

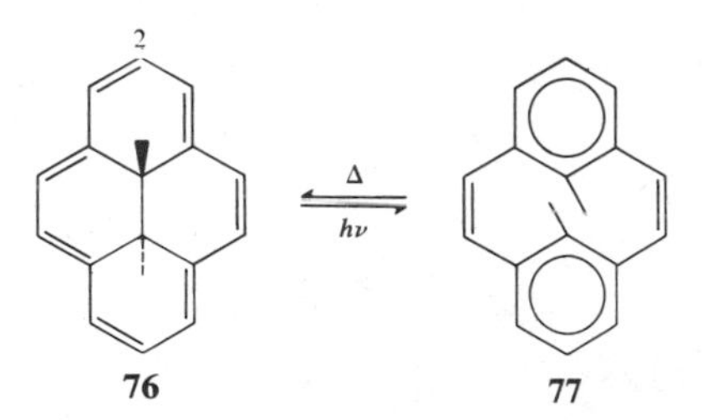

Figure 4.43

76 and the metacyclophane **77**; the metacyclophane is produced by photoirradiation of **76** and reverts thermally to **76** at room temperature.

Müllen and co-workers prepared *trans*-15,16-dimethyl-1,4:8,11-ethanediylidene[14]annulene (**79**), a compound based on the dicyclopentaheptalene structure, as dark red needles by treatment of **78** with lithium in tetrahydrofuran (THF) to give the dianion that was then methylated with dimethyl sulphate. The ^{1}H NMR spectrum shows the ring protons at δ 8.04–8.77 and the methyl protons at δ −4.53, and the molecule is strongly diatropic. The electronic spectrum resembles that of **76** and structurally is related to **76** (n = 0) by a 90° rotation of the ethane bridge. An x-ray crystallographic analysis indicates that the perimeter is reasonably flat and that the peripheral bond lengths are essentially equal (133–143 pm). Treatment with copper(II) nitrate in acetic anhydride gave the 6-nitroderivative.

1,4:8,11-Bisamino[14]annulene (**80**) has been prepared; it is closely related to **79** and, although unstable, is diatropic.

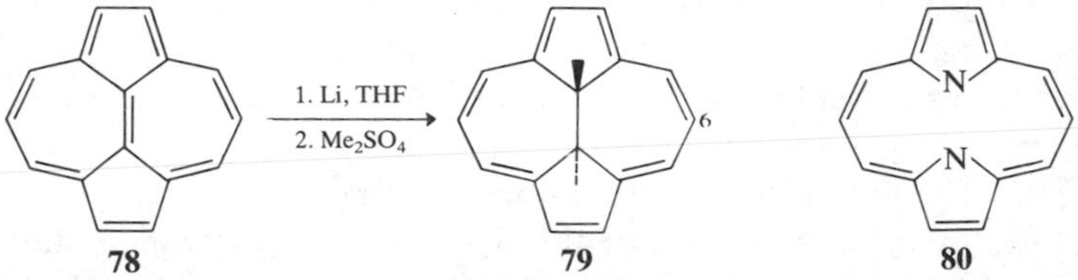

1. Li, THF
2. Me_2SO_4

78 79 80

Figure 4.44

[16]Annulene dioxide (**81**) is a paratropic compound, the inner protons appearing at δ 17.18 in the ^1H NMR spectrum, with the outer protons between δ 4.41 and 4.93 (Fig. 4.45).

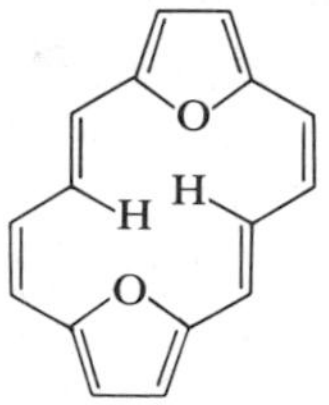

81

Figure 4.45

A variety of bridged [18]annulenes have been synthesized. The cycloheptatriene route (Fig. 4.40) has been used to prepare *syn,syn,syn*-1,6:8,17:10,15-trismethano[18]annulene (**82**). This compound is diatropic, and a comparison of the electronic spectrum with those of the 10 and 14 π-electron homologues, **42** and **69**, shows a remarkable similarity in the shape of the curves that are progressively shifted to the red with increasing ring size. Boekelheide and co-workers have prepared a number of very rigid bridged [18]annulenes in which the shape and length of the peripheral π system has been systematically altered. Haddon ring current calculations were carried out on these compounds and the relative ring currents compared. Table 4.4 shows values for the fraction of maximum calculated ring current and the degree of bond alternation for a variety of bridged annulenes

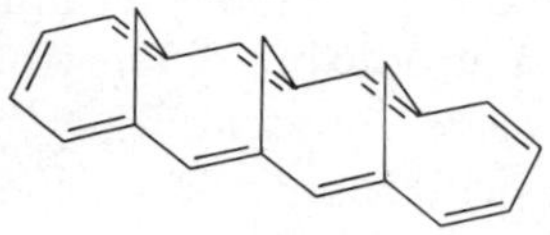

82

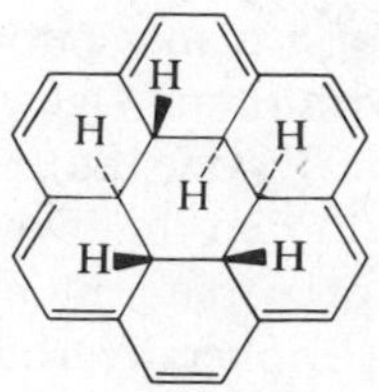

83

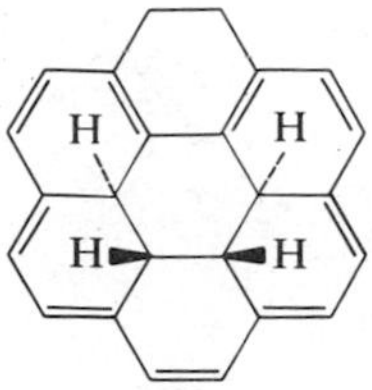

84

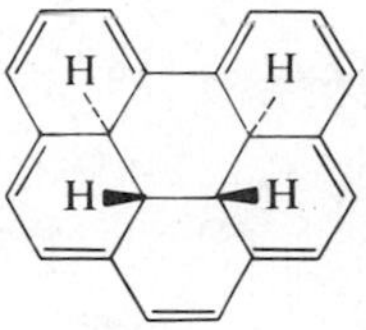

85

Figure 4.46

and [18]annulene itself, taking 15,16-dimethyldihydropyrene as the model compound (see also Table 2.6).

Some of the earliest bridged annulenes were those described by Badger and co-workers in the [18]annulene series. Heteroatoms were used as 1,4-bridging groups, which also facilitated synthesis from five-membered heterocycles. Of the systems shown in Figure 4.47, the compounds with three bridging oxygens or two oxygens and one sulfur are diatropic peripherally delocalized systems, whereas those with three sulfur or two sulfur and one oxygen resemble the constituent heterocycles. This difference in behavior may arise from the larger size of the sulfur atom, precluding

TABLE 4.4

Compound	Ring Current RC ($cm^2\ t^{-t}$)	k	Bond Alternation β_1/β_2
76	−1.6178	1.00	1.00
71 $n = 1$	−1.402	0.83	0.89
[18]Annulene	−1.2043	0.56	0.83
84 (or **85**)	−1.0295	0.59	0.83
83	−1.9403	0.88	0.91

SOURCE: Data taken from T. Otsubo, R. Gray, and V. Boekelheide, *J. Am. Chem. Soc.*, 1978, **100**, 2449.

attainment of a planar periphery, or the greater aromaticity of thiophene compared with furan. Ogawa and co-workers have more recently prepared [18]annulene dioxide (**86**), which is also diatropic but differs from Badger's compounds in that the two Kekulé structures are identical.

Elix has prepared a number of larger bridged annulenes containing 24, 30, and 36 peripheral carbon atoms. The trioxido[24]annulenes are paratropic, but the larger compounds are atropic.

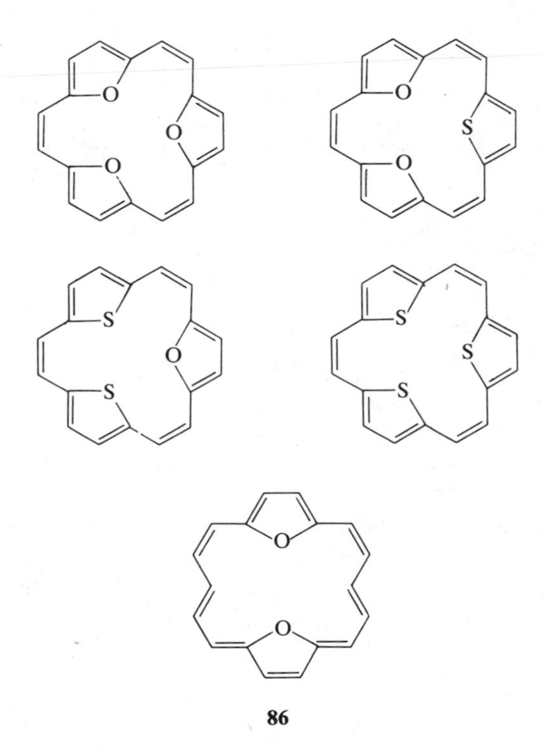

Figure 4.47

4.5. CONCLUSIONS

There are thus, besides benzene, a number of monocyclic conjugated compounds that have closed electronic shells and aromatic spectral properties. As would be predicted on theoretical grounds, these compounds are much more reactive and show less tendency to retain the closed electronic shell than does benzene. Part of this increased reactivity may be attributed to the less rigid σ framework and the availability of transannular reactions, and this is borne out by the greater aromaticity and lesser reactivity of the dehydroannulenes and bridged annulenes, which more closely resemble benzene in properties. Parallel with this series of molecules is a series of $4n$ compounds that deviate from planarity and are paratropic. Both series are interrupted by the medium ring compounds, which are nonplanar and atropic, largely owing to nonbonded repulsion and ring strain since the related bridged and dehydro systems exhibit diatropic or paratropic behavior. Several bridged systems involve transannular orbital overlap and are best considered as homobicyclic or homopolycyclic compounds rather than annulenes. As the ring size increases, the series merge so that the large ring compounds have properties independent of the number of π electrons, being atropic and polyolefinic.

FURTHER READING

For general reviews of the synthesis of annulenes with the emphasis on the work of Sondheimer's group, see F. Sondheimer, *Proc. Roy. Soc.*, 1967, **A297**, 173; *Proceedings Robert A. Welch Foundation*, XII Organic Synthesis, 1969, 125; *Pure Appl. Chem.*, 1971, **28**, 331; *Accounts Chem. Res.*, 1972, **5**, 81.

For the ^{1}H NMR spectra of the annulenes, see R. C. Haddon, V. R. Haddon, and L. M. Jackman, *Top. Curr. Chem.*, 1971, **16**, 103; for the ^{13}C NMR spectra of annulenes, see H. Günther and H. Schmickler, *Pure Appl. Chem.*, 1975, **44**, 807.

For a discussion of the fluxional properties of the annulenes, see J. F. M. Oth, *Pure Appl. Chem.*, 1971, **25**, 573.

For a discussion of the structure of [18]annulene in its historical context, see H. Baumann and J. F. M. Oth, *Helv. Chim. Acta*, 1982, **65**, 1885.

For a discussion of the dehydroannulenes, see M. Nakagawa in, T. Nozoe, R. Breslow, K. Hafner, Shô Itô, and I. Murata (Eds.), *Topics in Nonbenzenoid Aromatic Chemistry*, Vol. 1, Hirokawa, Tokyo, 1973, p. 191.

E. Vogel, *Chemical Society Special Publication*, 1967, **21**, 113; *Proceedings of the Robert A. Welch Foundation*, XII Organic Synthesis, 1969, 215; *Pure Appl. Chem.*, 1969, **20**, 237; *Israel J. Chem.*, 1980, **20**, 215; *Pure Appl. Chem.*, 1982, **54**, 1015—facets of Vogel's extensive work in the area of bridged annulenes.

V. Boekelheide, *Proceedings of the Robert Welch Foundation*, XII Organic Synthesis, 1969, 83; in T. Nozoe, R. Breslow, K. Hafner, Shô Itô, and I. Murata (Eds.), *Topics in Nonbenzenoid Aromatic Chemistry*, Vol. 1, Hirokawa, Tokyo, 1973, p. 47; *Pure Appl. Chem.*, 1975, **44**, 751—facets of Boekelheide's extensive work in the area of bridged annulenes.

For general accounts of the annulenes, see H. P. Figeys, in D. Lloyd (Ed.), *Topics in Carbocyclic Chemistry*, Logos Press, London, 1969, p. 269; P. J. Garratt and K. Grohmann, in E. Müller (Ed.), *Houben-Weyl Methoden der Organischen Chemie*, Band 5/1d, Georg Thiele Verlag, Stuttgart, 1972, p. 527; P. Skrabal, *Int. Rev. Sci. Org. Chem. Ser.* 2, 1976, **3**, 229; P. J. Garratt, in J. Frazer Stoddart (Ed.), *Comprehensive Organic Chemistry*, Sir Derek Barton and W. D. Ollis (series Eds.), Vol. 1, Pergamon, Elmsford, N.Y., 1979, p. 361; D. Lloyd, *Non-benzenoid Conjugated Carbocyclic Compounds*, Elsevier, Amsterdam, 1984.

For a recent theoretical account of the stabilization of bridged annulenes based on rehybridization to allow for the geometry of the σ framework, see R. C. Haddon and L. T. Scott, *Pure Appl. Chem.*, 1986, **58**, 137.

For recent experimental work appertaining to this chapter see the articles by D. Leaver, W. Flitsch, and K. Müllen, in *Pure Appl. Chem.*, 1986, **58**, 143, 153, 177.

5

MONOCYCLIC AROMATIC IONS

The Hückel Rule predicts that conjugated monocyclic ions with $(4n + 2)$ π electrons will be aromatic, whereas those with $4n$ π electrons will not. A considerable amount of work had been carried out prior to the enunciation of this rule, and many subsequent studies were made without reference to its predictions.

Thiele, in 1901, observed that when cyclopentadiene (**1**) is heated under nitrogen with a dispersion of potassium in benzene, potassium cyclopentadienide (**2**) is formed. In the same year he also noted that the similar treatment of cycloheptatriene (**3**) did not yield potassium cycloheptatrienide (**4**). The difference in acidity of the two methylene groups can now be

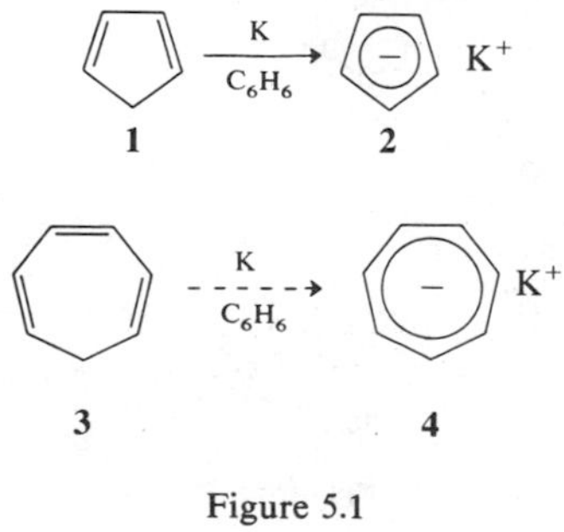

K, C_6H_6 ; K^+

1 **2**

K, C_6H_6 ; K^+

3 **4**

Figure 5.1

accounted for in terms of the stability of the resulting anions, since cyclopentadiene yields the aromatic 6 π-electron cyclopentadienyl anion, whereas the cycloheptatrienyl anion has an unfilled shell of 8 π electrons and is not aromatic (Fig. 5.2).

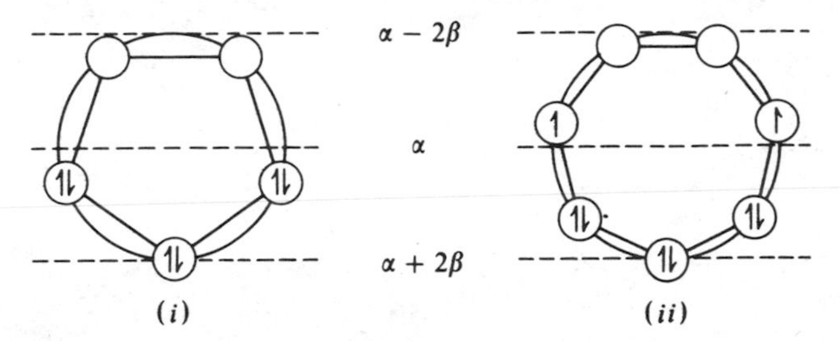

Figure 5.2. HMO energy levels of the cyclopentadienyl anion (i) and the cycloheptatrienyl anion (ii).

At the time, Thiele suggested that the difference in behavior of the two hydrocarbons might be due to a nonbonded interaction between the 1,6-carbon atoms of the cycloheptatriene giving an aromatic structure and thus deactivating the methylene group of **3** compared to that of **1**. Although this interaction probably contributes little to the deactivation of cycloheptatriene, it is a preelectronic concept corresponding to homoaromaticity, a topic discussed further in Chapter 10.

Prior to these experiments of Thiele, Merling, while carrying out an investigation of tropine, observed that cycloheptatriene on bromination gave an oily dibromide, which, on attempted distillation, decomposed to give hydrogen bromide and a crystalline solid. With the formation of this solid, Merling had prepared the first stable carbocation, cycloheptatrienium bromide (**5**), but, not surprisingly, this was not recognized until much later when Doering and Knox repeated these experiments in 1954. The cycloheptatrienium cation is predicted by the Hückel Rule to be aromatic (Fig. 5.3).

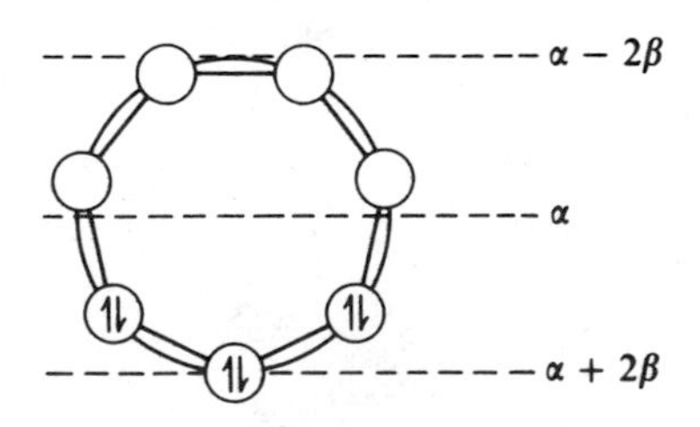

Figure 5.3

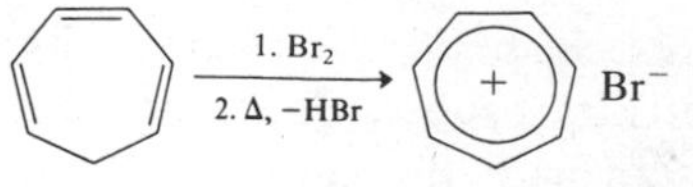

1. Br_2
2. Δ, −HBr

Br^-

3 5

Figure 5.4

Between 1901 and 1945, the main advances in this area were of a theoretical nature, culminating in Hückel's prediction that the cycloheptatrienium cation would be a stable aromatic system. Pfau and Plattner did prepare azulene (**6**), but this aroused more interest as the parent hydrocarbon of a number of natural products rather than as an aromatic system. Little attention was paid to the Hückel predictions until Dewar, in 1945, postulated structures for stipitatic acid and colchicine, which were based on a seven-

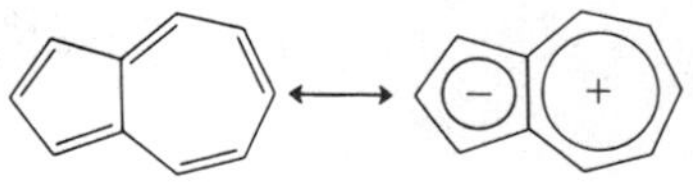

6a **6b**

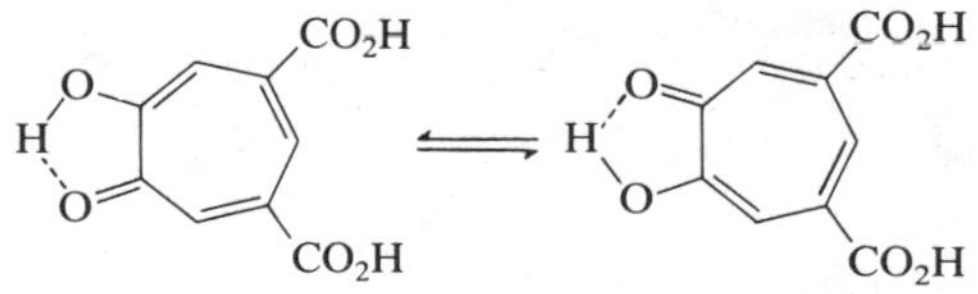

CO_2H CO_2H

CO_2H CO_2H

7a **7b**

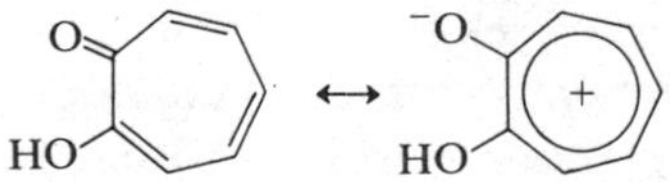

O HO $^-$O HO

8a **8b**

MeO MeO NHCOMe OMe MeO MeO NHCOMe O^- OMe

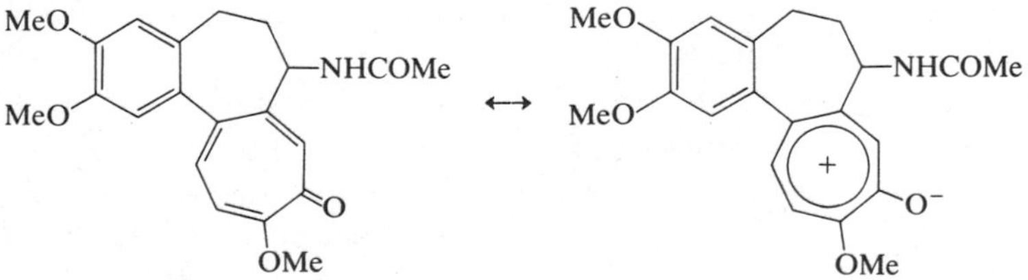

9a **9b**

Figure 5.5

membered aromatic ring. In his first communication, using the published experimental results of Birkinshaw, Chambers, and Raistrick, Dewar suggested that stipitatic acid had the structure **7**. He further postulated that there might be a family of aromatic systems in which hydrogen bonding gave structures analogous to azulene, and on this basis he gave the trivial name tropolone to the parent compound cycloheptatrienolone (**8**). However, in the structure subsequently suggested for colchicine (**9**),* the tropolone ring occurs as the methyl ether, and hydrogen bonding can thus not be important in this structure. The stability of the tropolone ring was now considered to arise from the contribution of the dipolar structure (e.g., **8b**), and the possibility was also advanced that tropolone might be a highly mobile tautomeric system.

At the same time as Dewar was subjecting the literature to this remarkable analysis, Nozoe and co-workers had arrived at similar conclusions regarding a natural product, hinokitiol, which they had isolated from the Formosan Cedar. The structure **10** had been assigned to hinokitiol, but an account of this work did not become generally available until 1951.

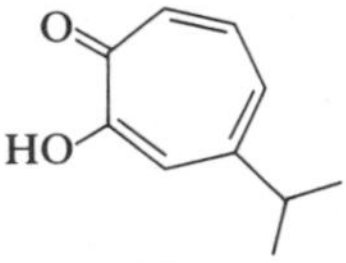

O HO 10

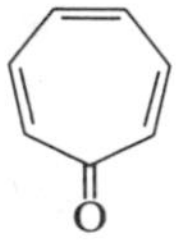

O 11

Figure 5.6

The structure suggested by Dewar for stipitatic acid was readily confirmed, and successful syntheses of the parent tropolone (**8**) were described a short time after. Subsequently, cycloheptatrienone (tropone)(**11**) was itself synthesized, and the importance of Hückel's Rule was finally clearly enunciated by Doering and Knox in their report of the synthesis of cycloheptatrienium (tropylium) bromide (**5**).

The cyclopentadienyl anion can be prepared by treatment of cyclopentadiene with either alkali metals or alkali hydroxides. The magnetic and electrochemical properties of the salts formed in this manner indicate that these are aromatic compounds. The IR and Raman spectra of the ions are simple, as expected for a molecule of D_{5h} symmetry, and the ^{1}H NMR spectrum shows a single resonance at δ 5.57. The chemical shift of the

*In his suggested structure for colchicine, Dewar left open the actual orientation of ring C with respect to ring A. The structure illustrated is that subsequently demonstrated to be that of colchicine.

protons is in reasonable agreement for a system with a diamagnetic current of the same magnitude as that of benzene having one-fifth excess electron density at each carbon atom. The electronic spectrum shows no absorption above 200 nm.

The cyclopentadienyl anion (**2**) reacts readily with electrophiles, is carboxylated with CO_2, and is alkylated or arylated with the appropriate halide. In all cases, dimeric dicyclopentadienides (**12–14**) are formed, although monomers can be prepared in the case of aryl and, less readily, alkyl cyclopentadienes. The cyclopentadienyl anion also reacts with aldehydes and ketones to give fulvene derivatives (**15**) (Fig. 5.7).

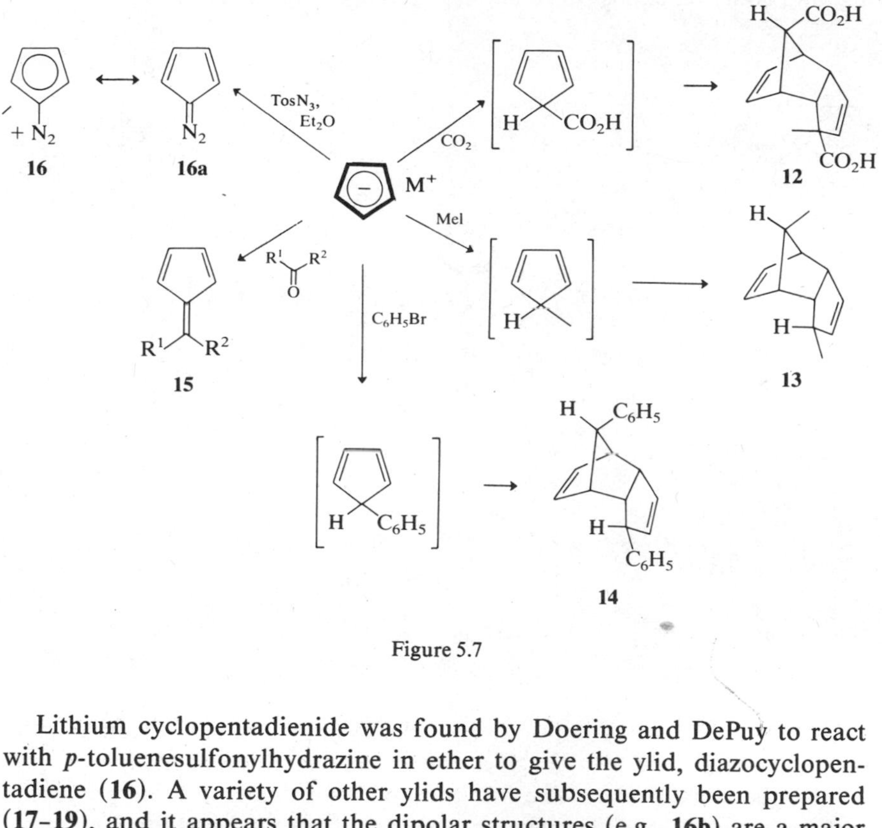

Figure 5.7

Lithium cyclopentadienide was found by Doering and DePuy to react with *p*-toluenesulfonylhydrazine in ether to give the ylid, diazocyclopentadiene (**16**). A variety of other ylids have subsequently been prepared (**17–19**), and it appears that the dipolar structures (e.g., **16b**) are a major factor in stabilizing these compounds. All of the ylids have large dipole moments, for example, that of triphenylphosphonium cyclopentadienylid (**19**) is 7.0 D, and these values can be ascribed to the contribution of the dipolar form.

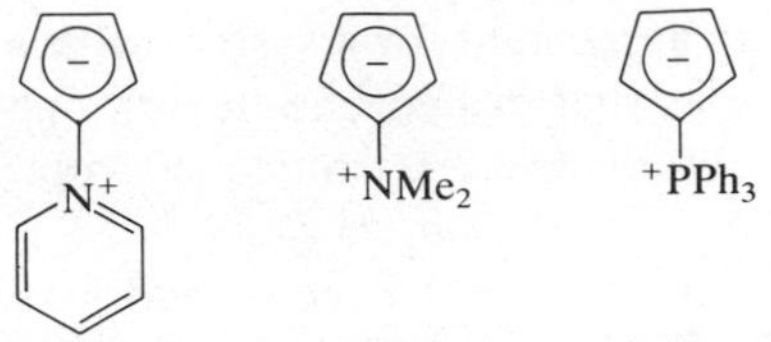

17 18 19

Figure 5.8

Both diazocyclopentadiene (**16**) and triphenylphosphonium cyclopentadienylid (**19**) undergo electrophilic substitution in the cyclopentadienyl ring, and some reactions of **16** are illustrated in Figure 5.9.

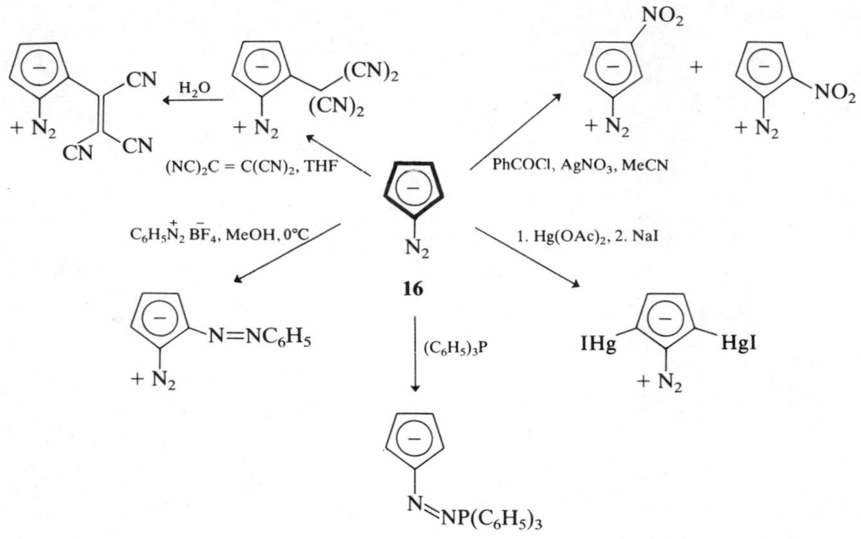

Figure 5.9

The other reaction of importance of the cyclopentadienyl anion is with transition metal salts, which gives the organometallic "sandwich" compounds, the metallocenes. The first compound of this type to be prepared was ferrocene (**20**), which was independently discovered by Miller, Tebboth, and Tremaine and by Kealy and Pauson. Both of these discoveries were fortuitous; the first group were investigating the reaction of alkenes with nitrogen over metal catalysts, and the second was attempting a synthesis of fulvalene. The sandwich structure for ferrocene was first proposed by

Wilkinson et al. a short time after the announcement of the synthesis; a number of analogous compounds were soon reported, and the aromatic nature of ferrocene was demonstrated. The IR spectrum shows only one type of C—H stretching mode, which is in accord with the symmetric structure. An electron count of the ferrocene system, assuming a contribution of 10 electrons from the cyclopentadienyl rings and 8 electrons from the iron, gives the krypton inert gas structure.

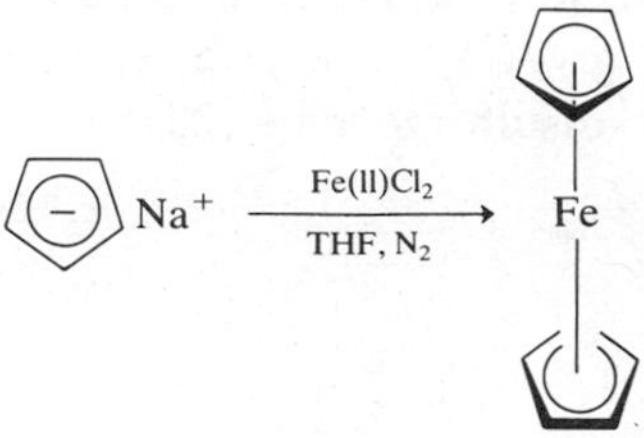

$$\textbf{2}\ \ (C_5H_5^-)\,Na^+ \xrightarrow[\text{THF, } N_2]{\text{Fe(II)Cl}_2} \text{Fe}(C_5H_5)_2\ \ \textbf{20}$$

Figure 5.10

As expected for such a structure with no unpaired electrons, ferrocene is diamagnetic. Alternative extreme descriptions of the ferrocene system have been made in which it is considered to be either two cyclopentadienyl anions bonded to a dipositive iron atom or two neutral cyclopentadienyl rings covalently bonded to a neutral iron atom. The latter description appears

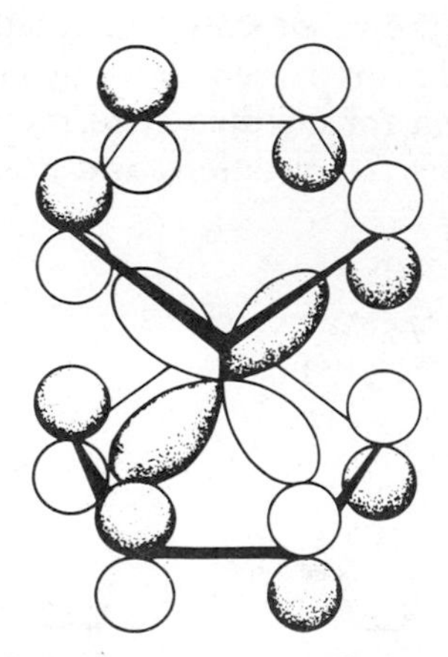

Figure 5.11. Overlap of the e_{1g} orbitals of the cyclopentadienyl rings (shown as p_x atomic orbitals) with the iron $3d_{xy,yz}$ orbitals in ferrocene.

to be more appropriate to the chemistry of ferrocene, and a satisfactory set of molecular orbitals for bonding can be provided. In the early MO treatments, most of the bonding was assumed to be derived from the lowering in energy caused by the mixing of the e_{1g} orbitals of the cyclopentadienyl rings with the $3d_{xy,yz}$ orbitals on the iron (Fig. 5.11). More sophisticated treatments suggest that twelve of the eighteen electrons are involved in the bonding but there is considerable disagreement on the relative energies of the orbitals and on the sign of the partial charge on the iron atom.

The chemistry of ferrocene reveals the aromatic nature of the compound. Friedel–Crafts acylation gives 1,1′-diacylferrocenes (**21**), whereas sulfonation gives ferrocene-1,1′-disulfonic acid (**22**).

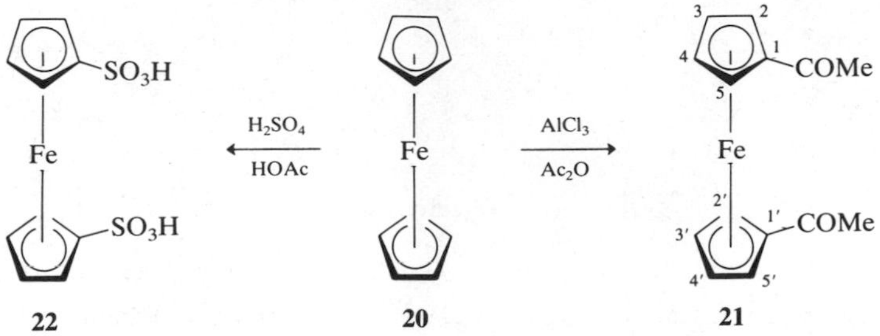

Figure 5.12

In the crystalline state, ferrocene has the cyclopentadienyl rings staggered as shown in formula **20**. The barrier to rotation of the rings is, however, low, and in solution or the vapor state free rotation occurs. This has been shown by both physical measurements and by classic substitution studies. Thus, the dipole moment for the diacetylferrocenes is in good agreement for a structure exhibiting free rotation, and acetylation of monoethylfer-

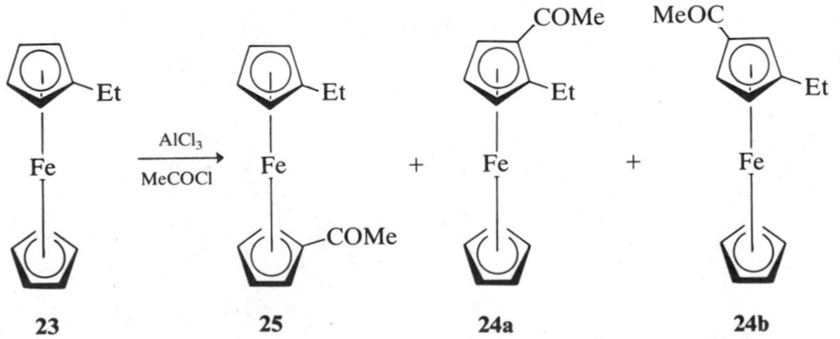

Figure 5.13

rocene (**23**) gives three isomeric acetylethylferrocenes, which were the three isomers expected if free rotation occurs. These isomers were **24a** and **24b**, with both substituents in the same ring, and **25**, with the substituent in different rings. In contrast to ferrocene, ruthenocene and most of the other metallocenes have eclipsed cyclopentadienyl rings.

Attempts to nitrate or brominate ferrocene directly are frustrated in that it is the iron atom that is attacked, being oxidized from the ferrous to the ferric state to give the blue-green ferricenium ion (**26**). The ferricenium ion is paramagnetic and water soluble in contrast to ferrocene, and the paramagnetic moment is consistent with the possession of an unpaired electron. The oxidation is reversible and the ferricenium ion is readily reduced to ferrocene.

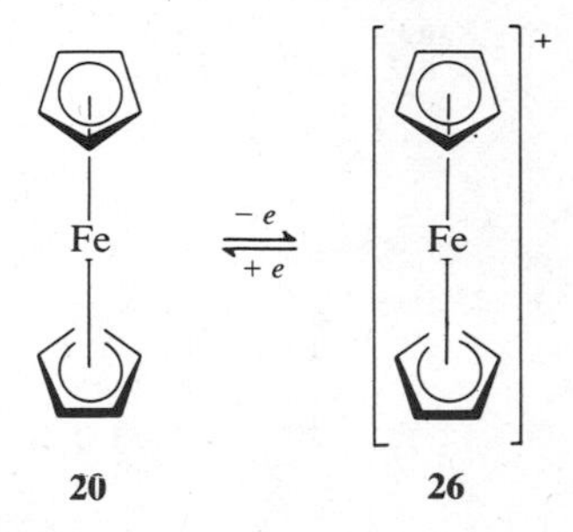

Figure 5.14

A voluminous chemistry of these types of sandwich compounds has been developed, which will not be discussed here, and the interested reader is directed to the references at the end of the chapter.

The chemistry of the tropylium cation (**5**) has been extensively developed. A summary of the methods of preparing **5** is shown in Figure 5.15. The preferred methods of synthesis appear to be that due to Dauben and co-workers in which cycloheptatriene (**3**) is reacted with the trityl cation [$(C_6H_5)_3C^+$], preferably as the perchlorate or tetrafluoroborate in sulfur dioxide or acetonitrile, and that due to Conrow in which **3** is treated with PCl_5 in carbon tetrachloride.

The cation **5** shows the simple IR and Raman spectra expected for a planar structure of D_{7h} symmetry, with no coincidence of the IR and Raman bands. The ^{1}H NMR spectrum shows a singlet at δ 9.28, the position of the chemical shift being in good agreement with that calculated on the basis of the benzene ring current and one-seventh of the positive charge on each carbon atom. The electronic spectrum has an absorption maximum at

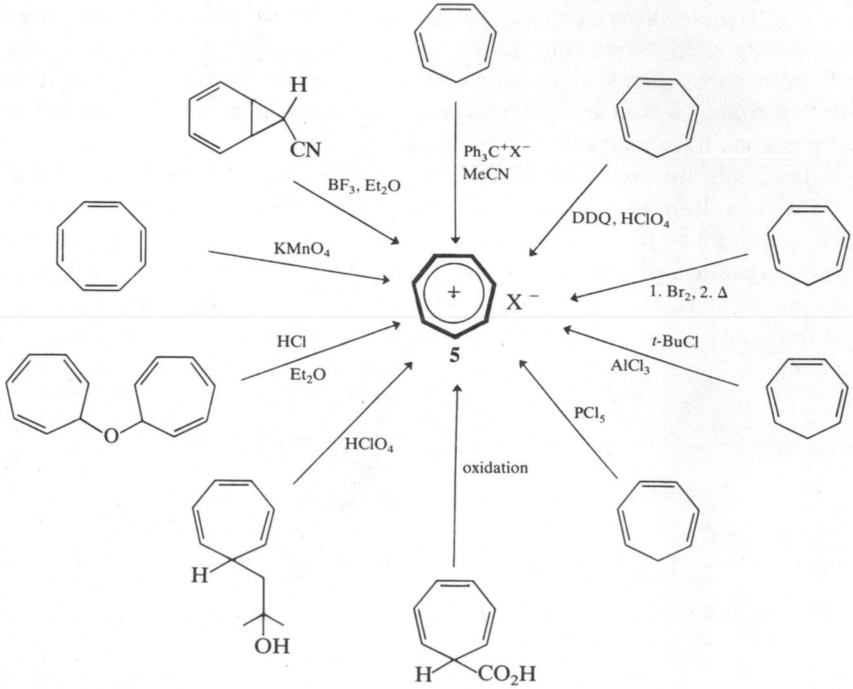

Figure 5.15

275 nm, which is in reasonable agreement with that calculated, 287 nm. The cation has a pK_a of 4.01 and thus has a comparable acidity in water to acetic acid.

The tropylium cation reacts as an electrophile, and its reactions with a variety of nucleophiles is shown in Figure 5.16. With water, hydrogen sulfide, and aqueous ammonia, ditropyl ether (**27**), ditropyl sulfide (**28**), and ditropylamine (**29**) were formed, respectively, whereas with ethereal ammonia tritropylamine (**30**) was formed. Succinimide gave N-tropylsuccinimide (**31**) and organometallic reagents gave 7-substituted cycloheptatrienes (e.g. **32**–**34**).

The cation is readily reduced by zinc to give bitropyl (**35**) and oxidized by chromic acid or silver oxide to benzaldehyde (**36**). The cation **5** can also act as an alkylating agent, and aromatic substitution occurs with phenols and tropolone. Thus, treatment of a mixture of phenol (**37**) and ditropyl ether (**27**) with HCl in ether gave 2-cycloheptatrienylphenol (**38**), whereas treatment of tropolone (**8**) with tropylium bromide (**5**) gave a mixture of **39**, **40**, and **41**. The compounds **39**–**41** are thermally labile and isomerize

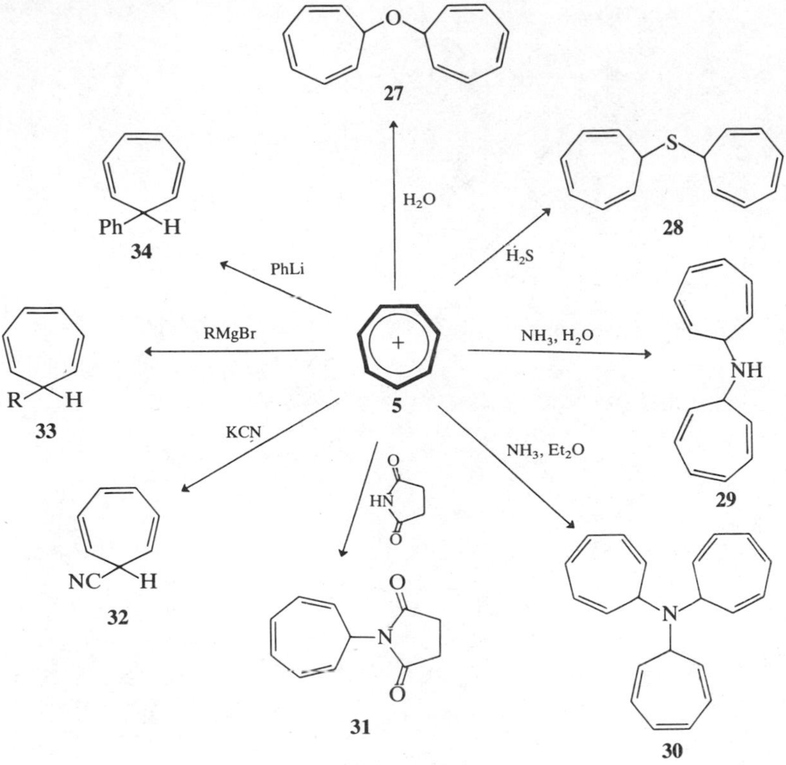

Figure 5.16

by 1,5-hydrogen shifts to give compounds with other arrangements of the double bonds in the cycloheptatrienyl rings.

The cation **5** can, under appropriate conditions, form sandwich compounds of the ferrocene type. Thus, treatment of cycloheptatrienyl molybdenum tricarbonyl (**42a**) with tritylfluoroborate in methylene chloride gave the tropylium complex **43a**, and the chromium complex **43b** can be prepared in an analogous manner. The true sandwich complexes of chromium and manganese (**45a, b**) were prepared by acylation of the corresponding complexes of benzene and the cyclopentadienyl anion (**44a, b**). The ring expansion presumably occurs because of the decreased tendency of the σ-transition state cation to isomerize when complexed to the metal.

In the case of the iron complexes, it appears that the tropylium cation acts as a 4 and not as a 6 π-electron donor. Thus, treatment of the iron tricarbonyl complex of cycloheptatrienyl methyl ether (**46**) with hydrofluoroboric acid gave the complex **47**. The structure **47**, in which the iron

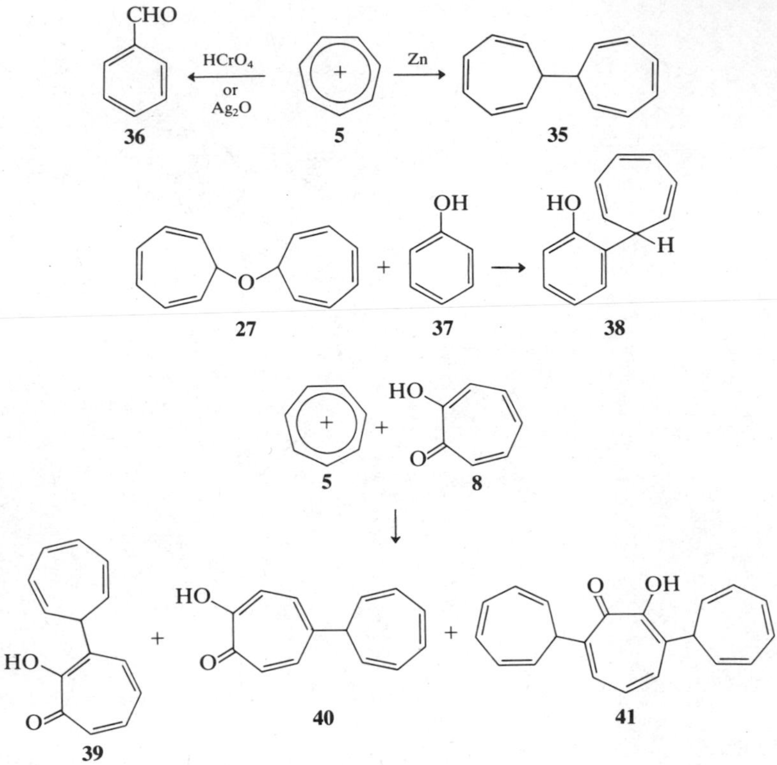

CHO
HCrO$_4$ or Ag$_2$O
Zn
36 5 35

OH HO H
27 37 38

HO
5 8
HO HO O OH
39 40 41

Figure 5.17

is complexed to two of the double bonds with a free allyl cation, was obtained by x-ray crystallographic and NMR studies. The structure is mobile, each of the double bonds in turn participating in bonding to the iron.

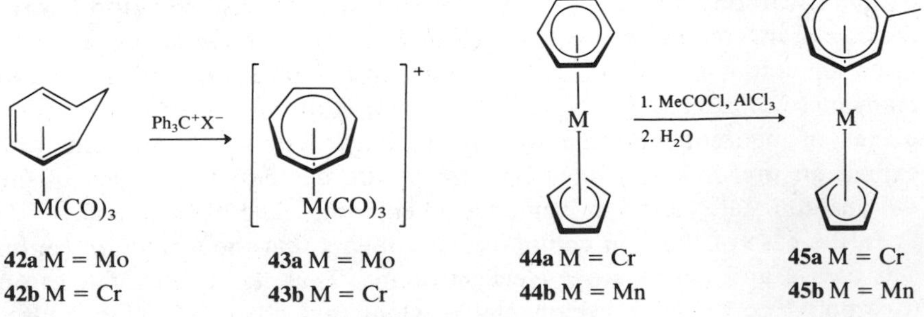

$Ph_3C^+X^-$
M(CO)$_3$ M(CO)$_3$
1. MeCOCl, AlCl$_3$
2. H$_2$O
M M

42a M = Mo
42b M = Cr
43a M = Mo
43b M = Cr
44a M = Cr
44b M = Mn
45a M = Cr
45b M = Mn

Figure 5.18

H OMe $\xrightarrow{HBF_4}$ + ⇌ +

$Fe(CO)_3$ $Fe(CO)_3$ $Fe(CO)_3$

46 **47a** **47b**

Figure 5.19

Numerous substituted and annelated tropylium cations have been prepared and the physical and chemical properties investigated, but these will not be considered here. The synthesis of the tropylium cation led to a search for other charged systems that could potentially be aromatic, and the simpler cyclic ions that contain $(4n + 2)$ π electrons are shown in Figure 5.20.

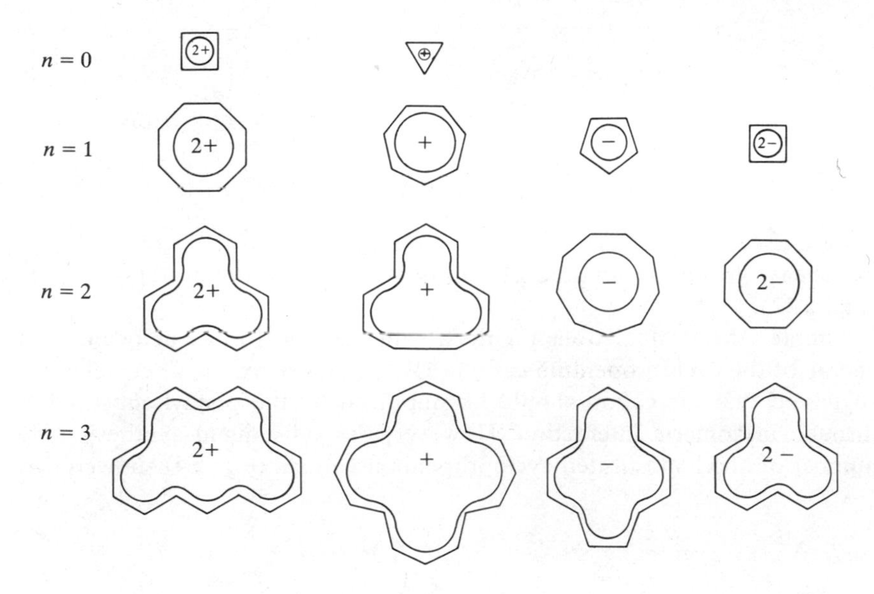

Figure 5.20 Potentially aromatic monocyclic ions containing $(4n + 2)$ π electrons.

The triphenylcyclopropenium cation (**51**), the first 2 π electron aromatic ion to be prepared, was synthesized by Breslow in 1957. Diphenylacetylene (**48**) reacts with phenyldiazonitrile (**49**) to give 1,2,3-triphenylcyclopropenyl cyanide (**50**), which, on treatment with boron trifluoride etherate containing water, gave the cation **51**. The cation reacts with methanol to give the

3-methoxy-1,2,3-triphenylcyclopropene (**52**), and **52** on treatment with HBr regenerates **51** as the bromide.

The ^{1}H NMR spectrum of the cation **51** shows that there is only one type of phenyl ring, and the electronic spectrum is similar to that of **52** except for a greatly increased intensity of absorption. An x-ray crystallographic analysis of **51** confirmed the symmetric structure and revealed that

C_6H_5—≡—C_6H_5 + $C_6H_5C(N_2)CN$ ⟶ **50** (cyclopropene with C_6H_5, C_6H_5, CN, C_6H_5)

48 **49** **50**

50 —(BF_3, Et_2O, H_2O)→ **51**

52 (C_6H_5, C_6H_5, C_6H_5, OMe) ⇌ (MeOH / HBr) **51** (C_6H_5, C_6H_5, C_6H_5, ⊕)

52 **51** X = BF_4, HBF
X = Br

Figure 5.21

the phenyl groups are twisted 21° out of the plane of the cyclopropyl ring (Fig. 5.22).

Simple HMO calculations predict a value of 2.0β for the delocalization energy of the cyclopropenium cation (**53**), and more extensive calculations suggested that the cation should be stabilized by the phenyl substituents through mesomeric interaction. However, the subsequent synthesis of a number of alkyl-substituted cyclopropenium cations (e.g., **54**) showed that

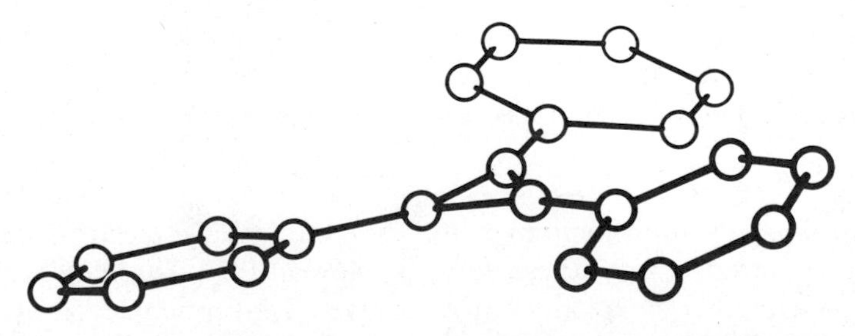

Figure 5.22. X-ray crystallographic structure of the triphenylcyclopropenium cation. Data from M. Sundaralingham and L. H. Jensen, *J. Am. Chem. Soc.*, 1966, **88**, 198.

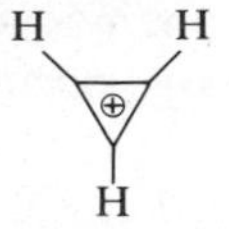

Figure 5.23

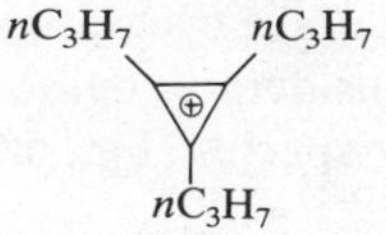

not only were the phenyl substituents unnecessary but that the alkylated ions were more stable than the arylated.

This conclusion was reached from a study of the pK_{R^+} values of the ions, and the results are shown in Table 5.1. Nevertheless, a study of the

TABLE 5.1

Ion	pK_{R^+}
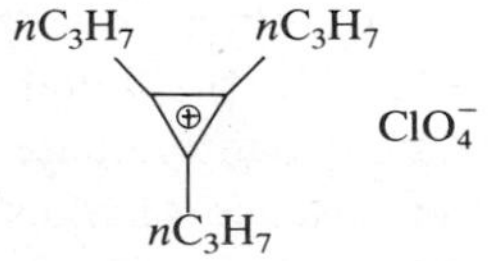 Tri-n-propylcyclopropenyl cation (nC_3H_7, nC_3H_7, nC_3H_7) ClO_4^-	7.2
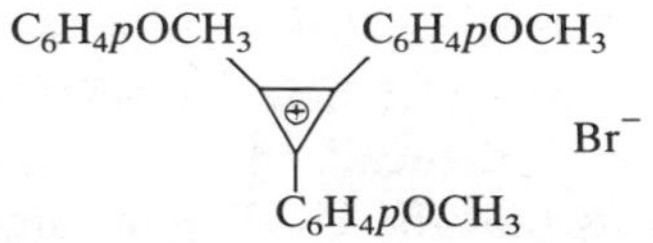 Tris(p-methoxyphenyl)cyclopropenyl cation ($C_6H_4pOCH_3$, $C_6H_4pOCH_3$, $C_6H_4pOCH_3$) Br^-	6.5
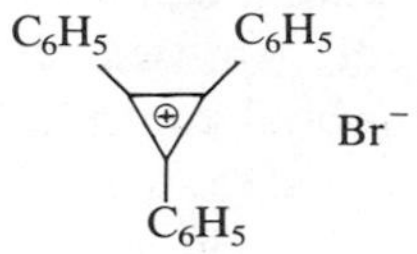 Triphenylcyclopropenyl cation (C_6H_5, C_6H_5, C_6H_5) Br^-	3.1
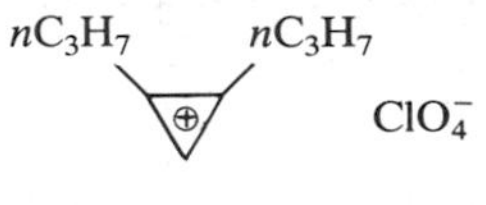 Di-n-propylcyclopropenyl cation (nC_3H_7, nC_3H_7) ClO_4^-	2.7
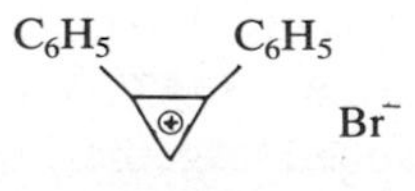 Diphenylcyclopropenyl cation (C_6H_5, C_6H_5) Br^-	− 0.67

SOURCE: Data from R. Breslow, H. Höver, and H. W. Chan, *J. Am. Chem. Soc.*, 1962, **84**, 3168.

^{13}C NMR spectra of the cyclopropenium and triphenylcyclopropenium ions suggests that a considerable delocalization of charge occurs into the benzene rings in the latter species (Fig. 5.24).

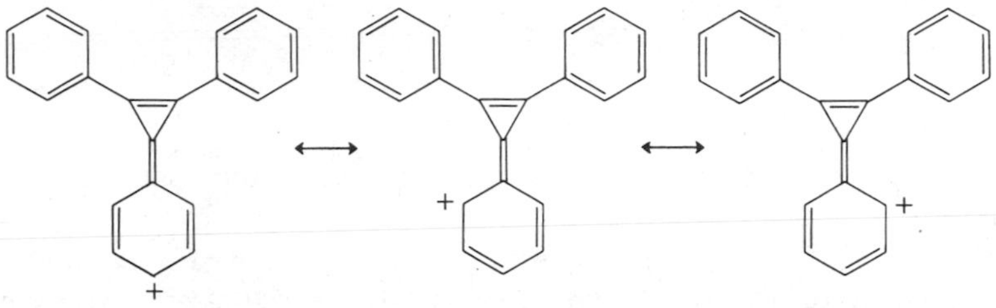

Figure 5.24

The parent cyclopropenium cation (**53**) has more recently been prepared. Reduction of tetrachlorocyclopropene (**55**) with tri-*n*-butyl tin hydride gave a mixture of mono- and dichloropropenes from which 3-chlorocyclopropene (**56**) was isolated. The ^1H NMR spectrum of **56** in SO_2 ionization occurs, and the chlorine migrates rapidly around the ring. When **56** is treated with antimony pentachloride in methylene chloride, a white precipitate is formed, which was shown to be cyclopropenium hexachloroantimonate (**53a**). The salt **53a** is stable at room temperature in the absence of moisture but is rapidly decomposed by water. The IR spectrum is simple, showing only four bands as expected for a molecule of D_{3h} symmetry, and the ^1H NMR spectrum showed only one resonance signal at δ 11.1. The ^{13}C NMR resonance signal is in the expected position for a delocalized cation, and the ^{13}C—H coupling constant, 265 Hz, suggests that the C—H bond has considerable *s* character. An orbital model of the cation can be formulated in which each carbon has an *sp* orbital to the hydrogen, two sp^3 orbitals for the bent cyclopropenyl C—C bonds, and a *p* orbital contributing to the π system (Fig. 5.25). A pK_{R^+} of −7.4 has been obtained for the cation, showing that it is much less stable than the substituted cyclopropenium cations (Table 5.1). However, this pK_{R^+} still indicates that the cyclopropenium cation is considerably thermodynamically more stable (about 75 kJ mol^{-1}) than the corresponding allyl cation.

The corresponding tetrafluoroborate **53b** and tetrachloroaluminate **53c** salts have been prepared, but these are only stable in solution. Evidence for the preparation of the cation **53** in solution has also been obtained when the pyrolysate from the thermal decomposition of **57** is treated with fluorosulfonic acid. Presumably, **58** is produced together with diethyl phthalate (**59**), and **58** subsequently decarboxylates under the influence of the strong acid.

$(nC_4H_9)_3SnH$

55 → 56 → 53

53a X = $SbCl_6$
53b X = BF_4
53c X = $AlCl_4$

CH_3O_2C, CO_2CH_3, CO_2CH_3 (**57**) $\xrightarrow{\Delta}$ [H, $COCH_3$] (**58**) + CO_2CH_3, CO_2CH_3 (**59**)

Figure 5.25

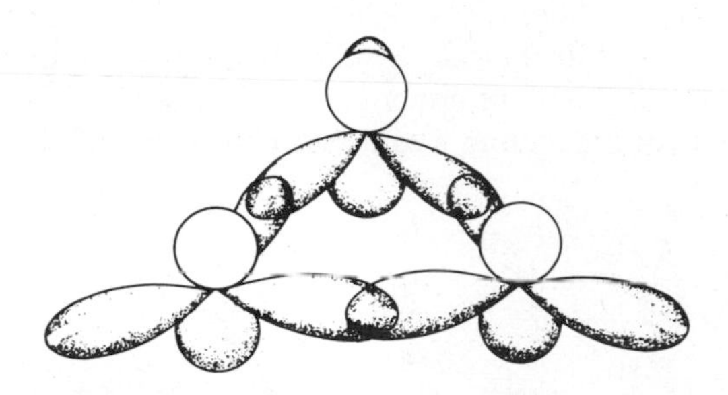

Figure 5.26. The bonding atomic orbitals in the cyclopropenium cation (the phase of the p_π atomic orbitals is indicated by shading).

The cyclobutadienediylium ion (**60**), a second 2 π-electron system, has not been prepared, but a number of derivatives has. Initial attempts to prepare the tetraphenyl **61** and tetramethyl **62** dications gave only the homoaromatic monocations **63** and **64**, respectively (see Chapter 10), but subsequent investigations by Olah and co-workers have led to the preparation of these and other cyclobutadiene dications.

Reaction of the dibromide **65** with SbF_5 in SO_2 at $-60°C$ gave the dication **61**, and reaction of the dichloride **66** with SbF_5–SO_2 gave, first, the previously observed monocation **64**, which was slowly transformed at $-75°C$

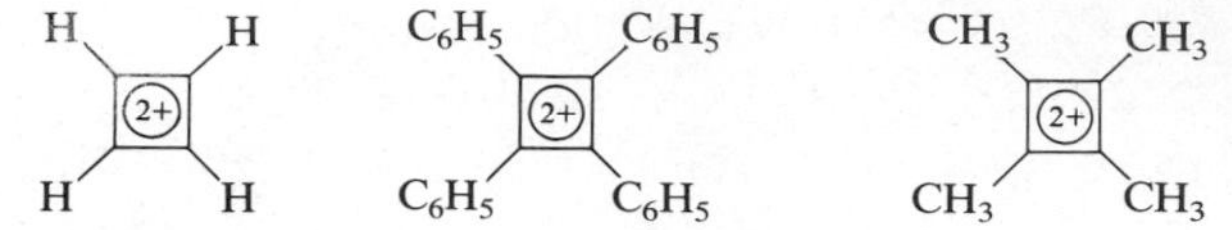

H H
2+
H H
60

C_6H_5 C_6H_5
2+
C_6H_5 C_6H_5
61

CH_3 CH_3
2+
CH_3 CH_3
62

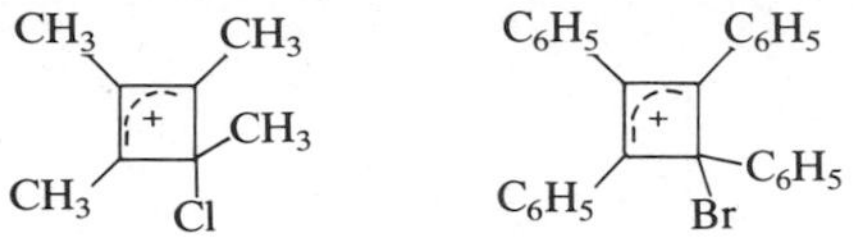

CH_3 CH_3
+ CH_3
CH_3 Cl
63

C_6H_5 C_6H_5
+ C_6H_5
C_6H_5 Br
64

Figure 5.27

to the dication **62**. The diphenyldifluoro **67** and triphenyl **68** dications were also prepared. The ^{13}C NMR spectra showed the ring carbons at low field [**61**, δ 173.4; **62**, δ 209.7; **68**, δ 190.9, 182.16 ($J_{^{13}C-H}$ = 209.6 Hz)] and the ^{1}H NMR spectrum of **68** showed the ring proton at δ 10.68. In the phenyl derivatives, the carbons *para* to the ring are also shifted to low field, indicating a substantial charge leakage to the phenyl rings.

Although in the tetraphenyl dication **61** the phenyl groups are all identical in the ^{13}C NMR, it is unlikely that the phenyl groups are all coplanar with the four-membered ring but are either equally twisted out of the ring plane or are in dynamic equilibrium at a rate that is fast on the NMR time scale.

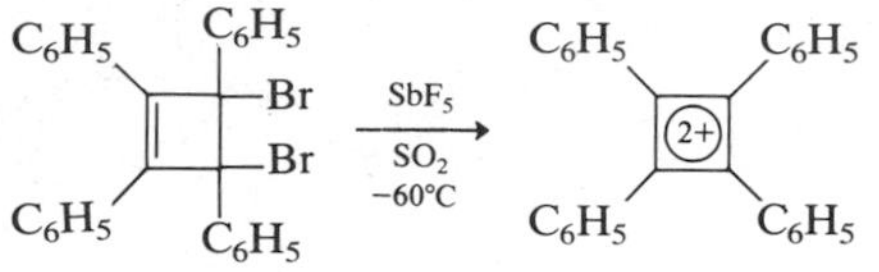

C_6H_5 C_6H_5
Br
Br
C_6H_5 C_6H_5
65

SbF_5
SO_2
−60°C

C_6H_5 C_6H_5
2+
C_6H_5 C_6H_5
61

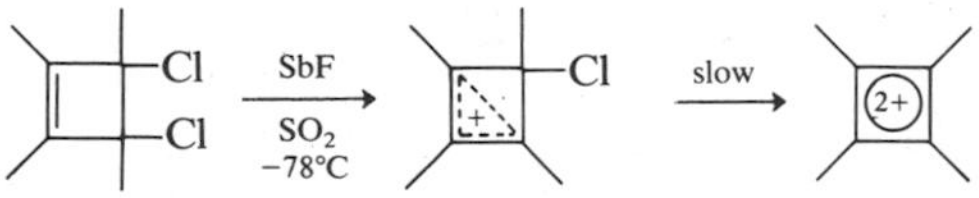

Cl
Cl
66

SbF
SO_2
−78°C

+ Cl
63

slow

2+
62

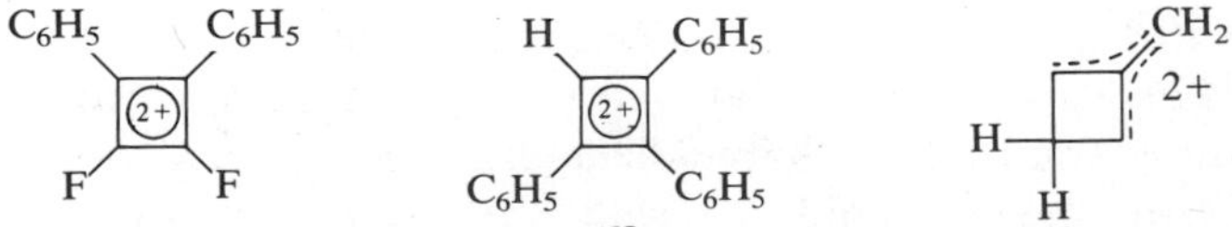

C_6H_5 C_6H_5
2+
F F
67

H C_6H_5
2+
C_6H_5 C_6H_5
68

CH_2
2+
H
H
69

Figure 5.28

Suggestions that the observed spectra arose from equilibrating monocations rather than dications appear to have been satisfactorily rebutted. Calculations suggest that the monomethyl cation would be more stable as the Y cation **69**.

Two theoretically possible 6 π-electron systems are the cyclobutadienyl dianion (**70**) and the cyclooctatetraenyl dication (**71**). Neither of the parent systems has been observed, but derivatives of both have been obtained.

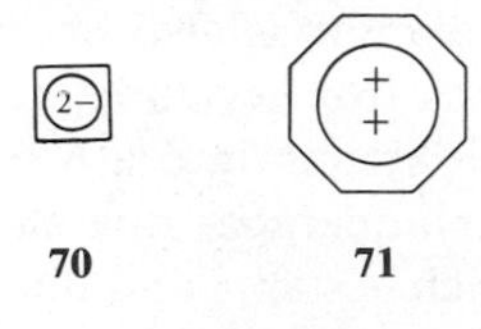

Figure 5.29

Pettit and co-workers produced some evidence for the intervention of the dianion **70** in the reaction of 3,4-dichlorocyclobutene with sodium naphthalide when quenching the product with MeOD gave 3,4-dideuteriocyclobutene. Since the dianion might be expected to be stabilized by ester groups as is the cyclopentadienyl anion, **72** was reacted with lithium di-isopropylamide (LDA) in tetrahydrofuran containing hexamethylphosphoric triamide when the dianion **73** was obtained. The ^{1}H and ^{13}C NMR spectra indicate that most of the charge is located on the ester groups and little is transferred to the nonsubstituted ring carbon atoms. The first pK_a is also similar to that of the saturated dimethyl cyclobutane-1,2-dicarboxylate dianion and to the acyclic dimethyl hex-3-enedioate dianion, and **73** thus does not appear to be a stabilized aromatic system.

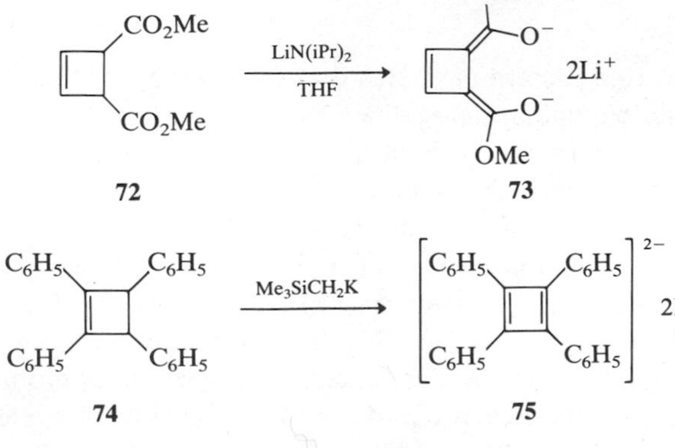

Figure 5.30

Attempts to generate the tetra-*t*-butylcyclobutadienyl dianion from tetra-*t*-butacyclobutadiene were unsuccessful, but treatment of tetraphenylcyclobutene (**74**) with the strongest known base, Me_3SiCH_2K, gave the tetraphenylcyclobutadienyl dianion (**75**). The ^{13}C NMR spectrum indicated that most of the charge was distributed on the phenyl rings and the low acidity of the monoanion, which could be isolated as a stable salt, again indicates that the dianion **75** has little aromatic stability. Calculations suggest that the Y dianion is preferred to the delocalized π dianion for methylcyclobutene, and the parent dianion is also predicted not to be stabilized.

Removal of two electrons from cyclooctatetraene should, in principle, convert the 8 π-electron $4n$ system into a 6 π-electron $(4n + 2)$ dication. However, cyclooctatetraene undergoes one-electron oxidation to form the homotropylium cation, which is stable and does not lose a second electron. Suitably substituted cyclooctatetraenes might destabilize the homotropylium structure and allow further oxidation. This concept was substantiated when it was found that treatment of 1,3,5,7-tetramethylcyclooctatetraene (**76**) with SbF_5 in SO_2ClF at −78°C gave the tetramethylcyclooctatetraenediylium ion **77**. The ^{1}H NMR spectrum shows signals at δ 4.27 and 10.80 for the methyl and ring protons, respectively, and in the ^{13}C NMR spectrum the ring carbons are at δ 182.7 (C—CH_3) and at 170.0 (C—H). On warming to −20°C, the dication rearranges to the bicyclic dication **78**.

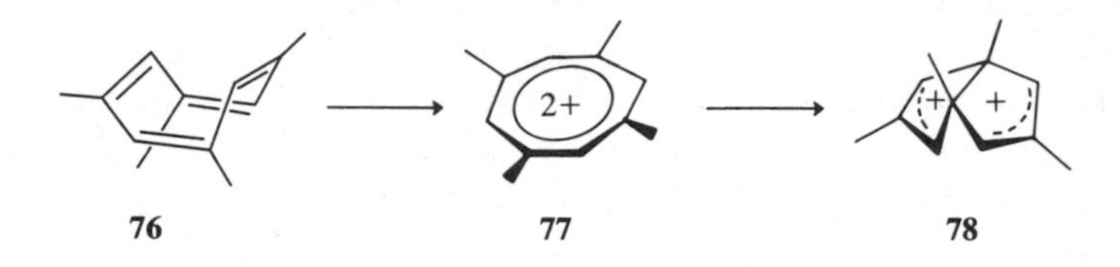

Figure 5.31

It is in the 10 π-electron systems that the greater propensity for charged systems to attain a planar delocalized state as compared to the isoelectronic neutral system is most clearly illustrated. Both the cyclooctatetraenyl dianion (**81**) and the cyclononatetraenyl dianion (**86**) have been prepared, the latter in two configurations, as planar delocalized diatropic systems, in sharp contrast to the nonplanar, atropic [10]annulenes (Chapter 4).

The cyclooctatetraenyl dianion (**81**) can be prepared by treatment of solutions of cyclooctatetraene (**79**) in ether, tetrahydrofuran, or liquid ammonia with alkali metals. The reaction was originally reported by Reppe and co-workers in their classic paper on the synthesis of cyclooctatetraene, and these investigators showed that a dialkali metal salt was formed since carboxylation gave a dicarboxylic acid. Quenching the solution of the salt with water gave a mixture of cyclooctatrienes. A number of authors suggested that the dianion might have aromatic character, but the first firm experimental evidence for this suggestion was provided by Katz in 1960, who studied the reaction of cyclooctatetraene with alkali metals by ESR and NMR spectroscopy. Reduction of cyclooctatetraene occurs by two one-electron processes to give first the radical anion **80** and then the dianion **81**. The

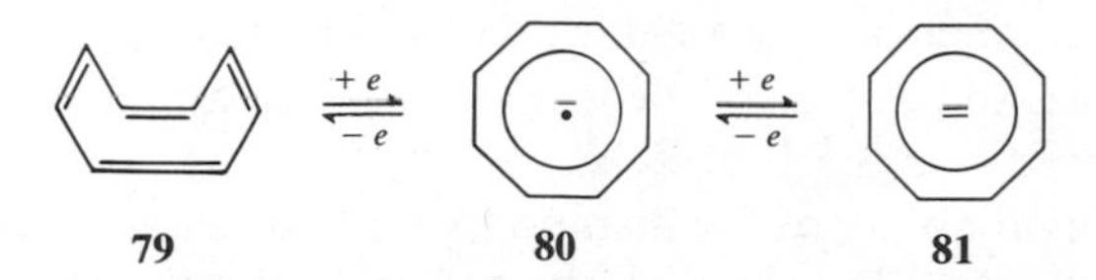

Figure 5.32

hyperfine structure of the ESR spectrum showed nine lines in the appropriate binomial ratio for eight equivalent protons, and the hyperfine splittings were consistent with a structure of D_{8h} symmetry. The ^{1}H NMR spectrum showed a single resonance at δ 5.7, and the IR spectrum was also simple. The chemical shift of the ^{1}H NMR signal is almost identical with that of cyclooctatetraene and indicates that the shielding effect of the one-fourth excess electron density on each carbon atom just balances the deshielding effect of the ring current. The calculated chemical shift of the *cis*-cyclononatetraenyl anion based on the value for **81** is in excellent agreement with that found (see later and Chapter 2).

The addition of the first electron to cyclooctatetraene leads to the formation of the radical anion **80** having a planar conformation. The pseudo-Jahn–Teller distortion that would have occurred in planar cyclooctatetraene (see Chapter 2) is removed by the addition of an electron that gives a nondegenerate state. The addition of the second electron fills the nonbonding orbitals and gives a closed shell configuration (Fig. 5.33). Unlike the other ions

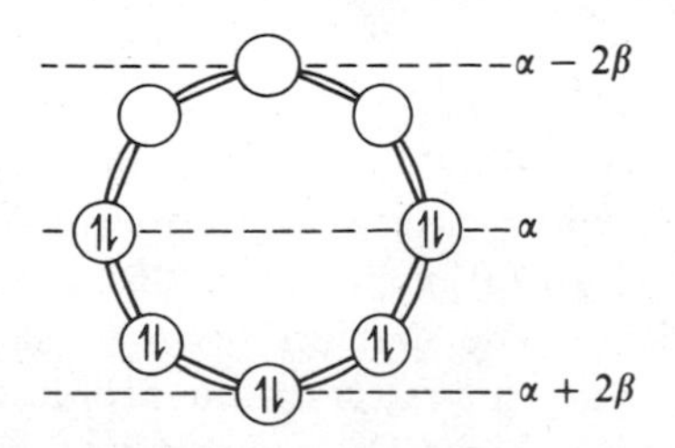

Figure 5.33

previously considered, the cyclooctatetraenyl dianion, in the simple HMO picture, does not gain in resonance energy on the addition of the electrons. The adoption of the planar configuration for the anion radical and the dianion can be attributed to the removal of the degenerate electronic state that would occur in planar cyclooctatetraene and to the delocalization of charge that can occur in the planar form. It appears that the main barrier to reduction occurs in the addition of the first electron when the ring must flatten. Early studies suggested that this radical anion **80** rapidly disproportionated to cyclooctatetraene and the dianion **81**, but this disproportionation has been shown to be strongly dependent on the nature of the counter ion and solvent used.

The ready availability of the dianion **81** has led to an extensive investigation of its chemistry. The dianion can act both as a reducing agent and as a nucleophile. Thus, tropylium bromide **5** is reduced by **81** to ditropyl (**82**), and **81** is alkylated by methyl iodide to give the dimethylcyclooctatetraenes **83** and **84**. The trienes **83** and **84** are thermally labile and rearrange readily to other products.

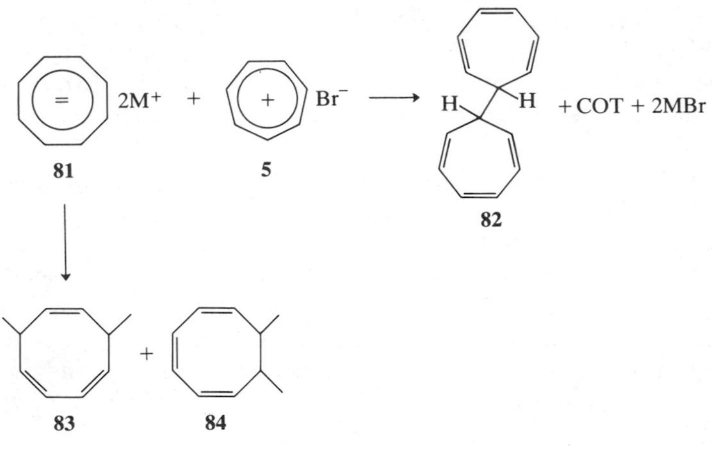

2M+ + Br⁻ → H H + COT + 2MBr

81 5 82

83 + 84

Figure 5.34

The dianion **81** reacts with aldehydes, ketones, and acid chlorides as a nucleophile to give a wide variety of products. With *gem*-dihalides the dianion gives *cis*-bicyclo[6.1.0]nonatrienes (**85**), a reaction that is preparatively useful (Fig. 5.35). The dianion forms a sandwich complex with uranium and thorium, and an x-ray crystallographic analysis of the uranium compound has shown it to be of D_{8h} symmetry with planar, eclipsed C_8H_8 rings (Fig. 5.36).

R^1
R^2
85
$R^1R^2CCl_2$
CO_2
CO_2H
CO_2H
CHO
CHO
81
C_6H_5COCl
H OH
H OH
O
C_6H_5
C_6H_5
O
+
O
OCC_6H_5
C_6H_5

Figure 5.35

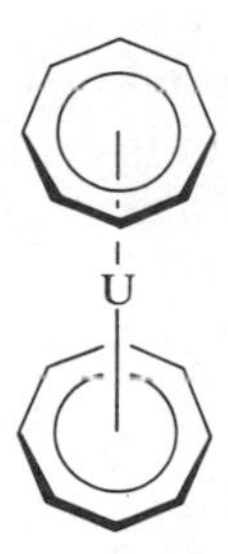

Figure 5.36

The physical and chemical properties of the cyclooctatetraenyl dianion support the view that this is a planar, diatropic, 10 π-electron aromatic system.

The cyclononatetraenyl anion, the next higher homologue of the cyclopentadienyl anion, has been synthesized both in the thermodynamically stable all-*cis* form **86** and in the mono-*trans* form **91**.

The all-*cis* anion **86** can be prepared by treatment of 9-chlorobicyclo[6.1.0]nonatriene (**87**) with lithium, by treatment of 9-methoxybicyclo[6.1.0]nonatriene (**88**) with potassium, and by deprotonation of bicyclo[6.1.0]nonatriene (**89**) or cyclononatetraene (**90**) with base

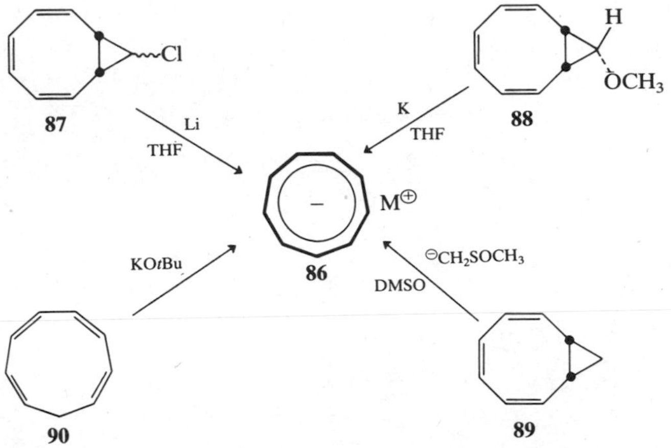

Figure 5.37

(Fig. 5.37). The formation of the all *cis*-anion **86** from anti-9-methoxy-bicyclo[6.1.0]nonatriene (**88**) with potassium, presumably via the cyclopropyl anion, is a process that is forbidden if concerted since conrotatory ring opening of the cyclopropyl anion is predicted and the mono-*trans* anion should result. This led to an investigation of this reaction at −40°C when it was found that the mono-*trans* anion **91** is, in fact, initially formed (Fig. 5.38).

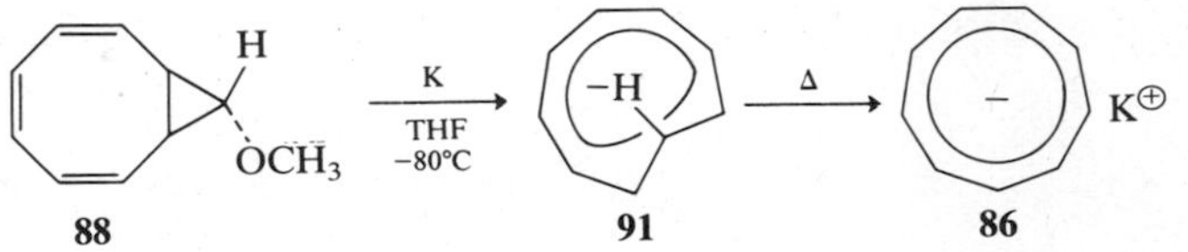

Figure 5.38

The anion **86**, which can be prepared as a crystalline tetramethylammonium salt, shows only one resonance signal in the ^{1}H NMR spectrum at about δ 7.0, depending on the nature of the counter ion and solvent. The position of this absorption is in good agreement with that calculated from the position of the proton signal in the cyclooctatetraenyl dianion, assuming that both systems have the same diamagnetic ring current. The electronic spectrum is simple, showing a strong absorption at 251 nm and a double, weaker maximum at about 320 nm, and a calculated spectrum for a planar

D_{9h} structure was in accord with that found experimentally. The simple infrared spectrum is also that expected for a molecule of D_{9h} symmetry.

Equilibrium studies with cyclopentadiene and indene indicated that **86** is thermodynamically more stable than the cyclopentadienyl anion (**2**) and that cyclononatetraene must have a pK_a between 16 and 21 on the Streitwieser scale. In accord with this view, cyclononatetraenyl trimethylstannate more readily forms an ion pair than does cyclopentadienyl trimethylstannate. The anion **86** reacts with water to give *cis*-cyclononatetraene (**90**), which is readily thermally rearranged to 8,9-dihydroindene (**92**). Hydrogenation of **86** over Pt/C gives a mixture of products of which cyclononane (**93**) is the major component. Carboxylation gives 1-carboxy-8,9-dihydroindene (**94**), and methylation with methyl iodide gives 1-methyl-8,9-dihydroindene (**95**). The anion **86** reacts with N,N,N′,N′-tetramethylchloroformadinium chloride (**96**) to give the nonafulvalene **97**. These reactions are illustrated in Figure 5.39.

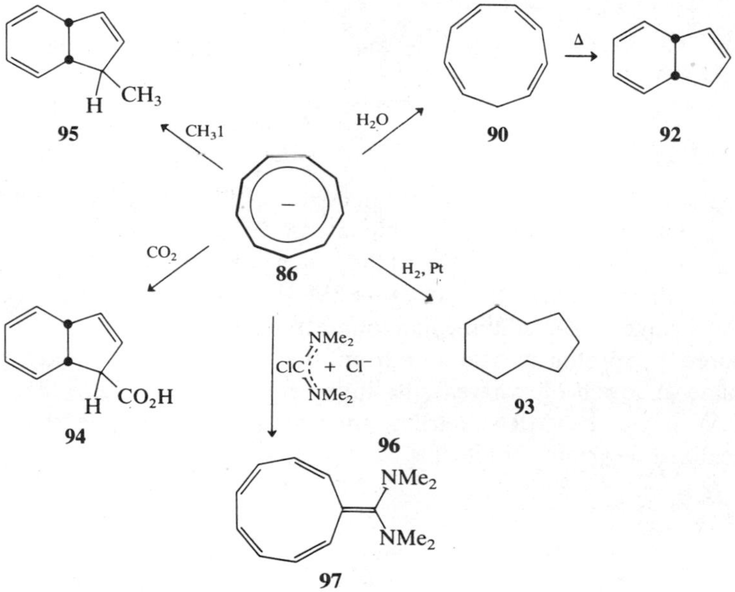

Figure 5.39

The physical and chemical properties of the *cis*-cyclononatetraenide ion are those expected for a planar 10 π-electron aromatic system.

Initially it was found that the mono-*trans* anion **91** was only stable below −10°C, rearranging to the all-*cis* anion above this temperature, but it was later shown that if the anion was removed from excess potassium it was thermally stable. The mono-*trans* anion **91** undergoes a topomerization similar to that found in the annulenes (**91**⇆**91a**⇆**91b**⇆, etc.), and for the pure anion the rate of topomerization is much faster than the rate of isomerization (Fig. 5.40).

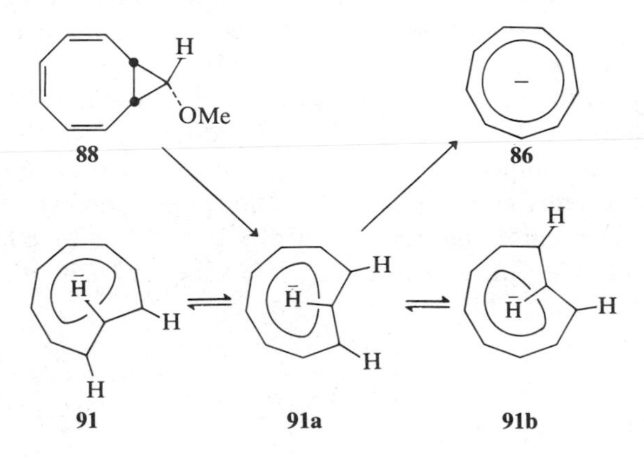

Figure 5.40

The low-temperature ^{1}H NMR spectrum of **91** shows the characteristic diatropic spectrum found in the higher $4n + 2$ annulenes, with the outer protons at low field (δ 7.3–6.4) and the inner protons at high field (δ −3.52). Like the all-*cis* isomer, **91** is an aromatic 10 π-electron system.

The bridged 1,6-methanocyclononatetraenyl anion (**99**) has also been prepared by proton abstraction from the triene **98**. The outer protons resonate at low field, whereas the inner protons are at high field (about δ −0.5) in the ^{1}H NMR spectrum, and this anion also appears to have a delocalized, aromatic structure.

$(CD_3)_2SO$

$Na^+\bar{C}D_2SOCD_3$

Na^+

98 **99**

Figure 5.41

Considerable progress has been made over the last few years in the preparation of macrocyclic aromatic annulenyl ions. The undecapentenium ion (**100**) remains unknown, but the bridged ion (**102**) has been prepared as its fluoroborate by treatment of the hydrocarbon **101** with trityl fluoroborate. The cation **102** is a diatropic system, the outer protons resonating at low field in the ^{1}H NMR spectrum, whereas the methylene protons appear as an AX system at δ −0.3, −1.8 (J = 10 Hz). The electronic spectrum is similar to the benzotropylium cation, and **102** is thermodynamically more stable than the tropylium ion.

BF_4^-

100 **101** **102**

Figure 5.42

The bridges in **99** and **102** remove the nonbonded interaction that would occur in the di-*trans* ions. The x-ray crystallographic data for **99** show the transannular C-1, C-6 distance to be long (229.9 pm), and this suggests that **99** is better represented as an open anion rather than a homobenzotropylium ion. This may reflect the availability of better charge delocalization in the open anion.

[12]Annulene, which is extremely unstable, can be reduced polarographically or with alkali metals to the 14 π-electron [12]annulenyl dianion (**104**), which is a stable, diatropic system. The dianion shows three signals in the ^{1}H NMR spectrum at δ 6.98, 6.23, and −4.6 and is temperature independent. The spectrum is unaltered on warming to 30°C, clearly showing that the dianion is much more stable than [12]annulene, and the temperature independence of the spectrum shows that there is no interconversion of inner and outer protons, again in complete contrast with **103** where the interconversion is rapid.

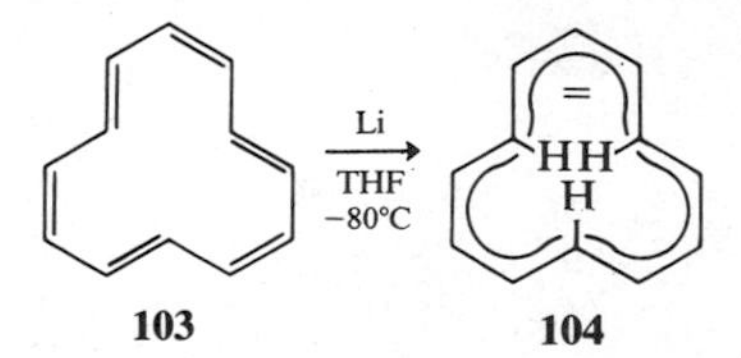

Figure 5.43

Both bisdehydro[12]annulene (**105**) and tridehydro[12]annulene (**108**) are reduced by potassium, first to the corresponding radical anions (**106**,

109) and then to the dianions (**107, 110**). Both the dianions are diatropic, the outer protons appearing at low field and the inner proton of **107** at very high field (δ −6.88). The bridged [12]annulenyl dianions **112** and **114** have been prepared by the reduction of 1,7-methano[12]annulene (**111**) and 1,6-methano[12]annulene (**113**), respectively, again with the radical anions intervening. In both anions the ring protons are at lower field in the ^{1}H NMR spectrum than the ring protons of the corresponding [12]annulene and the methylene bridge protons are at much higher field (**112**, δ −6.44; **114**, δ −5.52, −6.08), and these ions are diatropic.

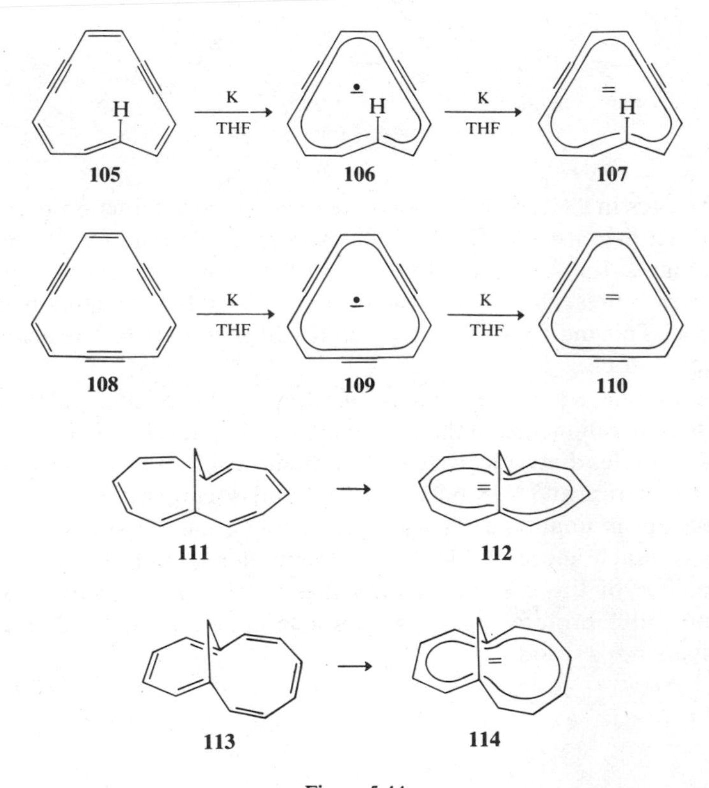

Figure 5.44

In all of these cases, reduction has converted a paratropic 12 π-electron $4n$ system into a diatropic 14 π-electron $(4n + 2)$ system, a change homologous with the conversion of [8]annulene to its dianion.

A bridged [13]annulenyl anion has been prepared by treatment of aceheptylene (**115**) with lithium in liquid ammonia followed by Me_2SO_4 to give the methylated derivative **116**, which was then deprotonated to give the anion **117**. The inner methyl protons in the ^{1}H NMR spectrum are at δ -3.75 and this appears to be a diatropic, 14 π-electron system.

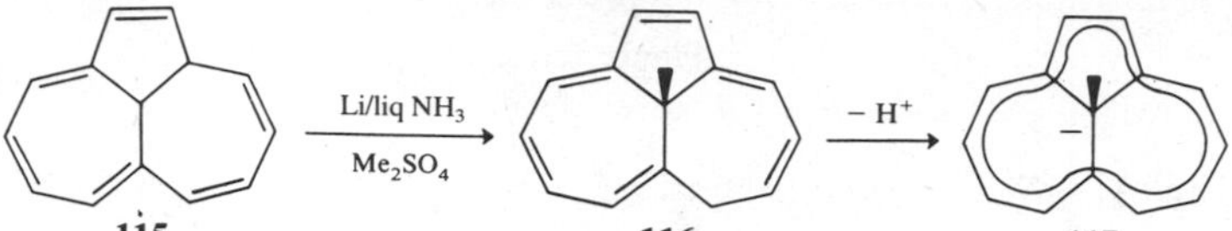

Li/liq NH_3, Me_2SO_4 ; $-H^+$

115 **116** **117**

Figure 5.45

The [15]annulenium ion is at present unknown, but a polycyclic cation with a 15C-14π periphery has been prepared. Treatment of **118a,b** with trityl fluoroborate gave the cation **119**. The ^{1}H NMR spectrum indicates that **119** is a diatropic, delocalized species and is best considered as having a 14 π periphery slightly perturbed by the central double bond. The pK_{R^+} value >8.4 reveals it to be the most stable hydrocarbon carbocation known.

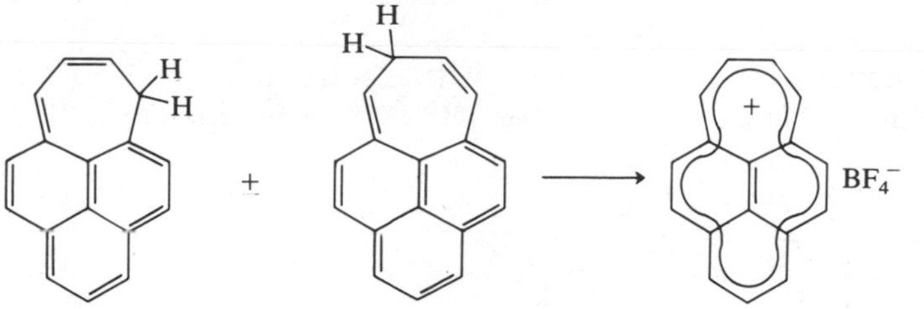

BF_4^-

118a **118b** **119**

Figure 5.46

The higher homologue of **102**, the bridged 15C-14π cation **120**, has recently been prepared and shown to be diatropic. The corresponding neutral radical and radical dianions of **102** and **120** have been prepared by the reduction of the respective cations, and the ESR spectra suggest that these are best described as peripheral delocalized systems lacking a substantial homoconjugative interaction across the bridges (e.g., **121**). This is in accord with the cations having a peripherally delocalized structure rather than being homoaromatic species.

The [16]annulenediylium ion (**123**), a 16C-14π system, has been prepared by treating [16]annulene (**122**) in SO_2–CD_2Cl_2 at -80°C with fluorosulfonic

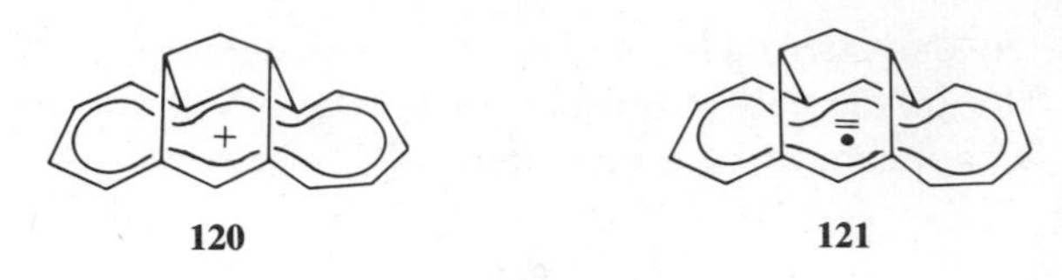

Figure 5.47

acid. The ^{1}H NMR spectrum shows 5 inner protons at high field and 11 outer protons at low field, and the dication clearly has the configuration shown. [16]Annulene thus differs from cyclooctatetraene in that it prefers two-electron oxidation to the dication rather than one-electron oxidation to the homoconjugated [15]annulenium cation.

The [17]annulenyl anion (**125**) has been prepared by deprotonation of cycloheptadecaoctaene (**124**), itself prepared from a dimer of cyclooctatetraene by addition of ethyl diazoacetate, photoinduced valence tautomerism, and decarboxylation. The ^{1}H NMR spectrum of **125** shows 12 outer protons at low field and 5 inner protons at very high field ($\delta -7.97$). Solutions of **125** are dark green, and the rich electronic spectrum extends to 673 nm (ε 7700). The anion is thermally stable, and the acidity of **121** is of the same order as cyclopentadiene. The anion **125** is clearly a diatropic, aromatic 18 π-electron system.

Reduction of [16]annulene with potassium in THF gave the [16]annulenyl dianion (**126**). The ^{1}H NMR spectrum shows the 12 outer

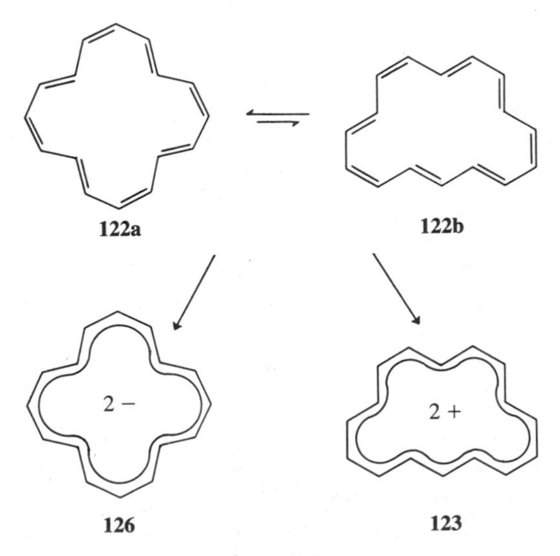

Figure 5.48

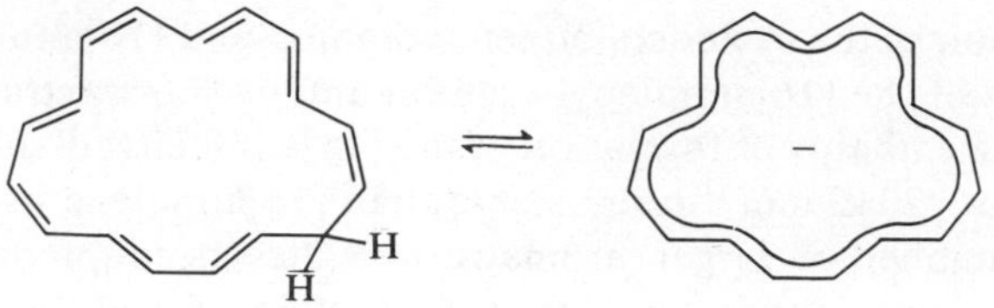

124 ⇌ 125

Figure 5.49

protons at low field and the four inner protons at very high field (δ −8.17), and **126** is thus diatropic. The dianion appears to exist in only one configurational form, and this is different to that adopted by the dication **123**. The ^{1}H NMR spectrum of **126** is temperature independent up to 140°C, which is in striking contrast to the behavior of the isoelectronic [18]annulene (see Chapter 4), and the delocalization energy of **126** must be about 40 kJ mol^{-1} greater than that of [18]annulene. A similar finding was made for the [12]annulenyl dianion (**104**), which, unlike [12]annulene or the isoelectronic [14]annulene, is conformationally rigid. These observations once more emphasize the greater resonance energy experienced by charged molecules, as compared to the corresponding isoelectronic neutral systems, on attaining a delocalized state.

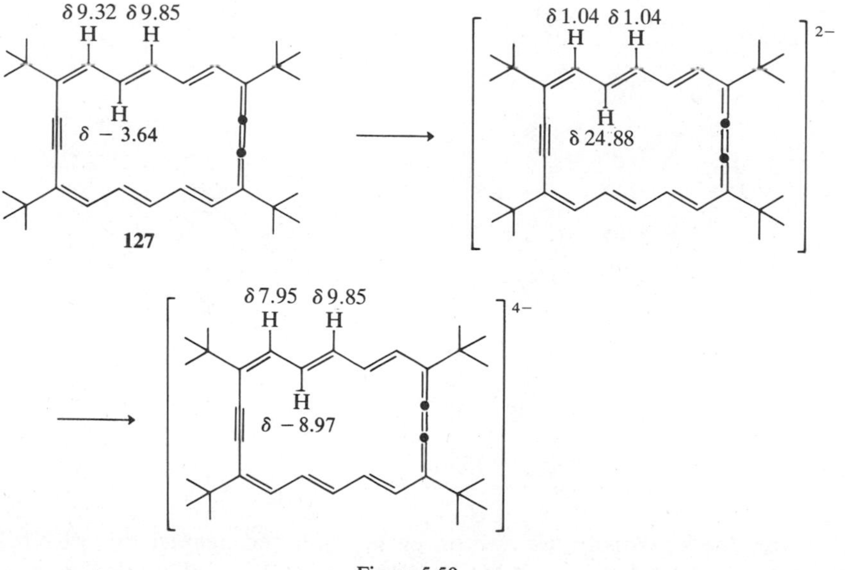

δ 9.32 δ 9.85 H H H δ − 3.64 **127** → [δ 1.04 δ 1.04 H H H δ 24.88]$^{2-}$ → [δ 7.95 δ 9.85 H H H δ − 8.97]$^{4-}$

Figure 5.50

Equilibration of the cyclooctatetraenyl dianion and [16]annulene favors the formation of the [16]annulenyl dianion and cyclooctatetraene, and an estimate of the enthalpy of formation of the [16]annulenyl dianion suggests that it is about 420 kJ mol^{-1} more stable than [16]annulene itself.

A small number of larger aromatic ions has been prepared, all of them derived from dehydroannulenes. Reduction of Nakagawa's $4n + 2$ π-electron dehydroannulenes first gave a series of $4n$ paratropic dianions, which could be reduced further to the corresponding $4n + 2$ diatropic tetra-anions. The ^1H NMR chemical shifts dramatically illustrate these changes in electronic configuration, and these are illustrated for 3,9,12,18-tetra-*t*-butyl-1,10-bisdehydro[18]annulene (**127**) and its dianion and tetra-anion in Figure 5.50.

Reduction of octadehydro[24]annulene (**128**) with potassium in THF first gave the anion radical **129**, which showed the expected nine-line ESR spectrum for a molecule with eight equivalent protons, that on further reduction gave the dianion **130**. Both **129** and **130**, unlike the precursor **128**, appear to be planar systems, with the electronic spectra indicating that the ions are delocalized.

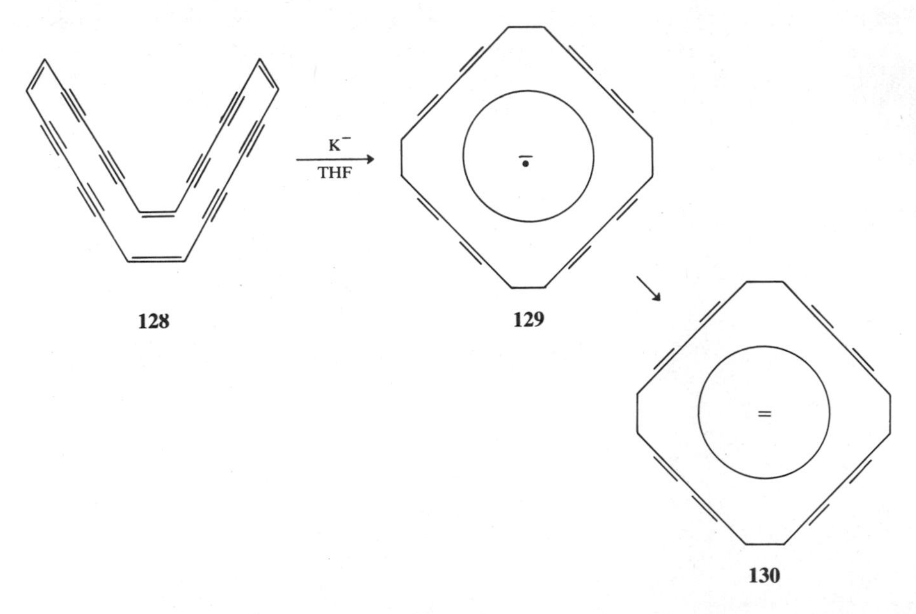

Figure 5.51

If the charge density per carbon atom is plotted against the ^1H NMR chemical shift for the series $(CH)_5^-$, $(CH)_6$, and $(CH)_7^+$, taking the chemical

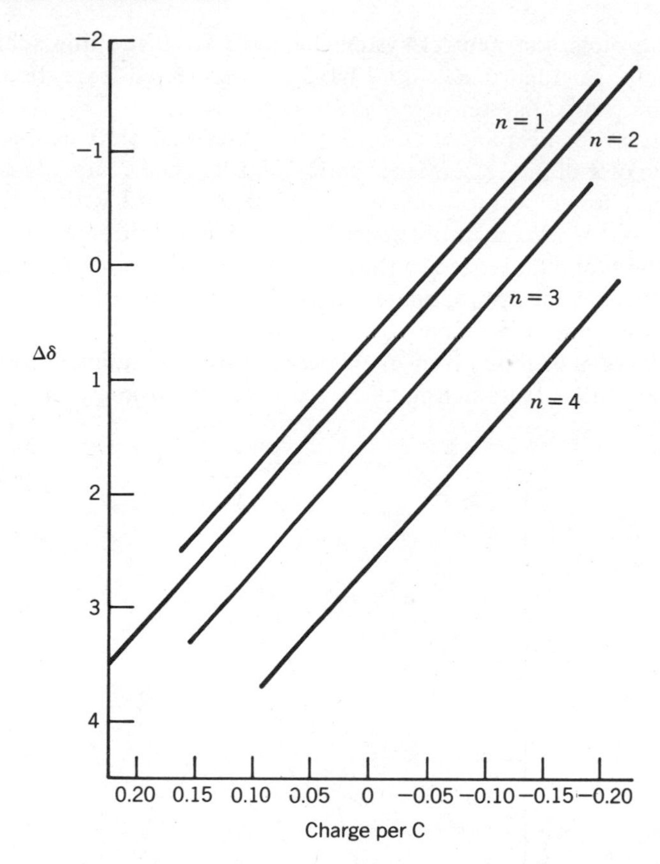

Figure 5.52. Proton chemical shift relative to benzene plotted against charge density per carbon atom for $4n + 2$ ions with different values of n.

shift of benzene as zero, then a straight line is obtained (Fig. 5.52). If the chemical shifts of the 10 π-electron ions $(CH)_8^{2-}$ and $(CH)_9^-$ are now added, they deviate from the line, showing smaller shielding effects than expected. However, if a plot of the ^1H NMR chemical shifts of the 10 π-electron ions $(CH)_8^{2-}$, $(CH)_9^-$, and $(C_{12}H_{11})^+$ (**102**) against charge density per carbon is made a second linear plot is obtained, slightly displaced from the first. This displacement can be attributed to a difference in the magnitude of the diamagnetic ring current in the two series, that in the 10 π series being greater. The slopes of the two lines are approximately parallel, as would be expected for an effect dependent upon charge density. Introducing the 14 π-electron and 18 π-electron ions provides two more parallel, linear plots, again the ring current increasing with the increasing value of n. For each of the plots, the neutral system deviated from the line such that it appears more shielded than would have been expected from the chemical shifts of the charged systems.

In contrast to the plot of the ^{1}H NMR chemical shift against charge density, the plot of the ^{13}C NMR chemical shift against charge density gives a linear plot for all values of *n* (Fig. 5.53). In accord with the Waugh–Fessenden and McWeeny profile of the induced magnetic field, the independence of the plot with regard to the value of *n* suggests that the magnitude of the induced field approximates to zero at the ring carbons, and consequently this plot only reflects the effect of charge.

The examples of ions given in this chapter nicely indicate the value of the Hückel Rule in predicting the occurrence of aromatic systems. The

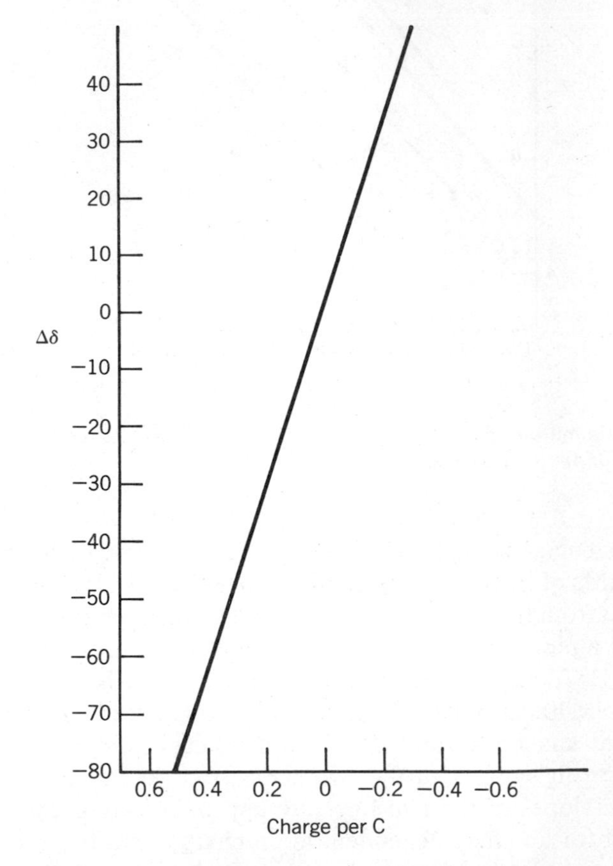

Figure 5.53. Carbon-13 chemical shift relative to benzene plotted against charge density per carbon atom for $4n + 2$ ions with different values of *n*.

smaller ions have been extensively investigated, and over the past 15 years much more has been learned about the larger members of the series. The limit at which bond alternation effects predominate has not been reached, and it may be expected to be at considerably larger ring sizes than the corresponding neutral systems. The magnitude of the resonance energy of the [16]annulenyl dianion and the preparation of macrocyclic polyanions suggest the possibility of synthesizing very large stabilized cyclic charged systems.

FURTHER READING

For a general review of these systems, excluding the cyclopentadienyl anion, the cycloheptatrienyl cation, and the smaller members of the series, see P. J. Garratt and M. V. Sargent, in J. P. Snyder (Ed.), *Nonbenzenoid Aromatics*, Vol. 2, Academic Press, New York, 1971.

See also D. Bethell, in J. Frazer Stoddart (Ed.), *Comprehensive Organic Chemistry*, Vol. 1, 411, Pergamon, Elmsford, N.Y., 1979, p. 411.

For early work on the tropylium cation see T. Nozoe, in Sir James Cook and W. Carruthers (Eds.), *Progress in Organic Chemistry*, Vol. 5, Butterworths, Woburn, Mass., 1961. For some aspects of later work on tropylium systems, see Shô Itô and Y. Fujise, in T. Nozoe, R. Breslow, K. Hafner, Shô Itô, and I. Murata (Eds.), *Topics in Nonbenzenoid Aromatic Chemistry*, Vol. 2, Hirokawa, Tokyo, 1977, p. 91.

Extensive reviews on the chemistry of ferrocene and related systems have appeared. See, for example, *Comprehensive Organometallic Chemistry*, Vols. 4, 5, and 6, Sir Geoffrey Wilkinson (Ed.), Pergamon, Elmsford, N.Y., 1982; P. L. Pauson, *Organometallic Chemistry*, Arnold, London, 1967; D. E. Bablitz and K. L. Rinehart, *Organic Reactions*, 1969, **17**, 1; G. Wilkinson and F. A. Cotton, *Progress in Inorganic Chemistry*, Vol. 1, Interscience, New York, 1959, p. 1; M. Rosenblum, *Chemistry of the Iron Group Metallocenes*, Part 1, Wiley, New York, 1965.

For a review of bridged ions containing 10 and 14 π-electrons, see E. Vogel, *23rd International Congress of Pure and Applied Chemistry*, Vol. 1, Butterworths, Woburn, Mass., 1971.

For macrocyclic annulene ions, see G. Schröder, *Pure Appl. Chem.*, 1975, **44**, 925.

See also P. J. Garratt in, J. Frazer Stoddart (Ed.), *Comprehensive Organic Chemistry*, Sir Derek Barton and W. D. Ollis (series Eds.), Vol. 1, Pergamon, Elmsford, N.Y., 1979, p. 361, D. Lloyd, *Non-benzenoid Conjugated Carbocyclic Compounds*, Elsevier, Amsterdam, 1984, and K. Müllen, *Pure Appl. Chem.*, 1986, **58**, 177.

6

MONOCYCLIC ANTIAROMATIC IONS

Besides the series of $(4n + 2)$ π-electron ions discussed in the preceding chapter, there is a corresponding series of ions with $4n$ π electrons. In the HMO theory, the $4n$ π-electron ions, like the neutral $4n$ annulenes, are predicted to have open, triplet, ground-state configurations. Thus, the simplest member of the group, the cyclopropenyl anion (**1**), has three molecular orbitals, two of which form a degenerate antibonding pair. These are shown in Figure 6.1.

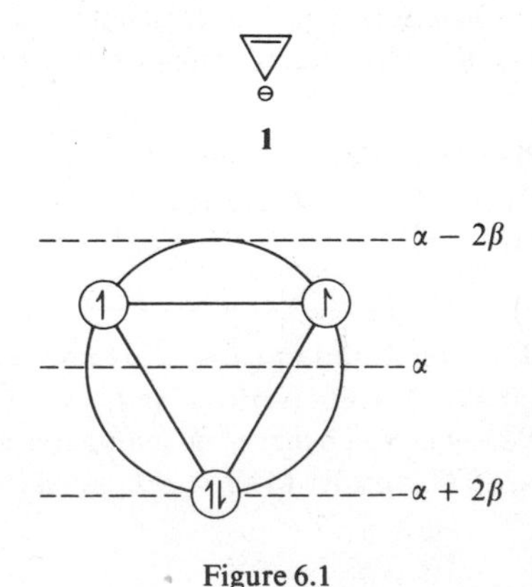

Figure 6.1

The simple HMO calculation suggests that the cyclopropenyl anion will have no delocalization energy (DE = 0) compared to a double bond and an isolated carbanion and that it will be of *higher* energy than the corresponding allyl anion, which has a DE = 0.83β. Breslow has suggested the use of the term *antiaromatic* to describe conjugated, cyclic systems that are thermodynamically less stable than the corresponding acyclic analogues. Dewar has made a similar suggestion on the basis of perturbational MO calculations, which predict that planar cyclooctatetraene would be 2β *less* stable than octatetraene itself. Using this definition, a further classification of cyclic conjugated systems can be made. Such a system would be *aromatic* if it has a larger DE than its acyclic analogue, *nonaromatic* if it has the same DE as its acyclic analogue, and *antiaromatic* if it has less DE than its acyclic analogue.

This type of calculation again suffers from the difficulties in determining these energies. Either the enthalpies of hydrogenation or combustion of real molecules can be found (e.g., cyclooctatetraene and octatetraene), or

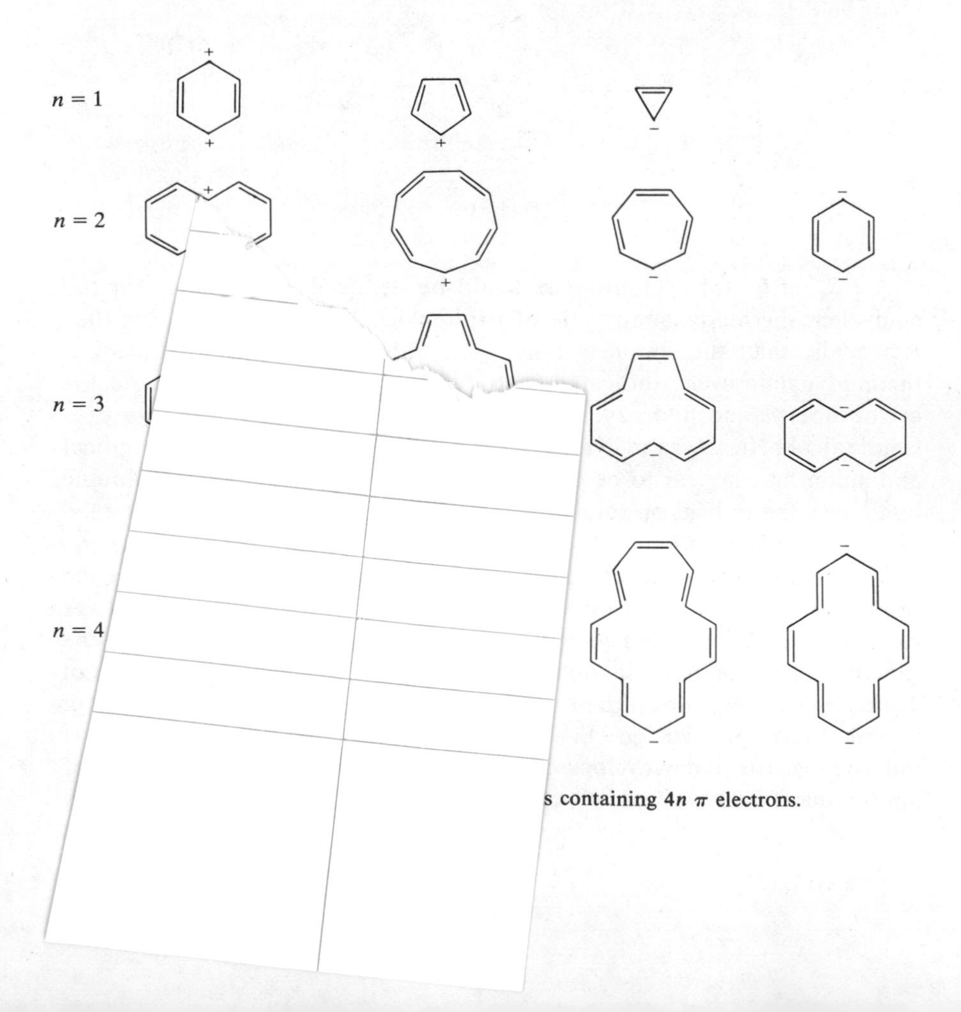

s containing $4n$ π electrons.

the DE of a nonreal system (e.g., planar cyclooctatetraene) can be estimated by direct calculation or by the use of some experimental data. Both of these methods are open to criticism of some kind. However, it seems likely that the term "antiaromatic" will be useful in describing molecules that are destabilized rather than stabilized by delocalization.

The concept of antiaromaticity is best illustrated for the ions, and in Figure 6.2 several smaller ions with $4n$ π electrons are shown.

The properties of a number of these ions or related derivatives are now known. The simplest system, the cyclopropenyl anion (**1**), has not been isolated, but some of its properties have been deduced from experiments involving the reduction of the cyclopropenium ion (**2**). This cation can be readily prepared as the fluoroborate salt in acetonitrile at −40°C, as was described in Chapter 5. Solutions so prepared are unstable at 25°C, so the electrochemistry was performed at −10°C. Using second harmonic AC voltammetry, Wasielewski and Breslow reduced the cyclopropenium ion first to the cyclopropenyl radical (**3**) and then to the cyclopropenyl anion (**1**) (Fig. 6.3).

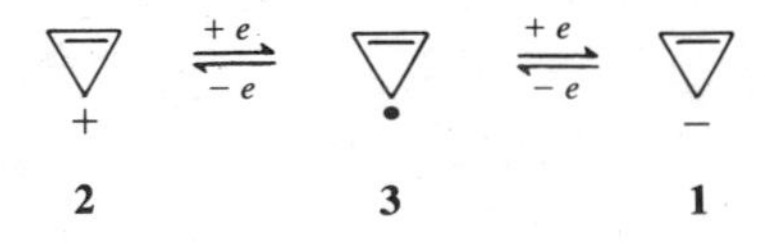

Figure 6.3

A pK_a of 61 for cyclopropene could be deduced (see Fig. 6.5 for the equivalent thermodynamic cycle of triphenylcyclopropene) and it is thus *less* acidic than the simple alkanes. A similar treatment, again using a thermodynamic cycle, indicated that cleavage of cycloprop-2-enol to the cyclopropenyl and hydroxyl radical is less favorable than cleavage of a simple alcohol to alkyl and hydroxyl radicals. Both the cyclopropenyl radical and anion thus appear to be destabilized by the interaction of the double bond with the radical or anion center; this phenomenon has been termed *resonance destabilization.*

These experiments had been preceded by a number of studies on the properties of the triphenylcyclopropenyl anion (**4**) and other derivatives carried out by Breslow and co-workers. Treatment of triphenylcyclopropene (**5**) with sodamide in liquid ammonia gave hexaphenylbenzene (**7**) and not **4**. The reaction was shown to proceed through the intermediacy of the dimer **6**, and **4** was not involved (Fig. 6.4). Comparison with triphenylmethane indicated that triphenylcyclopropene was much less acidic, and a pK_a of approximately 40 was adduced. More convincing evidence for the low

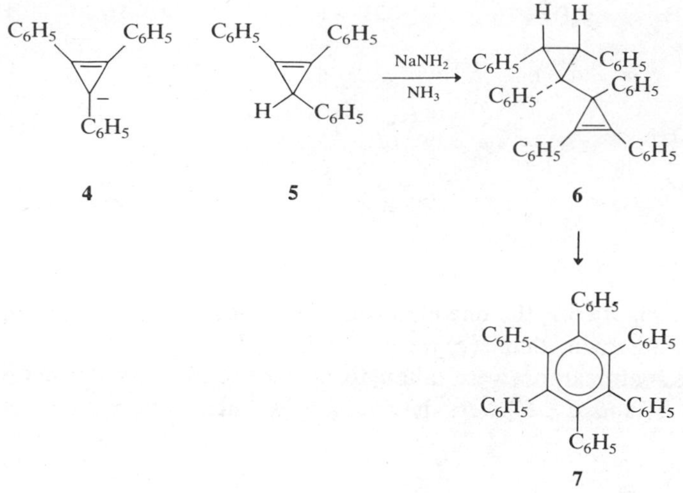

Figure 6.4

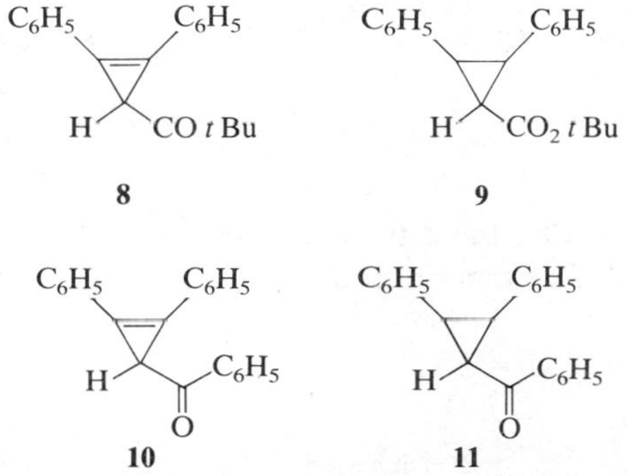

Figure 6.5

acidity of triphenylcyclopropene was obtained from an examination of the enolate anions derived from a number of derivatives. Thus, both **8** and **10** are much less acidic than the corresponding saturated analogues **9** and **11**. The base catalyzed ionization of **10** is 6000 times slower than that of **11**, which shows that the double bond, rather than increasing the acidity, decreases it by more than 7 pK units through resonance destabilization (Fig. 6.6).

A more direct estimate of the acidity of triphenylcyclopropene has been made using linear scan voltammetry. For the thermodynamic cycle (a)

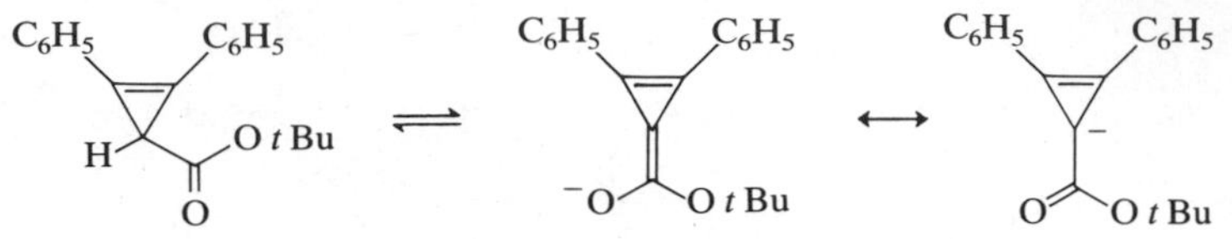

10 ⇌ 12

Figure 6.6

shown in Figure 6.7, the one-electron reductions of triphenylcyclopropene (b) and triphenylmethane (c) were examined. The formation of the alcohols from the hydrocarbon were taken to be independent of the nature of R, and E^1, E_1^1 and E_2, E_2^1 were determined by linear scan voltammetry.

$$(a)\ \mathrm{RH} \rightleftharpoons \mathrm{ROH} \overset{K_1}{\rightleftharpoons} \mathrm{R}^{\oplus} \overset{E_1}{\rightleftharpoons} \mathrm{R}^{\odot} \overset{E_2}{\rightleftharpoons} \mathrm{R}^{\ominus} \overset{K_2}{\rightleftharpoons} \mathrm{RH}$$

$$(b)\ (C_6H_5)_3CH \rightleftharpoons (C_6H_5)_3COH \overset{K_1}{\rightleftharpoons} (C_6H_5)_3C^{\oplus} \overset{E_1}{\rightleftharpoons} (C_6H_5)_3C^{\odot} \overset{E_2}{\rightleftharpoons} (C_6H_5)_3C^{\ominus} \overset{K_2}{\rightleftharpoons} (C_6H_5)_3CH$$

$$(c)\ (C_6H_5)_3C_3H \rightleftharpoons (C_6H_5)_3C_3OH \overset{K_1^1}{\rightleftharpoons} (C_6H_5)_3C_3^{\oplus} \overset{E_1^1}{\rightleftharpoons} (C_6H_5)_3C_3^{\odot} \overset{E_2^1}{\rightleftharpoons} (C_6H_5)_3C_3^{\ominus} \overset{K_2^1}{\rightleftharpoons} (C_6H_5)_3C_3H$$

Figure 6.7

Knowing these values, the difference in energy of these reactions $(E_1 + E_2) - (E_1^1 + E_2^1)$ can be determined. In this case,

$$(E_1 + E_2) - (E_1^1 + E_2^1) = 1.7\ \mathrm{V}$$

$$= 164\ \mathrm{kJ\ mol^{-1}} = 28\ \mathrm{p}K \text{ units at } 25°\mathrm{C} \qquad (6.1)$$

Since $\mathrm{p}K_{\mathrm{R}^+}$ of triphenylcyclopropenol is +3.1 and that of triphenylmethane is −6.6,

$$\frac{K_1^1}{K_1} = 10^{9.7} \qquad (6.2)$$

And thus knowing from equation (6.1) that the overall difference is 28 pK units,

$$\frac{K_2^1}{K_2} = 10^{18.3} \qquad (6.3)$$

Since triphenylmethane has a pK_a of 33, triphenylcyclopropene must have a pK_a of about 51.

This experiment was later repeated using second harmonic voltammetry as for the cyclopropenyl anion itself, and the same value for the pK_a was obtained. Similar experiments gave a pK_a of 74 for trimethylcyclopropene and 73 for tri-*t*-butylcyclopropene, and substitution of alkyl groups thus *destabilize* both the cyclopropenyl radical and anion, presumably in part through hyperconjugative encouragement of interaction of the double bond with the radical or anionic center.

The trimethylcyclopropenyl radical (**14**) has been generated by γ radiation of the cyclopropenium ion **13** at 77 K. The ESR spectrum indicates that **14** is best described as a localized radical and a double bond, but isomerization between the three equivalent structures is rapid (10^8 sec^{-1} at 292 K) with a low activation barrier (15–30 kJ mol^{-1}).

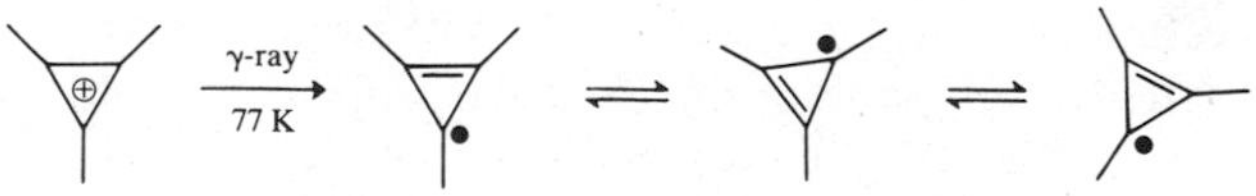

γ-ray
77 K
13 **14a** **14b** **14c**

Figure 6.8

Cleavage of the cyclopropenyl silyl derivative **15** labeled with ^{13}C at the carbon bearing the silyl group gave triphenylcyclopropene (**16**) in which the label is completely scrambled, indicating a similar pseudorotation in the triphenylcyclopropenyl anion.

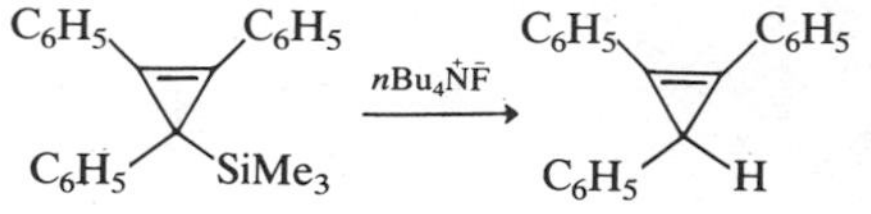

C_6H_5 C_6H_5 C_6H_5 $SiMe_3$ —$n\text{Bu}_4\overset{+}{\text{N}}\overset{-}{\text{F}}$→ C_6H_5 C_6H_5 C_6H_5 H

15 **16**

Figure 6.9

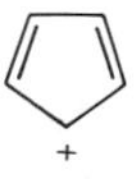

+

17

Figure 6.10

The cyclopentadienium cation (**17**), like the cyclopropenyl anion, is a $4n$ π-electron system with four π electrons. Again, the HMO calculation predicts that the two degenerate orbitals will only be half filled, but in this case a positive DE (1.24β) is predicted (Fig. 6.11).

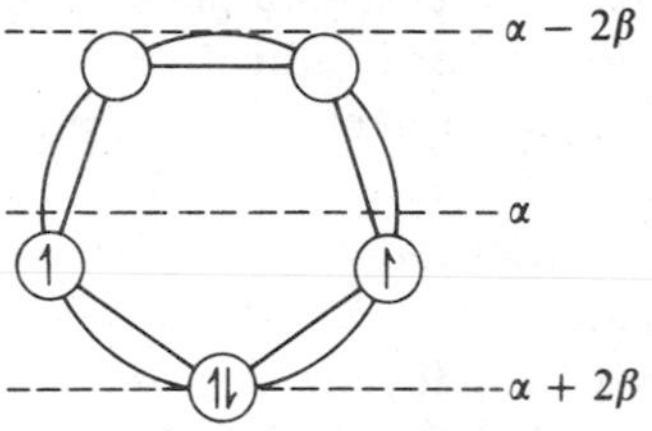

$\alpha - 2\beta$

α

$\alpha + 2\beta$

Figure 6.11

Initial investigations were carried out on substituted derivatives. Treatment of pentaphenylcyclopentadienol (**18**) with boron trifluoride in methylene chloride at $-60°C$ gives a deep blue solution of the cation **19**. This cation could exist in either the singlet **19a** or triplet **19b** states. The Jahn–Teller theorem suggests that the system should distort from the regular pentagonal structure, but the magnitude of this effect is not known.

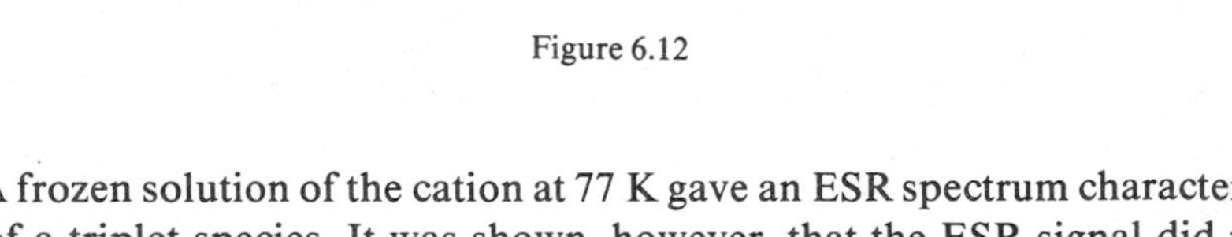

C_6H_5 C_6H_5 C_6H_5 C_6H_5 OH C_6H_5 — BF_3 / H_2O ⇌ — C_6H_5 C_6H_5 C_6H_5 C_6H_5 C_6H_5 ⟷ C_6H_5 C_6H_5 C_6H_5 C_6H_5 C_6H_5

18 **19a** **19b**

Figure 6.12

A frozen solution of the cation at 77 K gave an ESR spectrum characteristic of a triplet species. It was shown, however, that the ESR signal did not obey the Curie Law (IT = constant, I = intensity), and the slope of the plot of signal intensity against temperature suggested that the triplet state lies about 5.5 kJ mol^{-1} above the singlet ground state.

Since the triplet state should be favored in smaller molecules, the pentachlorocyclopentadienium cation (**21**) was prepared by treatment of hexachlorocyclopentadiene (**20**) with antimony pentafluoride. The ESR spectrum again indicated the presence of a triplet species, and in this case the signal intensity obeyed the Curie Law. The pentachlorocyclopentadienium cation (**21**) thus has the triplet ground state **21b**.

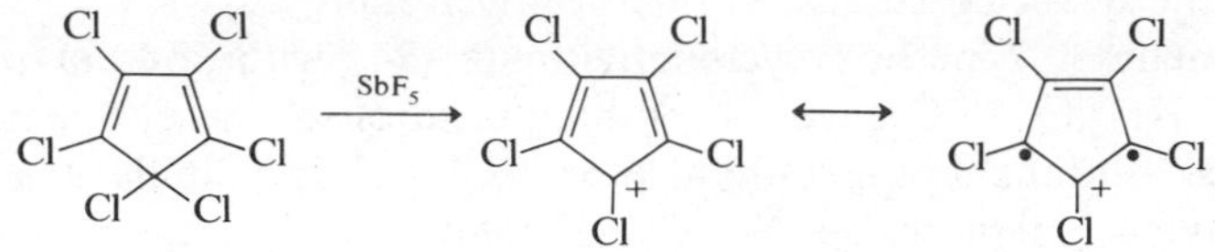

Cl Cl Cl Cl Cl Cl
SbF_5
Cl Cl Cl Cl Cl Cl
Cl Cl + Cl + Cl
20 **21a** **21b**

Figure 6.13

Finally, after numerous unsuccessful attempts, it was found that treatment by a molecular beam technique of 5-bromocyclopentadiene (**22**), prepared by the reaction of thallium cyclopentadiene with N-bromosuccinimide, with SbF_5 in di-*n*-butylphthalate at 78 K gave a species that showed triplet signals in the ESR spectrum. These signals were consistent with a planar D_{5h} species, as expected for a $C_5H_5^+$ cation, and they obeyed the Curie Law. The cyclopentadienium cation (**17**) is thus a ground-state triplet and is best depicted by **17b**. The same ESR spectrum could be obtained by similar treatment of 5-chlorocyclopentadiene.

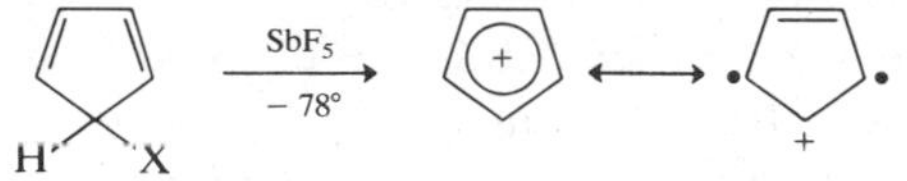

SbF_5
− 78°
H X
22 X = Br **17a** **17b**
X = Cl

Figure 6.14

Highly stabilized crystalline salts of substituted cyclopentadienium cations, such as **23**, have been prepared.

The 8 π-electron cycloheptatrienyl anion (**25**) was prepared by Dauben and Rifi in 1963. The HMO theory predicts that the two electrons will

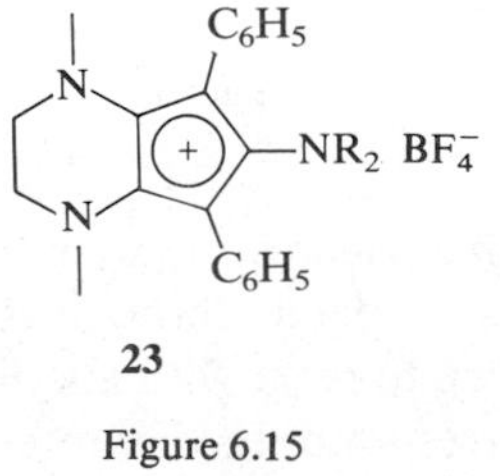

C_6H_5
N
+ $-NR_2$ BF_4^-
N
C_6H_5
23

Figure 6.15

occupy a pair of degenerate, antibonding orbitals and that the anion will have a DE of 2.10β, considerably less than the tropylium cation (2.99π). Semiempirical MO calculations suggest that the D_{7h} anion will be a triplet but that the distorted singlet will be of lower energy.

Treatment of 7-methoxycycloheptatriene (**24**) with sodium–potassium alloy in THF at −20°C gave a deep blue solution of the cycloheptatrienyl anion (**25**), and the anion could also be prepared by a similar treatment of 7-triphenylcycloheptatriene (**26**). Cycloheptatriene itself has been shown to undergo deuterium exchange under basic conditions, the cycloheptatrienyl anion (**25**) being the most probable intermediate.

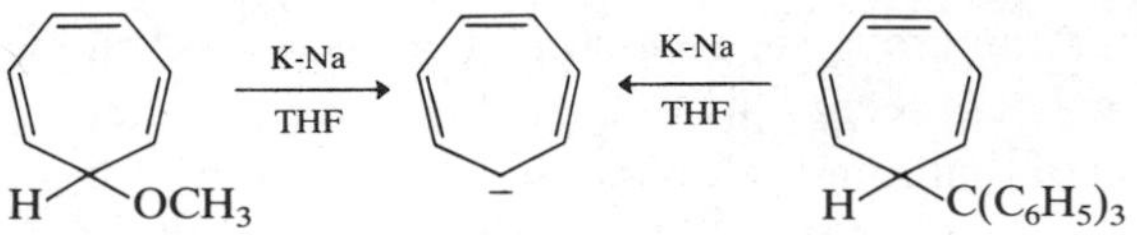

Figure 6.16

The heptaphenylcycloheptatrienyl anion (**28**) was prepared by treatment of heptaphenyltropylium bromide (**27**) with potassium in ether. The blue solution of the ion **28** yields heptaphenylcycloheptatriene-7-*d* (**29**), and the solution has an NMR but *no* ESR spectrum at low temperature. The heptaphenylcycloheptatrienyl anion thus appears to exist in an unsymmetric singlet rather than the symmetric triplet ground state.

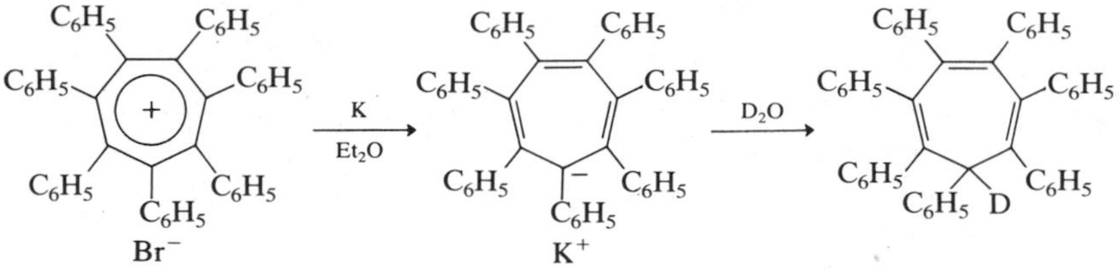

Figure 6.17

Some of the larger $4n$ π-electron systems depicted in Figure 6.2 have been prepared, often as bridged derivatives. A number of 15,16-dihydropyrenes has been shown to react with alkali metals to first give the anion radical and then the corresponding dianion. Thus, 15,16-dimethyl-15,16-

dihydropyrene (**30**) gives the anion radical **31** and then the dianion **32**. If **30** is considered to be a 14 π-electron system in which the central ethane bridge is a minor perturbation, then the HMO theory predicts that the added electrons will enter a pair of degenerate antibonding orbitals. However, the HMO theory is unlikely to be reliable for these molecules, and the finding that **32** is a singlet state with a well-resolved NMR spectrum is not too surprising. The NMR spectrum of **32** shows the expected *reversal* of the proton chemical shifts, the inside methyl protons appearing at very low field, δ 21.0, compared with the high field position in **30** (δ −4.25), whereas the outer protons are at high field (δ −3.2 to −4.0) rather than at low field as in the hydrocarbon (δ 8.67–7.95), and **32** is a paratropic, 16 π-electron system. The dianion **32** is more paratropic than [16]annulene, which probably reflects both the charge and the greater rigidity of the σ framework of **32**, leading to more mixing of excited states.

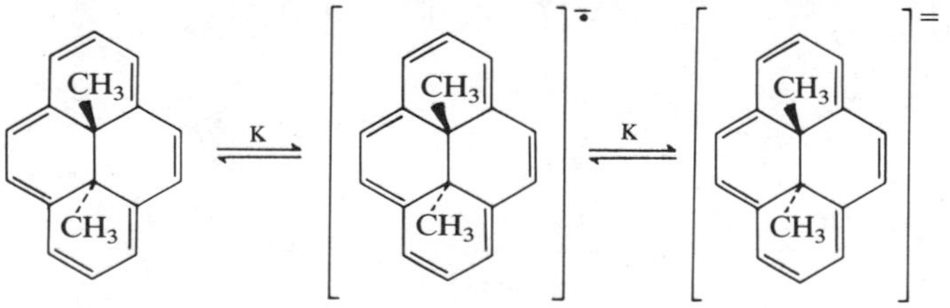

CH_3 CH_3 K **30** CH_3 CH_3 K **31** CH_3 CH_3 **32**

Figure 6.18

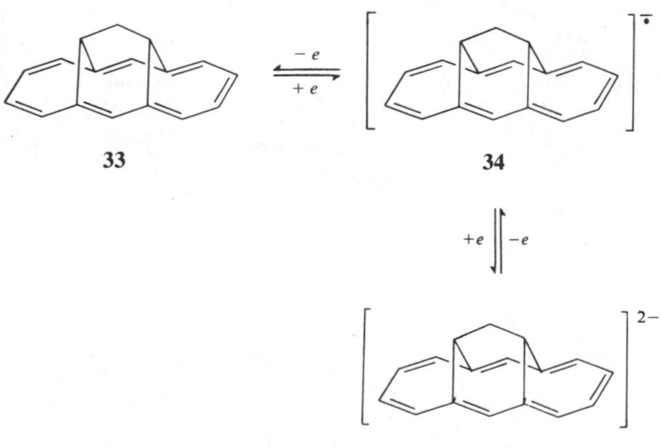

−e +e **33** **34** +e −e 2− **35**

Figure 6.19

A second 16 π-electron dianion **35** has been prepared by the two-electron reduction of the bridged [14]annulene **33**, the reaction proceeding via the radical anion **34**. Unlike the changes observed in the ^{1}H NMR spectra on going from **30** to the dianion **32**, the dianion **35** shows only small changes in chemical shift compared with **33** (ring protons, **33**, δ 7.6; **35**; δ 5.5; bridge protons, **33**, δ −0.9; **35**, δ 2.2) which indicates that **35**, unlike **32**, is not strongly paratropic. This difference in paratropicity has been attributed to the greater deviation from planarity of **35** compared with **32**.

Reduction of the series of $(4n + 2)$ π-electron bisdehydroannulenes (**36**, $n = 1, 2, 4$) prepared by Nakagawa gave the corresponding paratropic dianions (**37**, $n = 1, 2, 4$) (see Chapter 5).

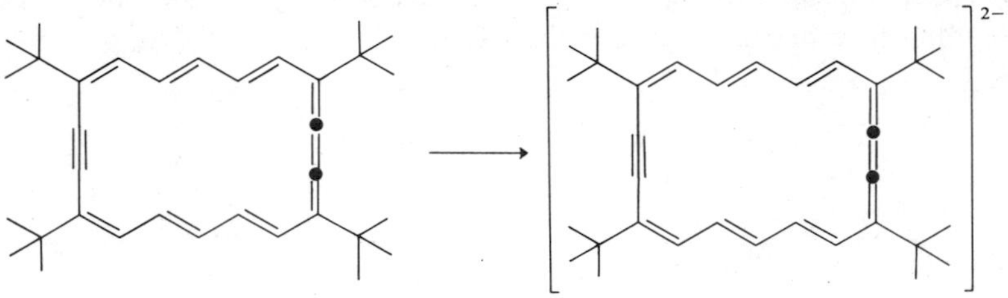

36 **37**

Figure 6.20

Oxidation of the 18 π-electron *syn,syn,syn*-1,6:8,17:10,15-tris-[18]annulene (**38**) in dichloromethane with fluorosulfonic acid in SO_2ClF at −80°C gave a violet solution of the dication **39**. The dication shows a seven line ^{13}C NMR spectrum with the expected downfield shift of the bridge carbons. The ^{1}H NMR spectrum shows signals at δ 4.35 and 7.58 (AX, 4H), 6.07 (s, 4H). 7.01–7.35 (A, A′, X, X′, 8H) and 8.25 (s, 2H); thus in comparison with **38** the bridge protons are shifted downfield and the ring protons are virtually unchanged. Presumably the paratropic shift of the ring protons has been offset by the downfield shift due to the positive charges.

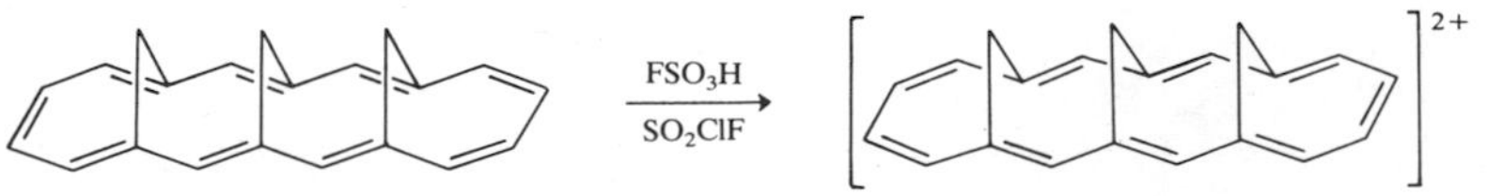

FSO_3H / SO_2ClF

38 **39**

Figure 6.21

[18]Annulene (**40**) is reduced by potassium in THF by two one-electron processes first to give the anion radical **41** and then the dianion **42**. The dianion **42**, with 20 π electrons, again shows the expected *reversal* of chemical shifts in the ^{1}H NMR spectrum compared to [18]annulene itself. Thus, the inner protons of **42** appear as two broad signals at δ 28.1 and 29.5 in the ^{1}H NMR spectrum at $-110°$, with the outer protons as a broad singlet at δ -1.13. The appearance of two sets of inner protons indicates that the dianion **42** exists in two configurations, probably **42a** and **42b**, whereas [18]annulene exists as only one detectable isomer. It may be remembered that in Chapter 5 it was shown that [16]annulene, which exists in two interconverting configurations, gives the dianion with only one configuration. In both cases, it seems that the energy difference between configurations is *less* for the 4*n* systems.

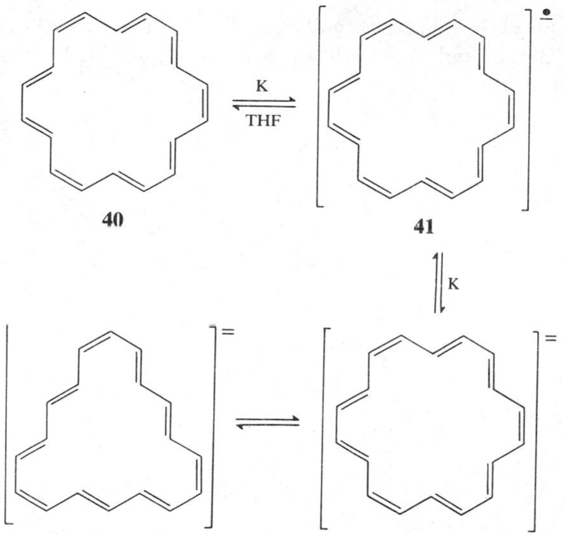

K
THF
40
41
K
42b
42a

Figure 6.22

It appears from the 4*n* π-electron ions that have so far been prepared that only in the small ions is it likely that the triplet electronic configuration will be more stable than an unsymmetric singlet ground state. Whereas in small systems it should be possible to determine if the ions are destabilized by conjugation, in the larger ions it will probably be difficult to find suitable model systems.

FURTHER READING

For a review of some of the early work described in this chapter see P. J. Garratt and M. V. Sargent, in E. C. Taylor and H. Wynberg (Eds.), *Advances in Organic Chemistry*, Vol. 6, Wiley-Interscience, New York, 1969, p. 1.

For a discussion of antiaromaticity, see R. Breslow, *Accounts Chem. Res.*, 1973, **6**, 393.

For a description of the work on the cyclopentadienium cation and other 4π systems, see R. Breslow, in T. Nozoe, R. Breslow, K. Hafner, Shô Itô, and I. Murata, *Topics in Nonbenzenoid Chemistry*, Vol. 1, Hirokawa, Tokyo, 1973, p. 81.

For a theoretical discussion of diradicals with four π-electrons, see W. T. Borden and E. R. Davidson, *Accounts Chem. Res.*, 1981, **14**, 176.

For a recent calculation suggesting that the cyclopropenyl anion has the hydrogens displaced from the ring plane (i.e., is nonplanar), see B. A. Hess, L. J. Schaad, and P. Čársky, *Tetrahedron Lett.*, 1984, **25**, 4721.

See also P. J. Garratt, in J. Frazer Stoddart (Ed.), *Comprehensive Organic Chemistry*, Sir Derek Barton and W. D. Ollis (series Eds.), Vol. 1, Pergamon, Elmsford, N.Y., 1979, p. 361.

7

ANNULENONES, FULVENES, AND RELATED SYSTEMS

The annulenones and fulvenes are cyclic systems composed of unsaturated, odd-membered rings in which the "odd" carbon atom is part of an exocyclic unsaturated group. The general formula for an annulenone is **1** and for fulvene is **2**. The polarization of the carbonyl bond in the annulenones leaves the odd carbon atom with a partial positive charge, and thus the annulenones with odd numbers of endocyclic double bonds are potentially aromatic, whereas those with even numbers of double bonds are potentially antiaromatic. However, for these effects to be manifest the contribution of the polarized structure to the ground state must be important.

The simplest system that can attain aromaticity by this form of cross conjugation is cyclopropenone (**3**), which may be considered to be a deriva-

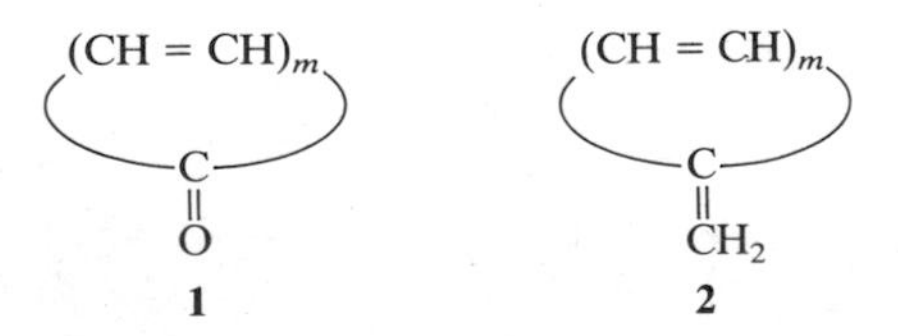

Figure 7.1

tive of the 2 π-electron cyclopropenium cation. Substituted cyclopropenones have been known for some time, but cyclopropenone (**3**), melting point −29 to −28°C, was more recently prepared and is much more stable than its saturated analogue cyclopropanone (**4**).

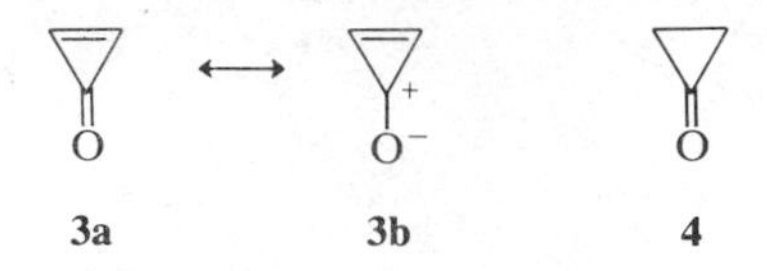

Figure 7.2

The synthetic route to **3** is outlined in Figure 7.3. The reduction of tetrachlorocyclopropene (**5**) with tri-*n*-butyl tin hydride gave a mixture of chlorocyclopropenes containing 3,3-dichlorocyclopropene (**6**), which was hydrolyzed with cold water to **3**. The ^{1}H NMR spectrum of **3** shows a single absorption signal at δ 9.0, and the ^{13}C NMR spectrum shows the carbonyl carbon at δ 155.1 and the double bond carbons at δ 158.3. The ^{13}C—H NMR coupling is large (217 Hz), which is characteristic of a cyclopropene

Cl Cl H H H H

Bu_3SnH

Cl Cl Cl Cl O

5 **6** **3**

Figure 7.3

or acetylene. The IR spectrum shows a strong doublet absorption at 1833, 1864 cm^{-1} for the carbonyl group, with the double bond stretch at 1641 cm^{-1}. Cyclopropenone is stable below its melting point but polymerizes at room temperature, rapidly at 80°C.

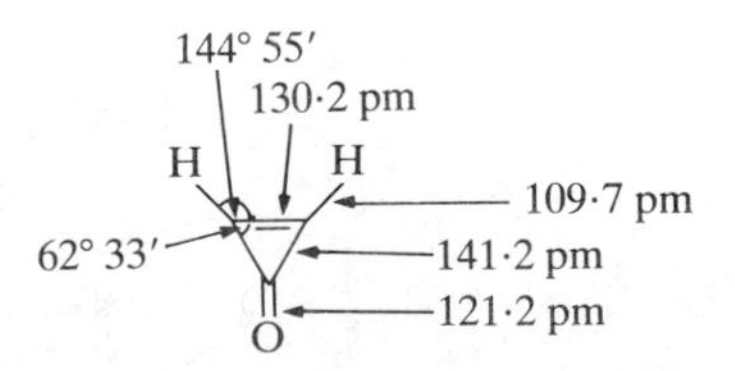

Figure 7.4. Structure of cyclopropenone from microwave data.

A microwave structural determination of cyclopropenone has been carried out and provides the structure shown in Figure 7.4. The dipole moment, 4.39D, was obtained from the molecular Stark effect and the magnetic susceptibility anisotropy, Δx, was found to be -17.8×10^6 erg mol^{-1} G^{-2}.* This latter value may be compared to those for methylenecyclopropane (−16.6 Fl) and cyclopropene (−17.0 Fl) and suggests that the diamagnetic ring current is similar to these systems. It would have been expected that if cyclopropenone is polarized, then the contribution from the cyclopropenium hybrid structure should increase the diatropicity. However, the cyclopropenium cation itself has only a small ring current compared to benzene (see Chapter 5), and in **3** this is presumably further reduced since all structures are not equivalent.

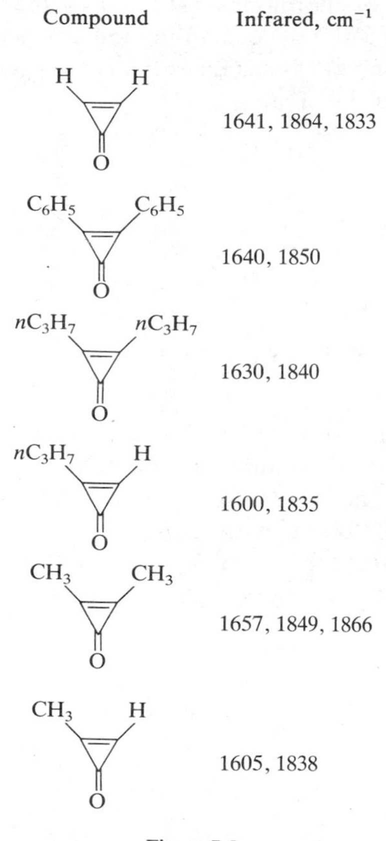

Figure 7.5

* 1×10^6 erg mol^{-1} G^{-2} = 1 Flygare = 1 Fl.

A large number of substituted cyclopropenones have been prepared since the first system, diphenycyclopropenone, was synthesized by Breslow and Vol'pin and their respective co-workers in 1959. The substituted cyclopropenones are more stable than the parent system and have characteristic IR spectra with bands in the 1800–1870 and 1600–1660 cm^{-1} regions, arising from the heavily mixed carbonyl and alkene stretching modes. The IR spectra of several cyclopropenes are tabulated in Figure 7.5.

Initial experimental measurements of the standard enthalpy of formation of diphenycyclopropenone suggested that it had a very large strain energy, but more recent measurements have shown the original values to have been in error, presumably owing to the presence of impurities in the sample. Using the later results, it is concluded that diphenylcyclopropene has a resonance energy of about 70 kJ mol^{-1} and is reasonably aromatic.

The ring protons of the monosubstituted cyclopropenones are acidic, and kinetic exchange with D_2O containing sodium bicarbonate is fast. Thus, 2-propylcyclopropenone (**7**) exchanges the ring proton under these conditions, presumably via the anion **8**.

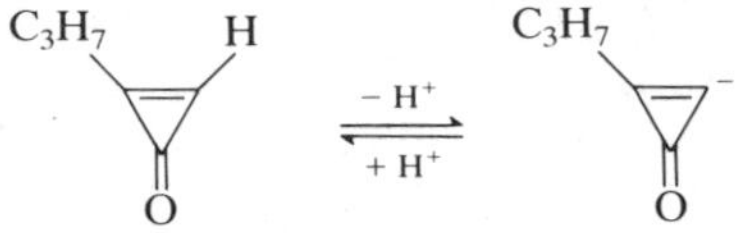

C_3H_7 H $\underset{+H^+}{\overset{-H^+}{\rightleftharpoons}}$ C_3H_7

7 **8**

Figure 7.6

Cyclopentadienone (**9**) has been observed as a transient intermediate in the pyrolysis of a number of compounds designed as precursors, including *o*-benzoquinone. The microwave spectrum of **9** has been observed, and the molecule is essentially planar with a dipole moment of 3.13D. Insufficient information was obtained for a complete structural analysis to be carried out.

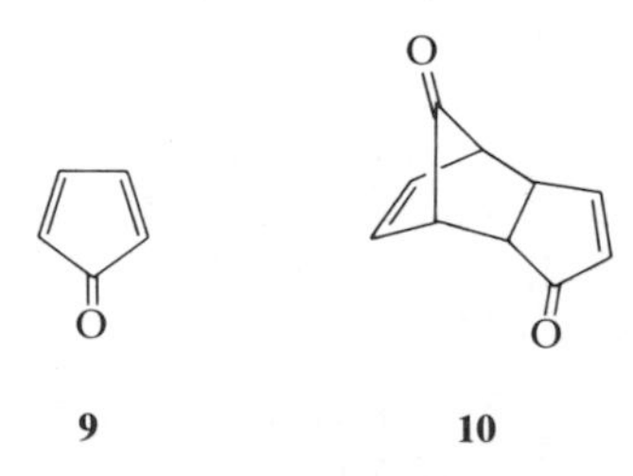

Figure 7.7

Earlier attempts to prepare cyclopentadienone had all given the dimer, dicyclopentadienone (**10**). For example, treatment of 5-bromocyclopentenone (**11**) with triethylamine gave a virtually quantitative yield of **10**. When the reaction was run in the presence of cyclopentadiene, the adduct **12** could be obtained, indicating the transitory formation of **9**.

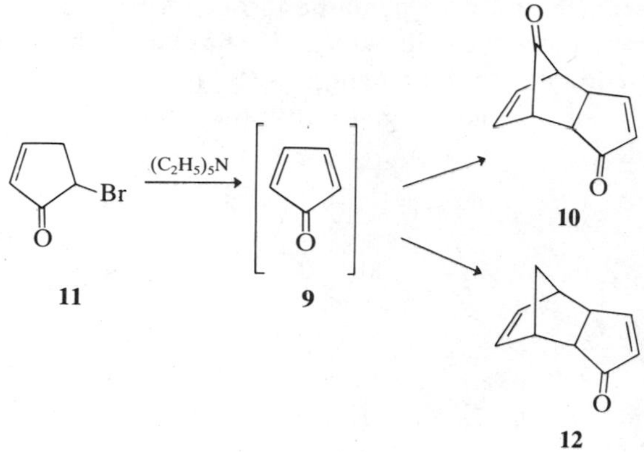
Br
$(C_2H_5)_3N$
O
O
O
O
O
11
9
10
12

Figure 7.8

Attempts to reduce the reactivity of cyclopentadienone by ketalization were unsuccessful, the corresponding dimers being obtained. When the diethyl ketal **14** was generated by treatment of the dibromoacetal **13** with

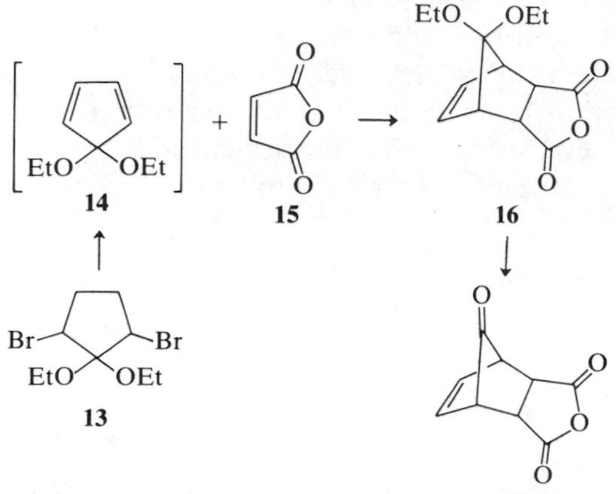
EtO
OEt
14
+
O
O
O
15
EtO
OEt
O
O
O
16
Br
Br
EtO
OEt
13
O
O
O
O
17

Figure 7.9

potassium *t*-butoxide in the presence of maleic anhydride (**15**), the Diels–Alder adduct **16** was obtained, which could be hydrolyzed to the 7-norbornenone derivative **17** (Fig. 7.9).

A number of sterically hindered cyclopentadienones have been prepared by Garbisch and Sprecher. 2,4-Di-*t*-butylcyclopentadienone (**18**) is a relatively stable compound that only dimerizes slowly at 25°C. The Diels–Alder reaction between **18** and cyclopentadiene occurs more rapidly at 25°C to give a mixture of products in which **18** has acted both as diene and dienophile. 3-*t*-Butylcyclopentadienone (**19**) is much more reactive than **18**, and with cyclopentadiene it gives only the adduct **20** for which it acted as a dienophile.

18 **19** **20**

Figure 7.10

The ^{1}H NMR spectrum of **18** has absorption signals for the H-3 and H-5 protons at δ 6.50 and δ 5.07, respectively. These values are at considerably higher field than would be expected for the α and β protons of an unsaturated ketone, and two explanations can be advanced. The high field position of these protons would be expected if the cyclopentadienone acted as a 4 π-electron system with the consequent paratropicity of a $4n$ annulene. The 4 π-electron nature of **18** would arise from the contribution of the polarized structure **18a**. Garbisch and Sprecher rejected this view on the grounds that the *t*-butyl protons resonate in the normal position and do not show an upfield shift. They considered that the upfield shift of the ring proton arises from a reverse polarization of the carbonyl group, as in **18b**, which increases

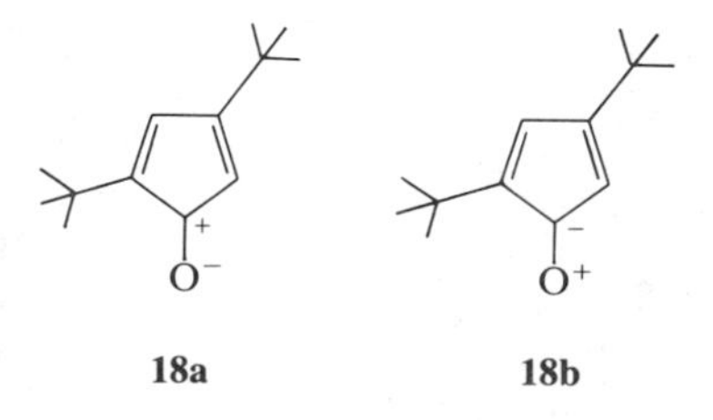

Figure 7.11

the charge density on the ring atoms. The present author is biased toward the paratropic view, believing that the diatropic effect in **18b** would offset any charge localization.

The properties of the two smallest members of the annulenone series appears to substantiate the view that the Hückel Rule also applies to these compounds. Cyclopropenone has its protons at low field in the ^{1}H NMR spectrum, whereas the substituted cyclopentadienones have the protons at high field. The high reactivity of cyclopentadienone may be seen as a reflection of its antiaromaticity, involving the cyclopentadienium cation as a contributing structure.

Tropone (**21**) was one of the earliest molecules to be synthesized as a test of the predictions of the Hückel theory. It is a stable compound in contrast to cyclopentadienone and for many years after its synthesis was generally accepted as an aromatic system. However, the basis of this assumption was more recently questioned, and it has been concluded from a study of its physical properties that tropone is not an aromatic system. The dipole moment of 4.30D was originally considered to be good evidence for a substantial contribution from the polarized structure **21b**. Calculation of the dipole moment by the CNDO/2 (complete neglect of differential overlap) method gave, however, a value of 3.88D.

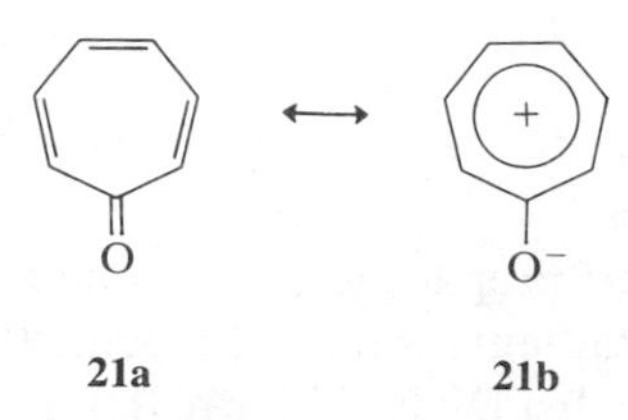

Figure 7.12

Similarly, the use of more recent values for the Pascal constants gave a value for the diamagnetic anisotropy of −47.4, which is close to the experimental value of −54. The magnetic susceptibility anisotropy exaltation, $\Delta\chi$, determined from microwave studies, was approximately zero and consistent with a nonaromatic nature for **21**. The microwave spectrum was interpreted to give the structure shown in Figure 7.13.

The fulvenes are systems related to the annulenones, the exocyclic carbonyl group having been replaced by an exocyclic methylene group. The simplest member of the series, methylenecyclopropene (**23**), has recently been prepared by Billups and Staley and their respective co-workers by the

120·8°
133·1°
128°
128·5°
O
124 pm
147 pm
133 pm
147·5 pm
134 pm

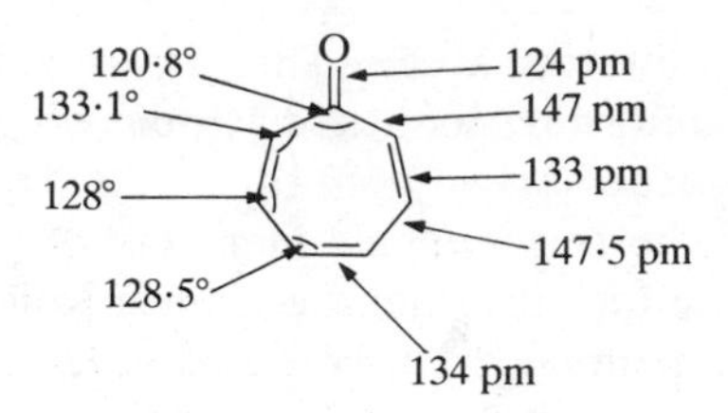

Figure 7.13. Structure of tropone from microwave data.

route shown in Figure 7.14. Halocarbene addition to allene gave the methylene halocyclopropane **22**, which underwent vapor-phase dehydrohalogenation to **23**. The ^{1}H NMR spectrum of **23** showed two triplets at δ 3.60 and 8.18, and the chemical shifts clearly indicate a considerable contribution from **23a**. The ^{13}C NMR spectrum shows the exocyclic carbon at δ 59.6 and the two equivalent ring carbons at δ 139.9. The electronic spectrum shows bands at 206, 242, and 309 nm, the long wavelength band being associated with a $\pi \rightarrow \pi^*$ charge transfer transition.

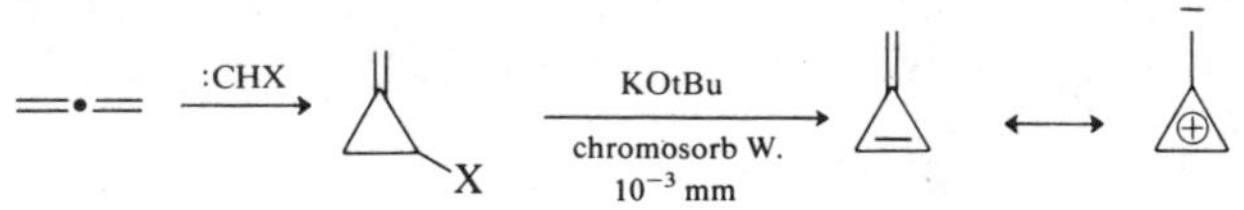

:CHX
X
KOtBu
chromosorb W.
10⁻³ mm

22 **23** **23a**

Figure 7.14

Other evidence for the preparation of **23** had earlier come from the reaction of the cyclopentadienyl anion (**24**) with the sulfonium salt **25** to give **26**, presumably through a Diels–Alder reaction of cyclopentadiene with **23**.

Numerous derivatives of **23** have been prepared. Thus, a Wittig reaction between diphenylcyclopropenone (**27**) and the ylid **28** gave the methyl-

BF4−
+SC6H5
Me
23
+

24 **25** **26**

Figure 7.15

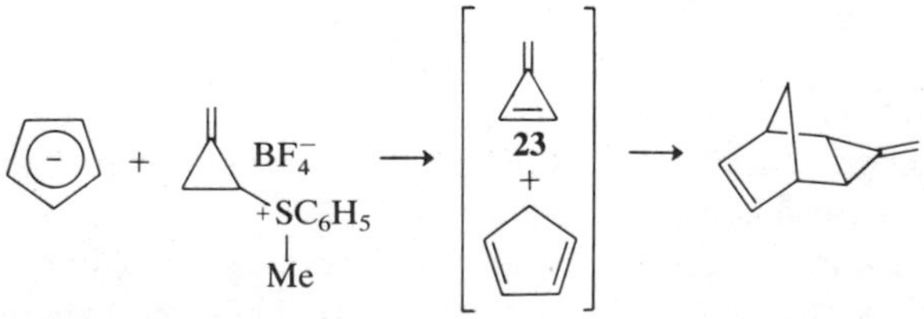

enecyclopropene derivative **29** as a yellow crystalline solid. In the ^{1}H NMR spectrum of **29**, the methylene proton appears at δ 5.02, which is at considerably higher field than the corresponding proton (δ 6.8) in the methylenecyclopropane **30**. This upfield shift may again by attributed to the diatropic contribution from the dipolar structure **29a**.

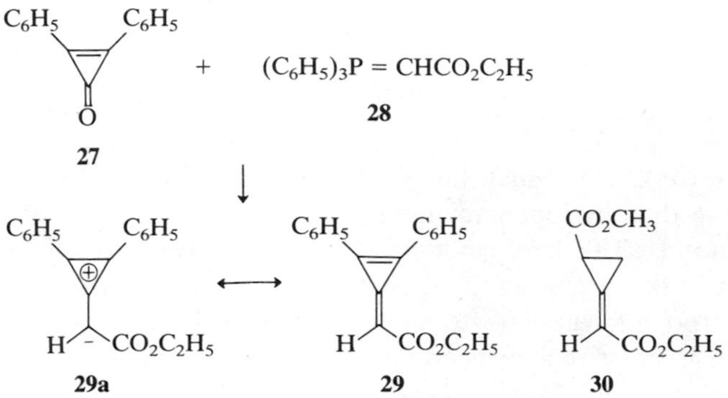

C_6H_5 C_6H_5 (cyclopropenone, O) **27** + $(C_6H_5)_3P = CHCO_2C_2H_5$ **28** → **29a** (C_6H_5, C_6H_5, ⊕, H, $CO_2C_2H_5$) ⟷ **29** (C_6H_5, C_6H_5, H, $CO_2C_2H_5$); **30** (CO_2CH_3, H, $CO_2C_2H_5$)

Figure 7.16

Reaction of malonitrile with diphenyl (**27a**) or di-*n*-propylcyclopropenone (**27b**) in the presence of base gives the corresponding methylenecyclopropenes **31a, b**. Whereas the aryl-substituted methylenecyclopropenes have electronic spectra with absorption maxima above 350 nm, the di-*n*-propyl derivative shows only absorption at 246 nm (ε 20,000), indicating that the phenyl group makes an important contribution to the long wavelength absorption.

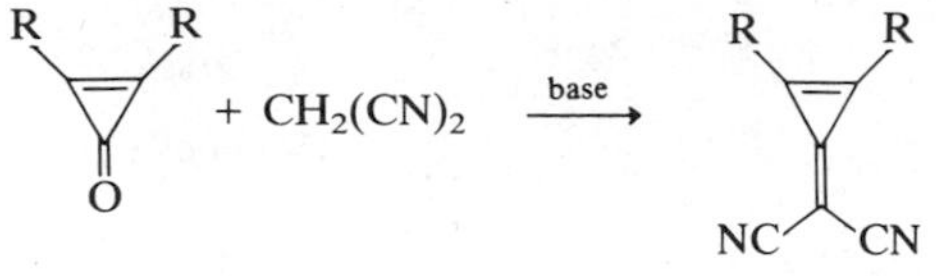

R R (cyclopropenone, O) + $CH_2(CN)_2$ —base→ R R (methylenecyclopropene, NC, CN)

27a R = C_6H_5
27b R = nC_3H_7

31a R = C_6H_5
31b R = nC_3H_7

Figure 7.17

The first methylenecyclopropene to be prepared was the orange quinonoid derivative **32** obtained by Kende. This molecule can be considered to be either a methylenecyclopropene or a vinylogous cyclopropenone.

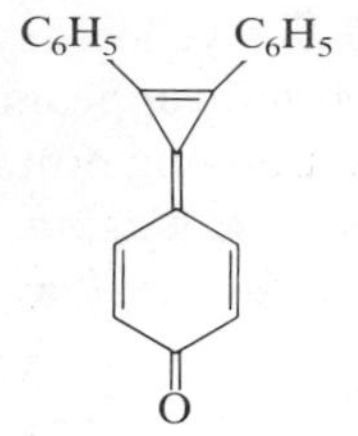

C_6H_5 C_6H_5

O

32

Figure 7.18

Fulvene (**33**), after which this series of compounds is named, is much less reactive than cyclopentadienone, and it can be isolated and distilled as the monomer. The dipole moment, 0.42D, has been determined from the Stark shift in the microwave spectrum and suggests that there is little contribution from the dipolar structure. The microwave spectra of **33** and its 6-deuteroderivative have been interpreted to give the structure shown in Figure 7.20.

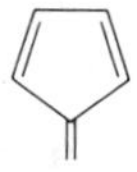

33

Figure 7.19

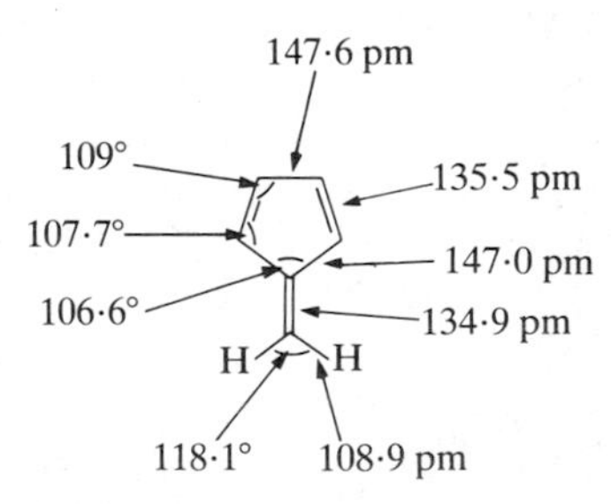

Figure 7.20. Structure of fulvene from microwave data.

Substitution at the exocyclic methylene position give 6-substituted fulvenes, which are more stable than the parent system. Although the microwave spectrum suggests little contribution of the dipolar structure to the ground state, nevertheless the exocyclic carbon atom behaves as if it is electron deficient with respect to the endocyclic carbon atoms. Addition of metal alkyls occurs across the exocyclic double bond to give the substituted cyclopentadienyl anion (e.g., **34** → **35**). The stabilization of the cyclopentadienyl anion makes the methyl hydrogens in 6,6-dimethylfulvene acidic, and **34** behaves in many reactions as though it were the tautomeric structure **34a**. The fulvenes are reduced with alkali metals first to the radical anion, which either dimerizes or is reduced further to the dianion.

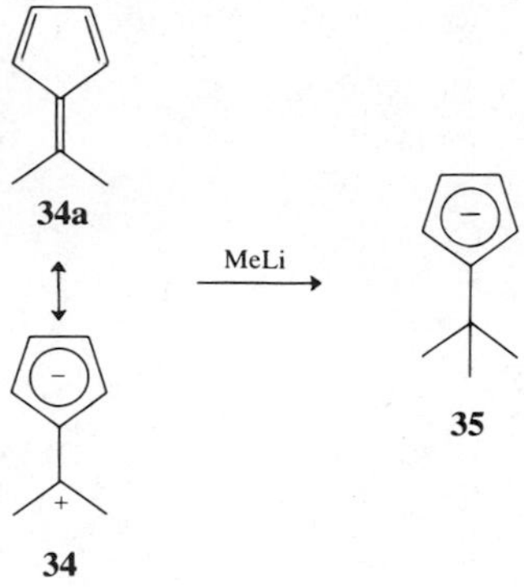

34a

MeLi

35

34

Figure 7.21

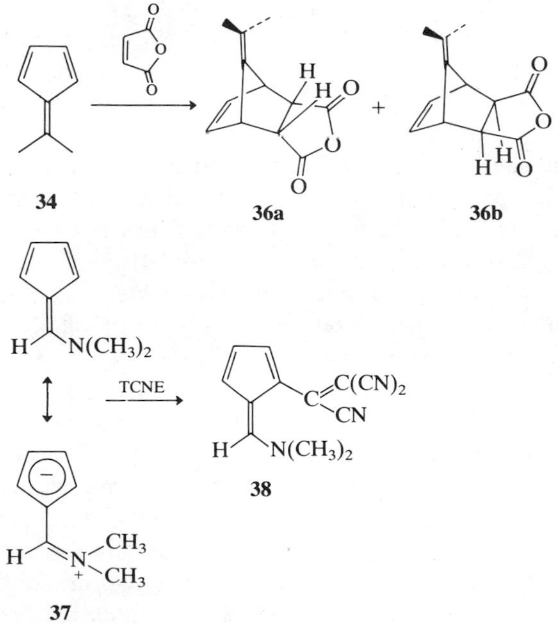

34

36a

36b

H N(CH$_3$)$_2$

TCNE

C(CN)$_2$ C CN

H N(CH$_3$)$_2$

38

H N CH$_3$ CH$_3$

37

Figure 7.22

6,6-Dimethylfulvene (**34**) reacts with maleic anhydride to give the Diels–Alder adducts **36a** and **36b**. By contrast, 6,6-(dimethylamino)fulvene (**37**), with electron withdrawing substituents at C-6, does not give Diels–Alder adducts with dienophiles. On reaction with tetracyanoethylene, the substitution product **38** and not the Diels–Alder adduct is obtained.

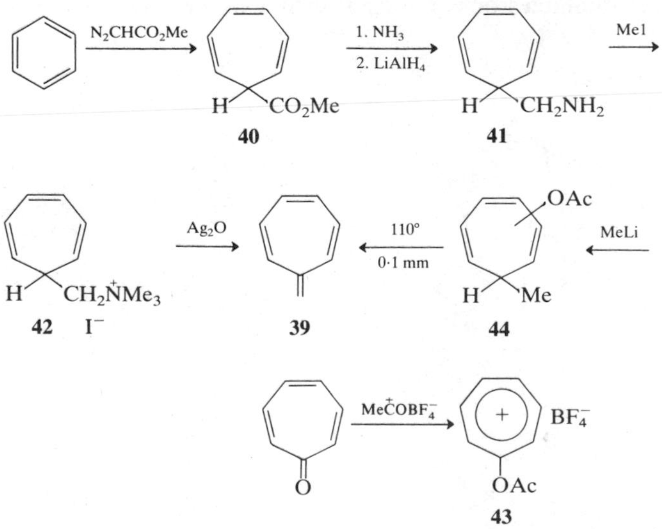

Figure 7.23

Heptafulvene (**39**) was originally prepared by von Doering and Wiley in 1954 (Fig. 7.23). Methyl diazoacetate adds to benzene to give 7-cycloheptarienyl carboxylate (**40**), which was converted into the amine **41** via the corresponding amide. Exhaustive methylation of **41** gave the trimethylammonium iodide **42**, which on treatment with silver oxide underwent a Hofmann elimination to give deep red solutions of **39**. More recently, Schenk, Kyburz, and Neuenschwander succeeded in preparing pure heptafulvene from tropone (Fig. 7.23). Tropone was converted to the tropylium salt **43**, which, on treatment with methyl lithium, gave a mixture of acetoxymethylcycloheptatrienes **44**. Pyrolysis of **44** at 100–110°C under vacuo gave **39**, melting point −43 to −42°C. The ^{1}H NMR spectrum showed signals at δ 5.95 (d, 2H), 5.6, 5.4 (m, 4H) and 4.45 (bs, 2H), the IR spectrum showed bands at 1647 and 1583 cm^{-1}, and the molecule had a rich electronic spectrum stretching into the visible region, which accounts for the red color of the solutions. A microwave spectrum gave a dipole moment of 0.48D, similar to fulvene, and the structure shown in Figure 7.24 was inferred.

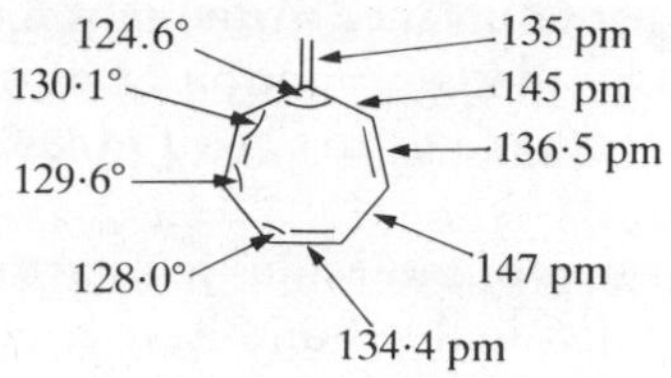

124.6° 130·1° 129·6° 128·0° 135 pm 145 pm 136·5 pm 147 pm 134·4 pm

Figure 7.24. Structure of heptafulvene from microwave data.

Heptafulvene reacts with dimethyl acetylenedicarboxylate (**45**) to give the Diels–Alder adduct **46**; involving an orbital symmetry allowed (8 + 2) cycloaddition (Fig. 7.25). Dehydrogenation with chloranil gave the disubstituted azulene **47**.

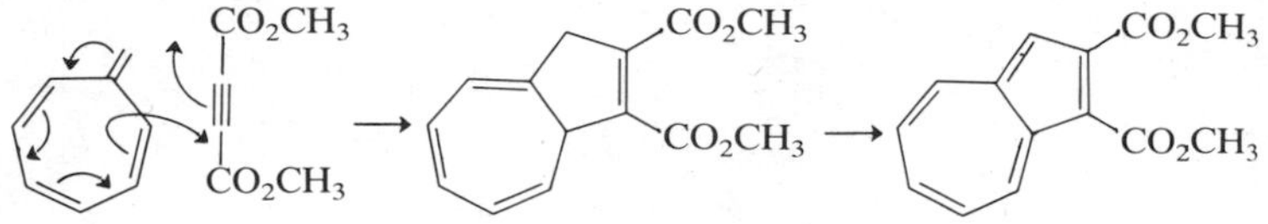

CO_2CH_3 CO_2CH_3 **39** **45** CO_2CH_3 CO_2CH_3 **46** CO_2CH_3 CO_2CH_3 **47**

Figure 7.25

The benzannelated derivatives are more stable than heptafulvene itself, and introduction of electron withdrawing groups at the exocyclic methylene position gives still more stable systems. The increased stability reflects the increased importance of the dipolar form, which involves the tropylium cation. 8,8-Dicyanoheptafulvene (**48**) is thus a stable compound with a large dipole moment (7.49D) due to the contribution of the dipolar forms **48a, b.**

The properties of methylenecyclopropene, fulvene, and heptafulvene do not suggest that there is much alternation of structural type between the compounds with odd and those with even numbers of double bonds. Whereas cyclopentadienone is readily dimerized and tropone is not, fulvene

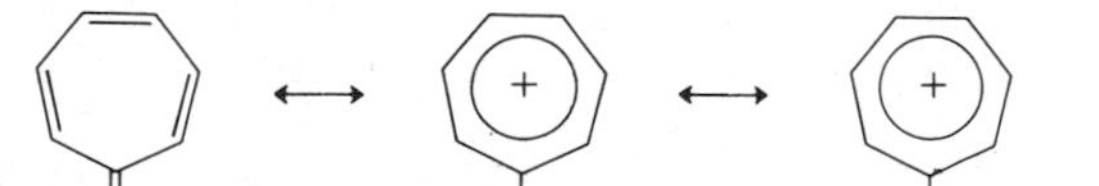

NC CN **48** NC CN **48a** NC N^- **48b**

Figure 7.26

is more stable than heptafulvene. Although 8,8-dicyanoheptafulvene appears to be stabilized by the contribution of dipolar structures, there is very little evidence to encourage us to classify the fulvenes as aromatic or antiaromatic.

The fulvalenes are a group of molecules related to the fulvenes, in which the protons on the exocyclic double bond have been replaced by a second "odd" unsaturated ring, giving the general formula **49**. The fulvalenes can be further classified into those systems in which *m* and *l* are both odd or both even integers and those in which *m* is odd and *l* is even. The simplest fulvalene, triafulvalene (**50**) is of the first type ($m = l = 1$), and neither it nor any of its derivatives has been prepared. The next member of the series,

(C=C)$_m$ C=C (C=C)$_l$

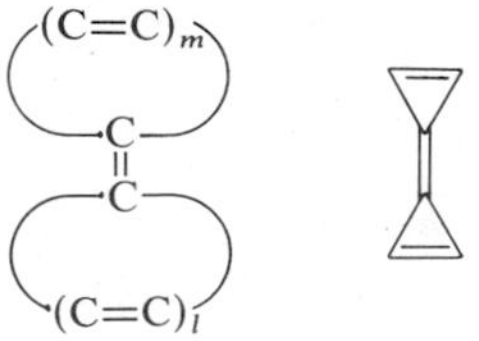

49 **50**

Figure 7.27

triapentafulvalene (**51**) belongs to the second type ($m = 1$, $l = 2$), and a number of derivatives of this system have been prepared. These compounds, which have also been called calicenes, appear to be stabilized by a significant contribution from the dipolar form **51a**. Thus, hexaphenyltriapentafulvalene (**52**) has a dipole moment of 6.3D. A polycyclic bicalicene is described in Chapter 9.

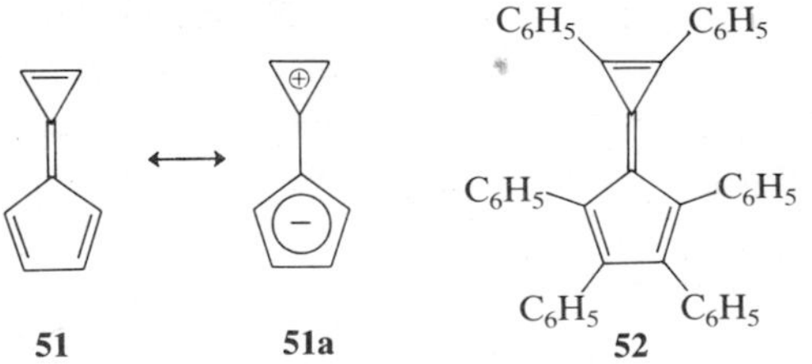

C_6H_5 C_6H_5 C_6H_5 C_6H_5 C_6H_5 C_6H_5

51 **51a** **52**

Figure 7.28

The next member of the series, fulvalene (**55**) itself, was prepared by von Doering and Matzner by the route shown in Figure 7.29. Sodium cyclopentadienide (**24**) was oxidized with I_2 at −80°C to the dimer **53**, which on treatment with *n*-butyl lithium was deprotonated to the dianion **54**. Oxidation of **54** with oxygen gives an orange solution of fulvalene (**55**). Attempts to concentrate this solution gave only polymeric products, but reaction with TCNE gave the Diels–Alder adduct from which fulvalene could be regenerated. The electronic spectrum of **55** is complex, showing absorption maxima at 266, 278, 289, 300, and 314 nm. The intensity of the absorption increases at longer wavelength, the 314-nm band having an extinction coefficient of 47,000. Fulvalene has also been prepared by the oxidation of dihydrofulvalene with oxygen in the presence of silver oxide and by photoirradiation of diazacyclopentadiene in a matrix at 77 K.

Na^+ **24** —I_2, −80 °C→ **53** —*n*BuLi, C_5H_{12}→ **54** $2Li^+$ —O_2, C_5H_{12}→ **55**

Figure 7.29

Partial oxidation of the dianion **54** gives the fulvalene radical anion **56**, the ESR spectrum of which indicates that the electron is delocalized over both five-membered rings.

56 ⟷ **56a**

Figure 7.30

Numerous annelated and substituted fulvalenes have been prepared that can be isolated as stable, polyenic compounds.

Sesquifulvalene (**57**), which has been known for some time in solution, was recently prepared in a pure form by Neuenschwander and co-workers by a route similar to that which they used for the preparation of heptafulvene (Fig. 7.31). The ^{1}H NMR spectrum shows multiplets at δ 7.05 (2H), 6.3 (4H), and 6.2 (4H), and the electronic spectrum shows maximum absorption at 393 nm (ε 25,900).

7,8,9,10-Tetraphenylsesquifulvalene (**58**) and a number of other sesquifulvalene derivatives had been prepared earlier. The physical properties of **58** indicate that it is a polyene with little contribution from the dipolar

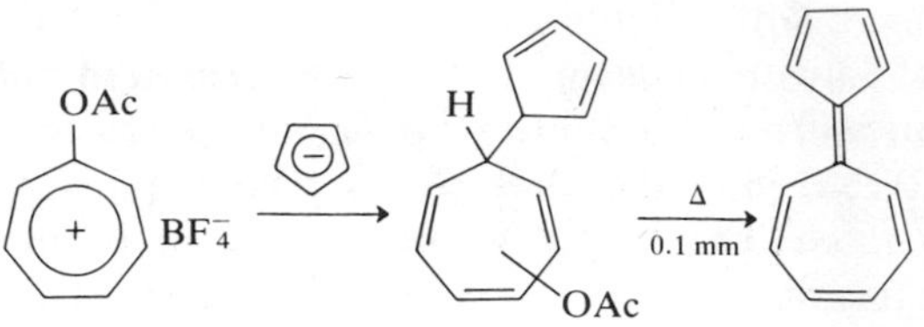

Figure 7.31

structure **58a**, and this is true for the other derivatives. The contribution of the dipolar form to all of the sesquifulvalenes appears to be small, and they are best represented as polyenes rather than aromatic systems.

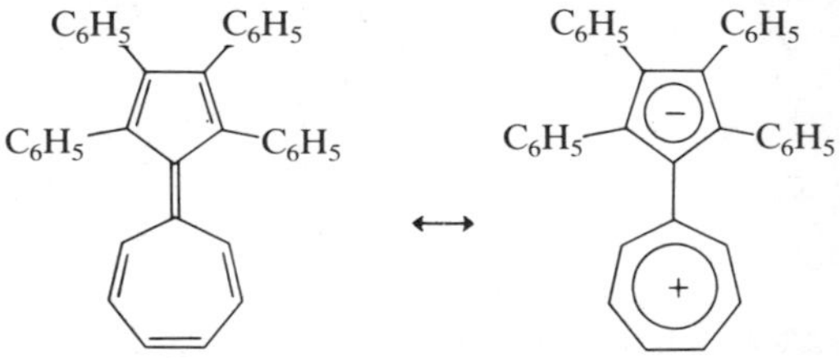

Figure 7.32

A variety of macrocyclic annulenones, fulvenes, and fulvalenes have been prepared over the last decade, usually as derivatives rather than as the parent systems.

The substituted [9]annulenones **59** and **60** have been synthesized, and both are nonaromatic rather than aromatic compounds. The nonafulvene **61** and the heptanonafulvalene **62** have also been prepared, and in both the nine-membered ring is nonplanar, the molecules behaving as polyenes.

Vogel and co-workers have prepared all of the possible 1,7-methano[11]annulenones (**63**–**67**).* All of these compounds are atropic; thus the ^{1}H NMR spectrum of **63** exhibits an AA′XX′ system at δ 7.01 and 6.90 and AB systems at δ 7.18, 6.02 and δ 1.68, 0.04, with ^{1}H coupling constants similar to those in the parent hydrocarbon bicyclo[5.4.1]dodeca-2,5,7,9,11-pentaene. Deuteronation of **63** produces a downfield shift of the

* Correct nomenclature, in which the carbonyl group takes precedence, would give different numbering to the bridge positions in the various compounds.

59

60

61

62

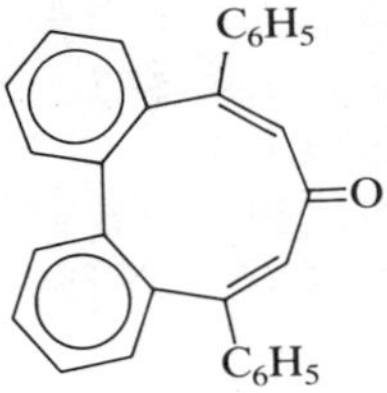

Figure 7.33

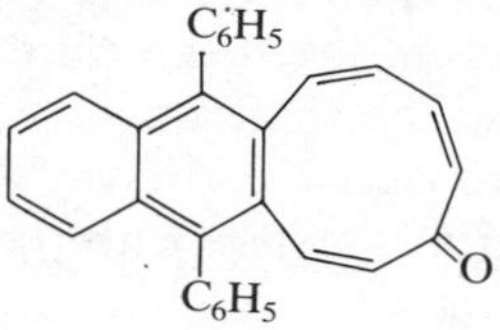

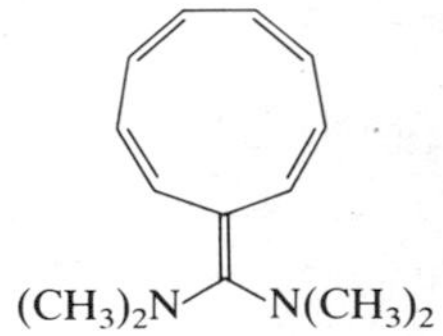

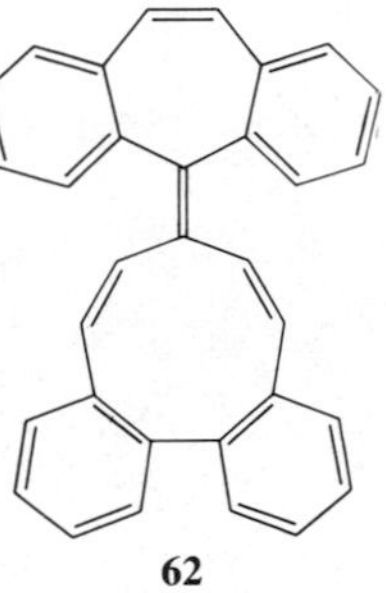

ring protons and an upfield shift of the bridge protons, the deuteroxy[11] annulenium ion **68** being diatropic.

A benzannelated 1,7-oxido[11]annulenone has also been synthesized, and this is again atropic.

63

64

65

66

67

68

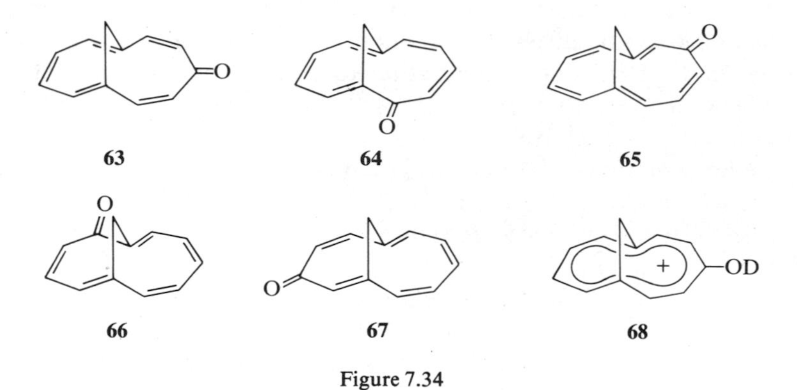

Figure 7.34

The undecapentafulvenes **69**, **70** have been prepared by the reaction of the 1,6-methanoundecapentaenium cation with either the fluorenyl anion or tetraphenylcyclopentadienide. In the ^{1}H NMR spectrum of both compounds, the protons on the 11-membered ring are at relatively high field, and this observation, together with the magnitude of the vinylic coupling constants, suggests that these are polyenic not aromatic systems. Protonation of **70** with trifluoroacetic acid gives the corresponding undecapentaenium cation **71**, a diatropic system with the ring protons at much lower field in the ^{1}H NMR spectrum.

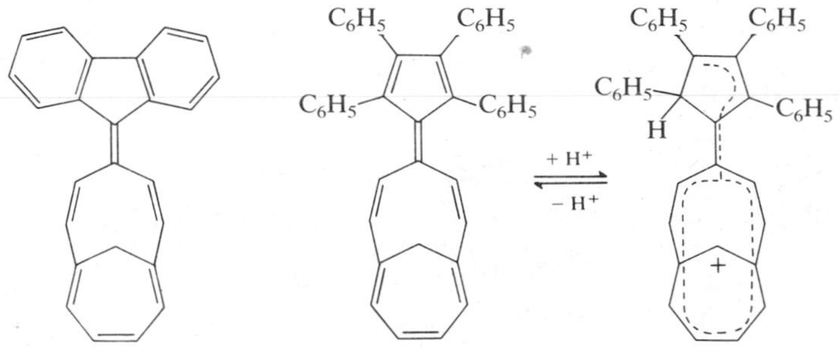

C_6H_5 C_6H_5 C_6H_5 C_6H_5 C_6H_5 C_6H_5 C_6H_5 C_6H_5 H $+H^+$ $-H^+$ +

69 **70** **71**

Figure 7.35

A variety of macrocyclic dehydroannulenones have been prepared by the method shown in Figure 7.36 for a [15]annulenone derivative. *cis,trans*-Octa-3,5-dien-1,7-diyne (**72**) was treated with a molar equivalent of ethylmagnesium bromide, and the resulting Grignard reagent **73** (a mixture of two stereomers) was then treated with the aldehyde **74**. The resulting mixture of alcohols was oxidatively coupled under the Glaser conditions (only the isomer **75** is of suitable configuration to undergo this reaction) to give the cyclic alcohol **76**. Oxidation with manganese dioxide gave the [15]annulenone derivative **77**.

The series of trimethylbisdehydro[*n*]annulenones (**78**) prepared by Ojima and co-workers (n = 15, 17, 19, 21, 23, 25) show an alternation in the position of the electronic spectrum absorption maximum similar to that observed for the annulenes, and in the ^{1}H NMR spectrum, **78e** (n = 23) has the inner protons at higher field than the outer, whereas **78f** (n = 25) has the outer protons at higher field than the inner. These latter differences are much smaller than those observed in the annulenes but do suggest that **78e** is diatropic and **78f** paratropic. This difference is emphasized on deuteronation when the tropicity of both types of ring is increased.

A number of annulenediones have been prepared, including both bridged and dehydro derivatives. The 1,6-methano[10]annulenedione **80**, (R = H) exists in the norcaradiene form **79** (R = H), whereas the 11,11-

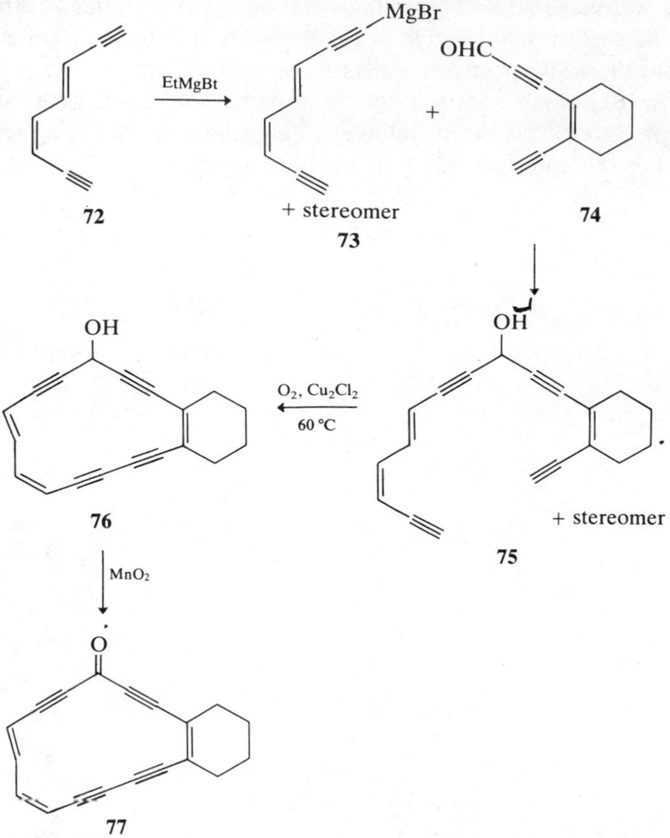

EtMgBt

MgBr

OHC

+

72

+ stereomer

73

74

OH

OH

O_2, Cu_2Cl_2

60 °C

76

+ stereomer

75

MnO_2

O

77

Figure 7.36

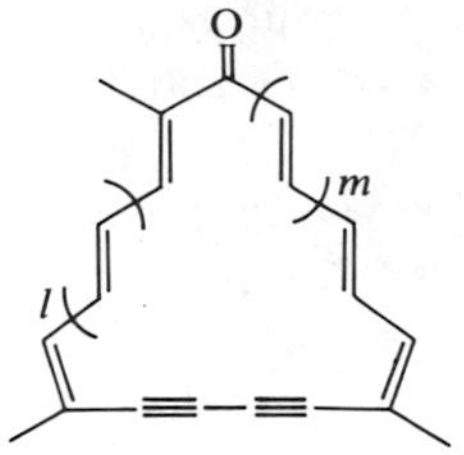

O

m

l

78a $n = 15$, $l = 0$, $m = 1$
78b $n = 17$, $l = 1$, $m = 1$
78c $n = 19$, $l = 1$, $m = 2$
78d $n = 21$, $l = 2$, $m = 2$
78e $n = 23$, $l = 2$, $m = 3$
78f $n = 25$, $l = 3$, $m = 3$

Figure 7.37

difluoroderivative **80** (R = F) exists in the open [10]annulene form. Mild reduction of the tricyclic **79** (R = H) converts it into the semiquinone, which, from the ESR spectrum, exists in the open form.

The dione **81**, which does not have a norcaradiene alternative form, has also been prepared. This dione behaves as a quinone, undergoing reductive acetylation to the diacetate **82**, a 10 π-electron analogue of benzoquinone diacetate.

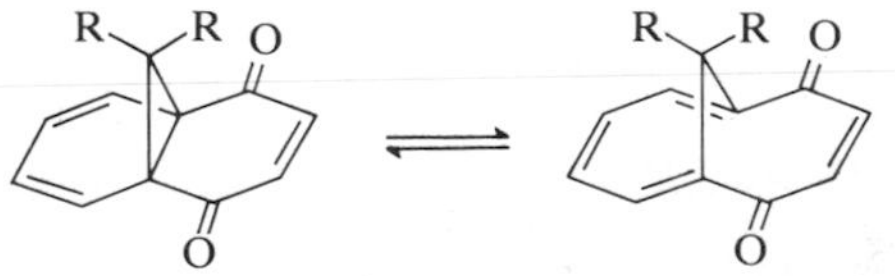

79 R = H
R = F

80 R = H
R = F

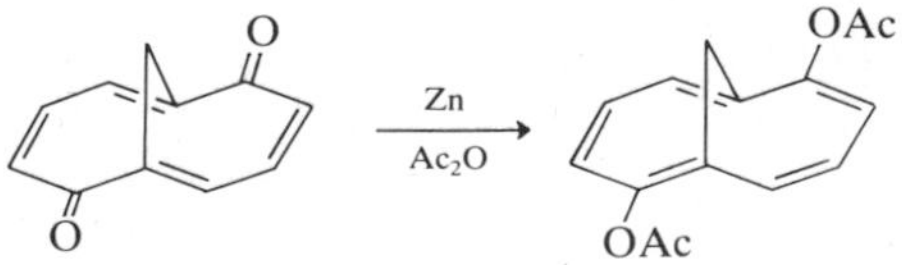

Zn
Ac_2O

81

82

Figure 7.38

A number of macrocyclic tetradehydro[18]annulenediones (**83–86**) have been prepared with the carbonyl groups in pseudo-*ortho* and pseudo-*para* positions. These quinones can be readily reduced to the corresponding dianions by electrochemical reduction. The electrochemical reduction potentials show that these compounds are more readily reduced than *p*-benzoquinone, suggesting a considerable stability for the dianions **87** and **88**.

The pentatridecafulvalene **89** was prepared by treating the corresponding ketone with cyclopentadiene in the presence of sodium methoxide. The fulvalene **89** was not protonated with trifluoroacetic acid and appears to have little dipolar character. Other related compounds have been prepared, and a number of macrocyclic dehydroannulenofulvenes has also been synthesized. None of these compounds shows a great deal of aromatic character.

The cross-conjugated systems described in this chapter exhibit little aromatic stabilization, although in the cases of both the small ring and macrocyclic annulenones there is some evidence for an alternation of magnetic properties. The fulvenes and fulvalenes are best described as

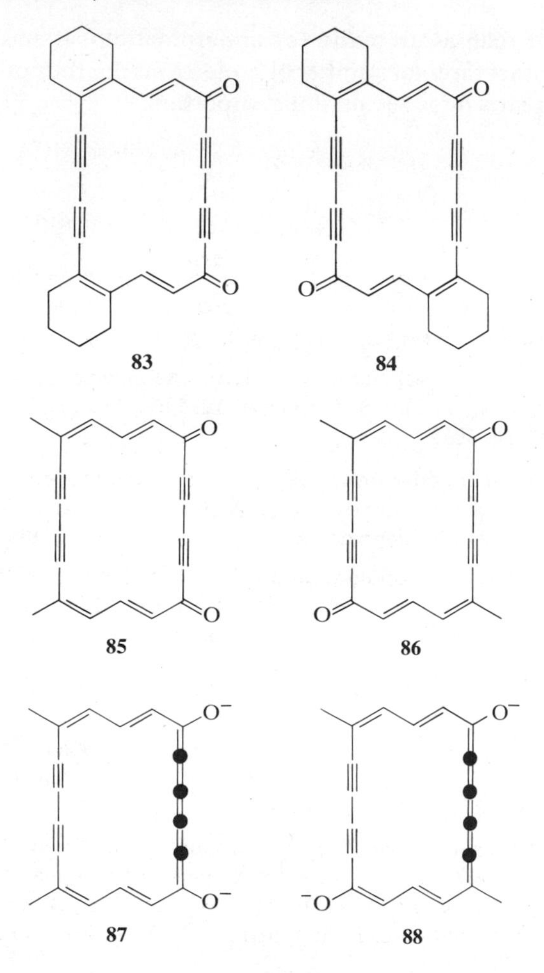

Figure 7.39

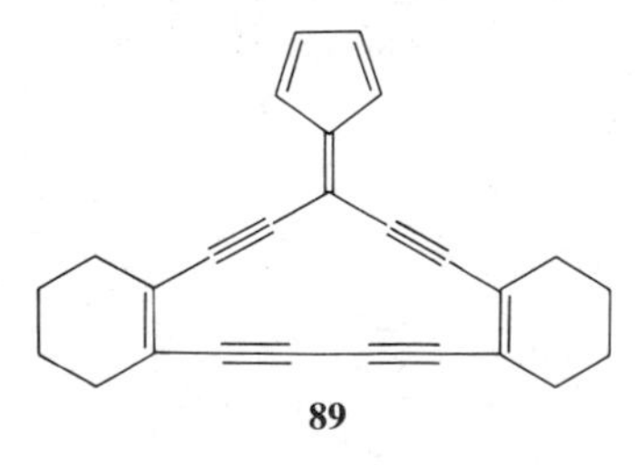

Figure 7.40

polyenes rather than as aromatic (or antiaromatic), systems and in these compounds neither the total number of π electrons nor the possible intervention of dipolar structures seems to be important.

FURTHER READING

For a general discussion of these systems, see P. J. Garratt and M. V. Sargent, in E. C. Taylor and H. Wynberg (Eds.), *Advances in Organic Chemistry*, Vol. 6, Wiley-Interscience, New York, 1969, p. 1.

For a discussion of cyclopropenone and triafulvene chemistry, see T. Eicher and J. L. Weber, *Topics in Current Chemistry*, 1975, **57**, 1; Z.-i. Yoshida, *Pure Appl. Chem.*, 1982, **54**, 1059.

For the more recent thermodynamic data on diphenylcyclopropenone and a succinct review of the relevant literature, see A. Greenberg, R. P. T. Tomkins, M. Dobrovolny, and J. F. Liebman, *J. Am. Chem. Soc.*, 1983, **105**, 6855.

For more recent work on tropolone and other annulenones, see Shô Itô and Y. Fujise, *Topics in Nonbenzenoid Aromatic Chemistry*, T. Nozoe, R. Breslow, K. Hafner, Shô Itô, and I. Murata (Eds.), Vol. 2, Hirokawa, Tokyo, 1977, p. 91.

For recent studies on fulvenes and fulvalenes see M. Neuenschwander, *Pure Appl. Chem.*, 1986, **58**, 55.

For 10 π-troponoid systems, see E. Vogel, *Topics in Nonbenzenoid Aromatic Chemistry*, T. Nozoe, R. Breslow, K. Hafner, Shô Itô, and I. Murata (Eds.), Vol. 2, Hirokawa, Tokyo, 1977, p. 243.

For the larger annulenones, see M. V. Sargent and T. M. Cresp, *Topics in Current Chemistry*, 1975, **57**, 111, and for more recent work J. Ojima, K. Wada, and M. Terasaki, *J. Chem. Soc., Perkin 1*, 1982, 51, and V. K. Shama, H. Sharihari-Zavarah, P. J. Garratt, and F. Sondheimer, *J. Org. Chem.*, 1983, **48**, 2379 and references in these papers.

For a review on heptafulvene see D. J. Bertelli, *Topics in Nonbenzenoid Aromatic Chemistry*, T. Nozoe, R. Breslow, K. Hafner, Shô Itô, and I. Murata (Eds.), Vol. 1, Hirokawa, Tokyo, 1973, p. 29, and for more recent work see W. K. Schenk, R. Kyburz, and M. Nuenschwander, *Helv. Chim. Acta*, 1975, **58**, 1099. See also H. Prinzbach and L. Knothe, *Pure Appl. Chem.*, 1986, **58**, 25.

8

HETEROCYCLIC SYSTEMS

8.1. INTRODUCTION

There are two types of heterocyclic systems that in principle could exhibit aromatic properties: type (i) in which the heteroatom replaces a carbon atom in a carbocyclic aromatic compound and provides one electron to the π system and type (ii) in which the heteroatom replaces a carbon-carbon double bond unit (C=C) and provides two electrons to the π system. Pyridine (**1**) is the classic example of type (i) and pyrrole (**2**) that of type (ii).

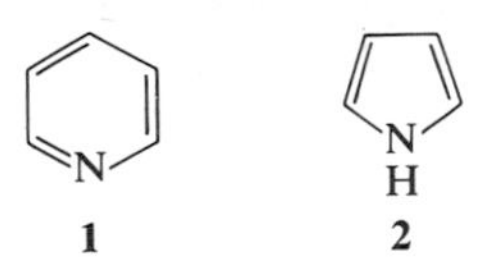

Figure 8.1

The properties of pyridine would be expected to be similar to those of benzene, the differences being due to the difference in electronegativity of nitrogen and carbon. Whereas benzene has D_{6h} symmetry with all the carbon atoms and C—C bonds equivalent, pyridine is a distorted hexagon with C-2 equivalent to C-6, C-3 equivalent to C-5, and C-4 unique. In the simple HMO treatment, the σ framework of pyridine is considered to be hexagonal and the nitrogen atom to be sp^2 hybridized with the lone pair of electrons occupying one of these orbitals. The remaining p orbital on nitrogen then combines with the five carbon p atomic orbitals to form a set of six molecular orbitals (Fig. 8.2). The six π electrons can then occupy the three bonding

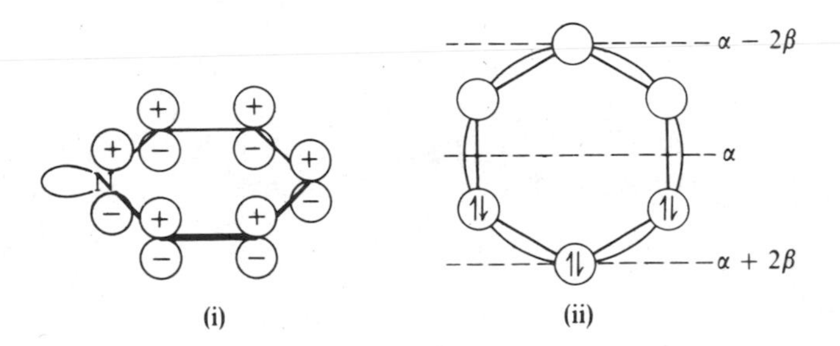

Figure 8.2. (i) p_π atomic orbitals of pyridine; (ii) molecular orbitals of pyridine assuming $\alpha_N = \alpha$, $\beta_{CN} = \beta$.

orbitals as in benzene. Two problems now arise: how to account for the alteration of the hexagonal σ framework of benzene by the substitution of nitrogen and how to estimate the values of the coulomb and resonance integrals for the C—N bond. One approach is to ignore the changes in the σ framework and allow all the changes to be absorbed in the coulomb and resonance integrals. Taking the empirical parameters α and β for the coulombic and resonance integrals of benzene, these can be modified so that

$$\alpha_{\mathrm{N}} = \alpha + h_{\mathrm{N}}\beta \tag{8.1}$$

and

$$\beta_{\mathrm{CN}} = k_{\mathrm{CN}}\beta \tag{8.2}$$

The problem then is to assign specific values to the h_{N} and k_{CN} parameters for nitrogen. For elements such as nitrogen, which are more electronegative than carbon, the value of h will be positive and α_{N} should be greater than α. However, the value of h presumably also depends on the number of electrons contributed by the heteroatom to the π system and should differ in pyridine and pyrrole. The value of k depends on the bond length. Streitwieser has given values for the h and k parameters for a number of heteroatoms, taking into account the number of electrons donated by the heteroatom and the bond order. The values of $h_N = 0.5$ and $k_{CN} = 1$ were suggested for nitrogen in pyridine. These values can now be introduced into equations (8.1) and (8.2) and the new values for α and β introduced in the appropriate place in the determinant (1.12).

A similar treatment can be applied to pyrrole (**2**). In this case, the nitrogen p atomic orbital combines with the four carbon p orbitals to form five molecular orbitals similar to those of the cyclopentadienyl anion (Fig. 8.3).

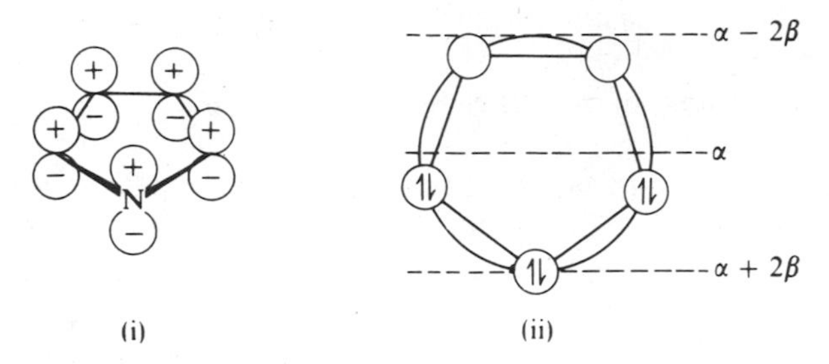

Figure 8.3. (i) p_π atomic orbitals of pyrrole; (ii) molecular orbitals of pyrrole assuming $\alpha_N = \alpha$, $\beta_{CN} = \beta$.

The values suggested for nitrogen in pyrrole are $h_N = 1.5$ and $k_{CN} = 1$, although calculations have been carried out using a wide variety of values for these parameters.

The general conclusions of the MO theory are perforce in general agreement with the earlier inductive ideas of the English school of chemists. Pyridine is thus found to be polarized so that nitrogen has an excess of π-electron density with the carbons being electron deficient. Pyridine is generally deactivated toward electrophilic substitution, with the C-3,5 atoms being less deactivated than the others. Pyrrole by contrast has the nitrogen atom electron deficient as compared to the carbon atoms and is thus activated toward electrophilic substitution. The dipole moments of pyridine and pyrrole nicely reflect this difference, pyrrole having a dipole moment (1.8D) with the nitrogen as the positive pole and pyridine a dipole moment (2.2D) with nitrogen as the negative pole (Fig. 8.4). Heterocycles like pyridine, in

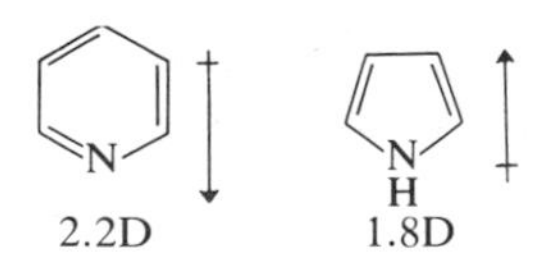

Figure 8.4

which the carbon atoms are electron deficient, are classified as π deficient, whereas those like pyrrole, in which the carbon atoms have excess electrons, are classified as π excessive.

The electronic spectrum of pyridine resembles that of benzene with three major absorption bands. The long wavelength bands are, however, more intense than those in benzene since, owing to the lower symmetry, these transitions are no longer forbidden. The spectrum of pyrrole is less readily explained, and its similarity to cyclopentadiene has been interpreted both for and against the supposition that pyrrole is a delocalized system.

The ^{1}H NMR spectrum of pyridine clearly highlights the effect of introducing nitrogen for carbon. The H-2,6 and H-4 protons all appear at lower field than those of benzene, whereas the H-3,5 protons appear at higher field, these changes being in accord with the changes in electron density (Fig. 8.5). In pyrrole all of the protons attached to carbon are at higher field than those in benzene, again indicating the difference in electron density. The ^{1}H NMR spectra of these compounds are consistent with the direction of the dipole moment and with their reactivity to electrophilic substitution.

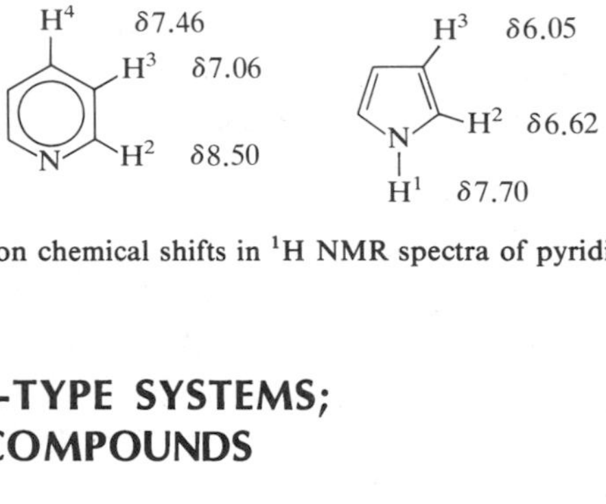

H^4 δ7.46
H^3 δ7.06
H^2 δ8.50
N

H^3 δ6.05
H^2 δ6.62
N
H^1 δ7.70

Figure 8.5 Proton chemical shifts in ^{1}H NMR spectra of pyridine and pyrrole.

8.2. PYRROLE-TYPE SYSTEMS; π-EXCESSIVE COMPOUNDS

Other electronegative elements besides nitrogen may be conceptually substituted for the methylene group in cyclopentadiene to form heterocyclic analogues of pyrrole. Further, the substitution of heteroatoms into other "odd" cyclic systems, such as cyclopropene, will give a series of molecules of the general type **3**.

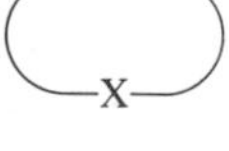

$(=)_m$
X

3

Figure 8.6.

Assuming a two-electron contribution from the heteroatom, then those systems in which m is odd will contain $4n$ π electrons and those in which m is even will contain $(4n + 2)$ π electrons. The first four members of this series are shown in Figure 8.7.

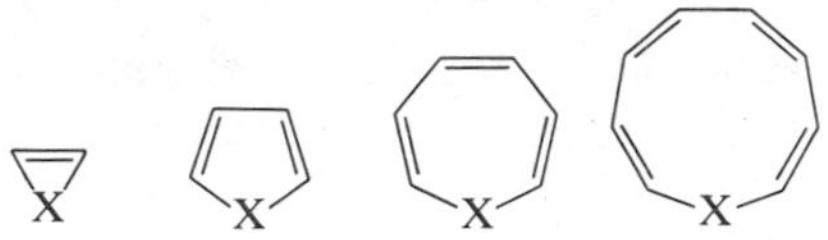

no of π-electrons	4	6	8	10

Figure 8.7. Heterocyclic homologues of the pyrrole type.

The most usual heteroatoms to have been introduced are oxygen, nitrogen, and sulfur, although a number of 6 π-electron compounds containing other heteroatoms have been prepared (vide infra). The three-membered ring compounds with $m = 1$ are $4n$ systems containing 4 π electrons and should not be aromatic. None of these compounds has been isolated, but evidence has been obtained for their intermediacy in reactions and, in one case, for formation in matrixes by low-temperature photolysis of appropriate precursors. Oxirene (**4a**, X = O) has been postulated as an intermediate in the Wolf rearrangement and other reactions on the basis of scrambling experiments. MO calculations predict oxirene to be less thermodynamically stable than ketene. Thiirene (**4b**, X = S) is produced by the photoirradiation of vinylene thiocarbonate (**5**) in an argon matrix at 8 K, and the IR spectrum shows bands at 3208, 3170, 1660, 912, 660, and 563 cm^{-1}, the C=C stretching frequency being very similar to that of cyclopropene (1656 cm^{-1}). Substitution by deuterium or other groups showed the expected changes in the IR spectral bands, and calculated spectra were in reasonable agreement with that observed. 2-Azirine (**4c**, X = NH) has, like oxirene, been implicated in a number of rearrangements through labeling and product studies, and the formation of a substituted 2-azirine in a low-temperature photoirradiation experiment has been claimed on the basis of a band at 1867 cm^{-1} in

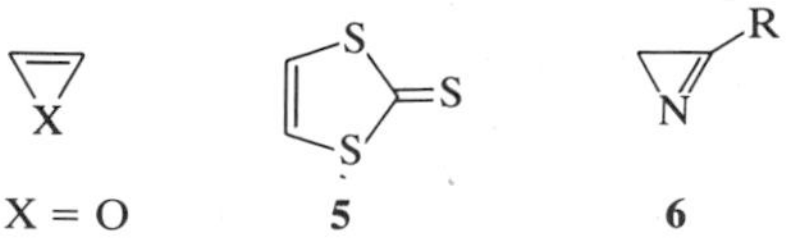

4a X = O
4b X = S
4c X = NH

5

6

Figure 8.8

the IR spectrum. A number of 1H-azirines (**6**) have been prepared, and these all appear to be thermodynamically more stable than their 2H-azirine isomers.

The compounds with $m = 2$ have six π-electrons and should be aromatic. This group is made up of the familiar compounds pyrrole (**2**), furan (**7**), and thiophene (**8**). These molecules have very different properties from those of cyclopentadiene, being far less reactive than would be expected for a fixed *cisoid* diene. In all of these systems, the heteroatom forms the positive end of the dipole in contrast to their saturated analogues, in which the heteroatom is the negative end. Pyrrole is a very weak base because the lone pair of electrons on nitrogen is involved in the delocalized π system, which would be destroyed on protonation. In fact, pyrrole protonates in strongly acid media not on nitrogen but on carbon. The importance of the delocalized π system is nicely demonstrated for thiophene, which on oxidation with peroxybenzoic acid gives the dimeric derivative **9** rather than the sulfoxide or sulfone. Thiophene dioxide, which has not been isolated, thus appears to be even more reactive than cyclopentadiene toward Diels–Alder addition in contrast to thiophene, which is largely inert to dienophiles.

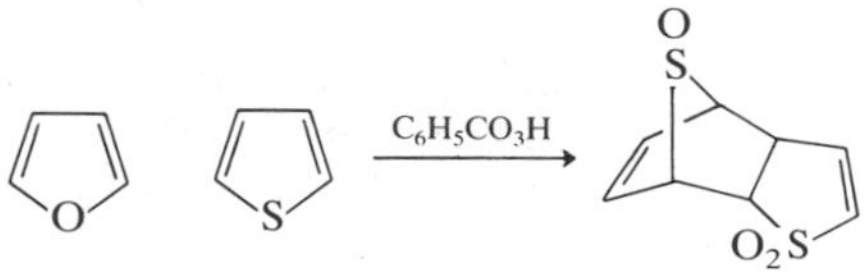

O S $C_6H_5CO_3H$ O_2S

7 **8** **9**

Figure 8.9

Methylation of thiophene gives the S-methylthiophenium salt **10** in which only one lone pair of electrons is used in the S—Me linkage, the other pair remaining available for contribution to the 6 π system.

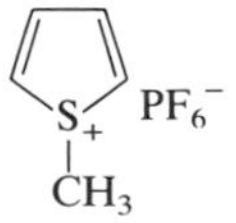

S^+ PF_6^- CH_3

10

Figure 8.10

Furan reacts readily with dienophiles to give Diels–Alder adducts, whereas thiophene, as mentioned above, reacts only with powerful dienophiles or under very forcing conditions. Pyrrole normally gives substitution products rather than adducts. All three heterocycles undergo electrophilic substitution, furan being the most reactive and thiophene the least. Due to the electron donation by the heteroatom, the carbon atoms have excess electron density, and all three compounds are more reactive than benzene toward electrophiles. This is in complete contrast with pyridine, which is much less reactive than benzene toward electrophiles.

The other members of groups V and VI can also form five-membered heterocycles. Arsole (**11**, X = As) and stibole (**11**, X = Sb) appear to be nonaromatic systems, the arsenic atom, for example, being pyramidal. Nevertheless, the barrier to inversion is lower than in other arsines, which suggests a lowering of the energy of the planar transition state in arsole.

X

X = As, Sb

11

Figure 8.11

The heterocyclic systems with $m = 3$ have eight π electrons and are isoelectronic with the tropylium anion. These compounds should thus not exhibit aromatic behavior. Oxepin (**12**) and 1H-azepine (**13**) have been prepared by Vogel and co-workers, but only derivatives of thiepin (**14**) have so far been isolated (Fig. 8.12). Oxepin was obtained by the route outlined

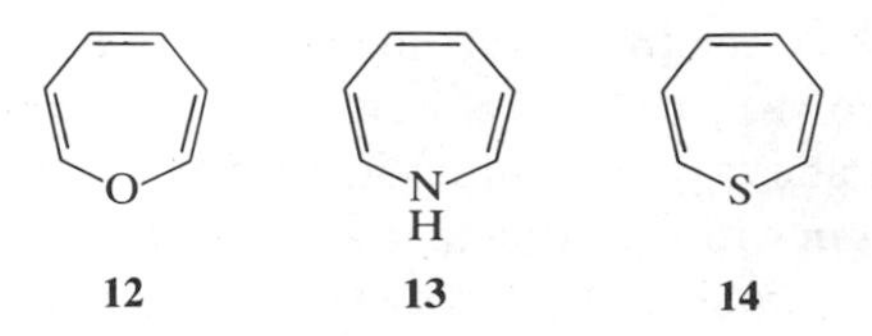

Figure 8.12

in Figure 8.13, which was subsequently modified to prepare 1,6-methano[10]annulene. Epoxidation of 1,4-cyclohexadiene gave the oxirane **15**, which on bromination gave **16**. Dehydrobromination of **15** gave a mixture of oxepin (**12**) and its valence tautomer benzene oxide (**17**). The ^{1}H NMR spectrum indicated that this was a 1 : 1 mixture of **12** and **17**, and the temperature dependence of the spectrum indicated that the interconversion of **17** and **12** is rapid.

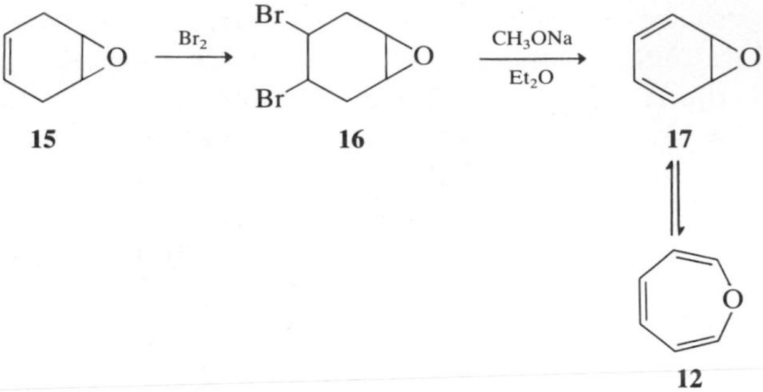

15 16 17 12

Figure 8.13

The oxepin molecule is nonplanar, existing in a pseudo tub conformation and rapidly equilibrating between the equivalent structures **12a** and **12b** (Fig. 8.14). Changes in the electronic spectrum indicate that the equilibrium between **12** and **17** is very solvent dependent, the amount of **17** increasing in more polar solvents.

12a 12b

Figure 8.14

1H-Azepine (**13**) was prepared by the sequence shown in Figure 8.15. Demethylation of the ester **18** with trimethylsilyl iodide gave the silyl ester **19**, which was hydrolyzed to the acid **20** with methanol. Decarboxylation of the acid at a little above room temperature gave **13**, which, unlike oxepin, was not in equilibrium with its valence tautomer **21**. The ^{1}H NMR spectrum of **13** is temperature independent between −120 and −30°C and shows absorptions at δ 5.57 (H-4, 5), 5.22 (H-2,7) and 4.69 (H-3,6). The ^{13}C NMR spectrum has signals at δ 138.0 (C-2,7), 132.3 (C-4,5) and 113.0 (C-3,6) and is very like that of oxepin [δ 141.8 (C-2,7), 130.8 (C-4,5), 117.6 (C-3,6)] and unlike that of benzene oxide [δ 128.7 (C-3,4,5,6), 56.6 (C-2,7)]. 1H-Azepine is only obtainable as a red solution and readily polymerizes, but it can be isomerized into 2H-azepine (**22**) by reaction with trimethylamine and converted into N-substituted derivatives by treatment with a number of reagents (Fig. 8.15).

A number of N-substituted azepines had previously been prepared and were all shown to have nonplanar, pseudo tub structures and showed little

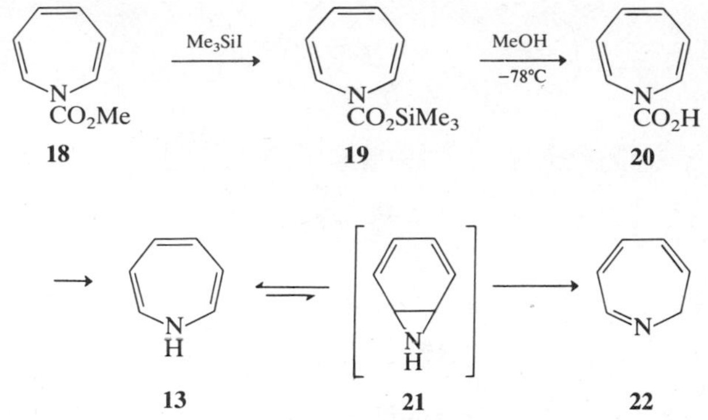

Figure 8.15

tendency to valence isomerize to the bicyclic tautomers. Both oxepines and azepines exhibit the properties of polyenes, but the oxygen atom shows a greater preference for the three-membered ring form.

Thiepin has so far resisted synthesis, largely because of the ease with which sulfur is extruded to give benzene derivatives. A number of monocyclic thiepins have been prepared in which bulky groups at the 2,7-positions destabilize the intermediate for sulfur extrusion. The earliest example was **23** prepared by Schlessinger and co-workers, but simpler derivatives have now been synthesized by Murata's group (e.g., **24**). Benzo[b]thiepin and a variety of benzannelated thiepins are also known.

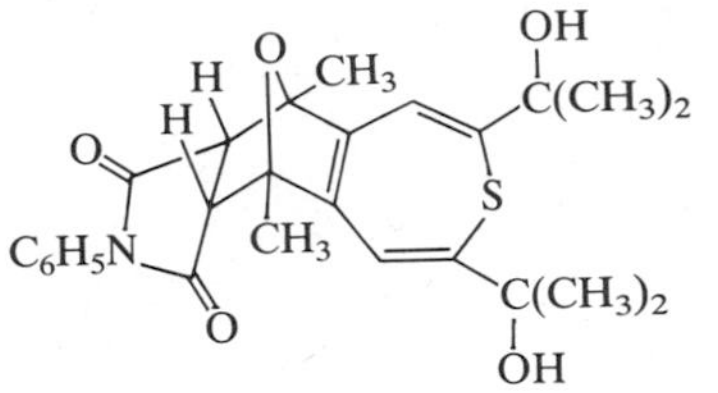

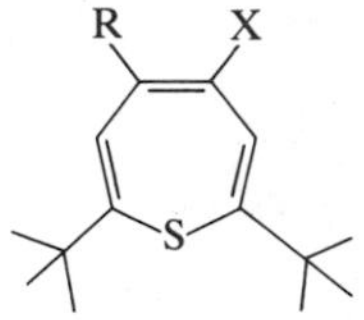

24a R = Me, X = CO_2Et
24b R = H, X = CO_2Et
24c R = H, X = CHO

Figure 8.16

The systems with $m = 4$ contain 10 π electrons, are isoelectronic with the cyclononatetraenyl anion, and should be aromatic. Both the parent oxonin (**25**) and azonine (**26**) have been prepared, but only substituted derivatives of thionin (**27**) are known. Oxonin (**25**) was prepared by the low-temperature photoirradiation of cyclooctatetraene epoxide (**28**). The

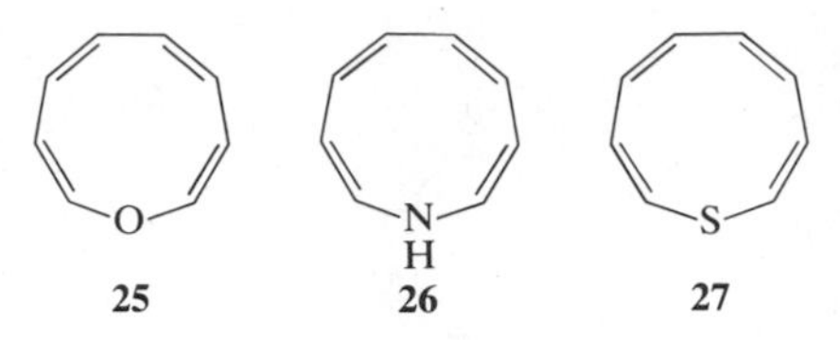

Figure 8.17

^{1}H NMR spectrum of oxonin has resonance signals in the region δ 6.21–5.08, and the spectrum is temperature independent. Oxonin is thermally unstable and rearranges to the *cis*-bicyclic derivative **29** at 30°C. Photoirradiation leads to the formation of an equilibrium mixture of **25** and an extremely labile compound that is probably mono-*trans*-oxonin (**30**). The latter isomer rearranges below −15°C to the *trans*-bicyclic compound **31** (Fig. 8.18). The assignment of the mono-*trans* structure to **30** was made on the assumption that the rearrangement **30** → **31** would be orbital-symmetry controlled. Oxonin is clearly a nonplanar, nonaromatic compound and resembles cyclononatetraene rather than the cyclononatetraenyl anion.

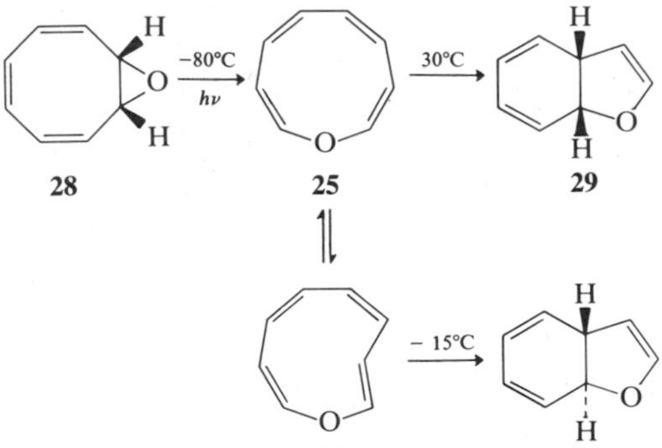

Figure 8.18

Azonine (**26**) was prepared by treatment of the N-carbethoxyazonine (**32**) with potassium *t*-butoxide in THF at low temperature. The ^{1}H NMR spectrum of **26** consists of a two-proton doublet at δ 7.15, a four-proton multiplet at δ 6.73, and a broad two-proton multiplet at δ 6.08–5.75. Hydrogenation of **26** over rhodium on charcoal at 0°C gave azacyclononane (**33**).

Azonine is thermally more stable than oxonin and is recovered unchanged after heating to 50°C. The chemical shift of the protons in the ^{1}H NMR

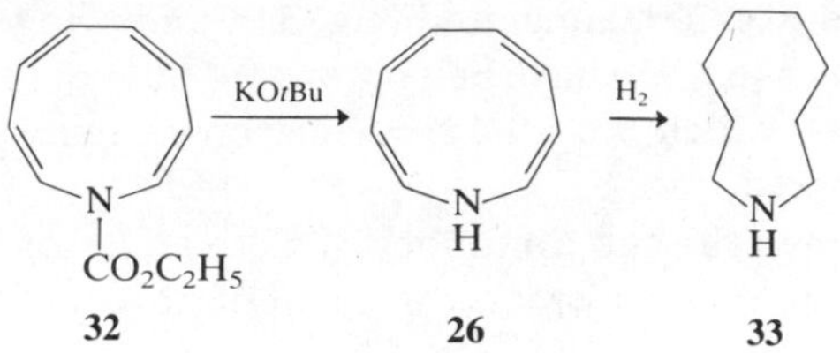

KO*t*Bu H_2

$CO_2C_2H_5$

32 **26** **33**

Figure 8.19

spectrum of **26**, at lower field than the corresponding protons of oxonin and the N-carboethoxy derivative **32**, together with the ^{13}C NMR spectral data and the greater thermal stability, was interpreted to indicate that azonine is diatropic with some aromatic character. As a solvent, azonine also gave solute chemical shifts typical of a diatropic medium.

A variety of benzannelated *cis*-heteronins (**34a–c**) are known, and all possess nine-membered polyenic rings, including the azonine anions. That this difference in property was skeletal in origin was nicely shown for *cis*-benzo[4,5]azonine (**34b**), which could be rearranged to the 6,7-*trans* derivative **35** via the anion. The mono-*trans* isomer **35**, unlike its all-*cis* isomer **34c**, was diatropic. The dibenzo derivatives **36a** and **36b** are also atropic with buckled nine-membered rings.

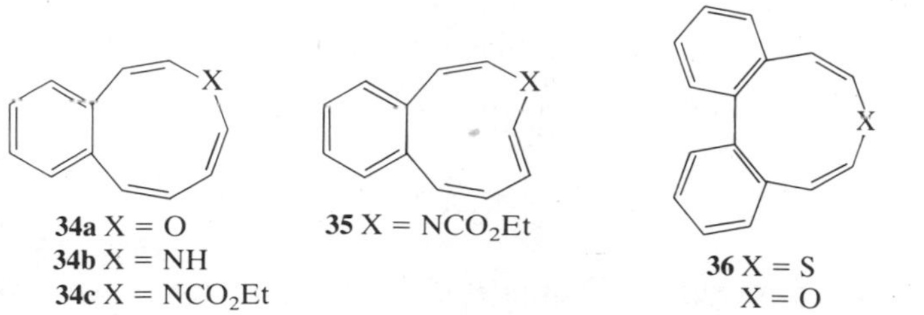

34a X = O
34b X = NH
34c X = NCO_2Et

35 X = NCO_2Et

36 X = S
X = O

Figure 8.20

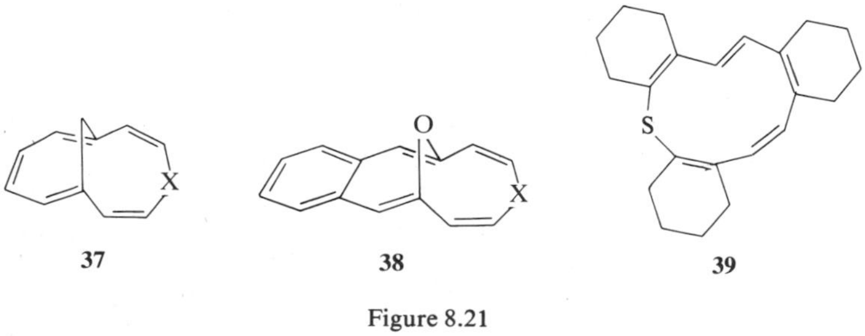

37 **38** **39**

Figure 8.21

Several bridged (e.g. **37**), annelated (e.g. **38**), and bridged and annelated hetero[11]annulenes (e.g. **39**) have been prepared but not the parent systems. Compounds **37–39**, which are all 12 π-electron systems, are atropic with buckled 11-membered rings.

Oxa[13]annulene (**40**) and a number of derivatives of aza[13]annulene (**41, 42**) have been prepared by rearrangement of polycyclic precursors (Fig. 8.22).

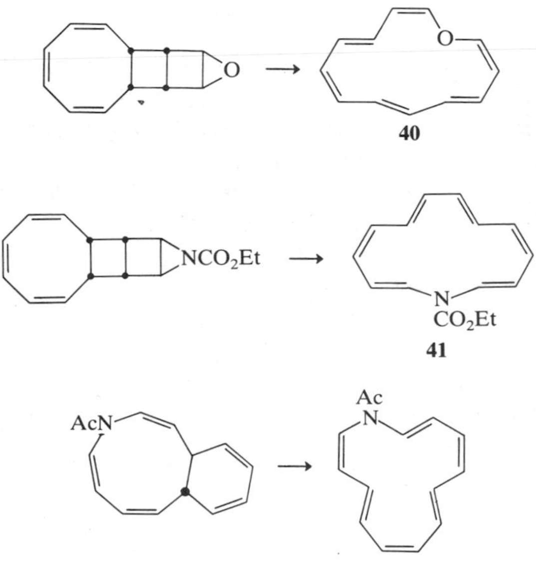

O → O

40

NCO_2Et → N CO_2Et

41

AcN → Ac N

42

Figure 8.22

Several heterodehydro[13]annulenes (e.g., **43**) are known, the thia-annulene **43b** being diatropic, the oxygen analogue **43a** weakly diatropic, and the sulfoxide and sulfone atropic.

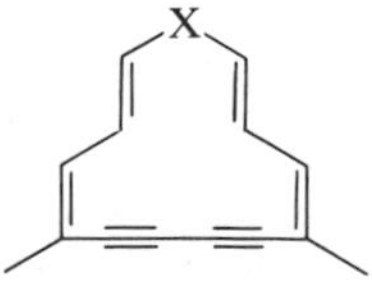

X

X = O, S, SO_2, SO

43

Figure 8.23

A variety of annelated or dehydrogenated oxa and thia[15]annulenes (e.g., **44**, **45**) are known but not the parent systems. Both **44** and **45** are paratropic.

44

45

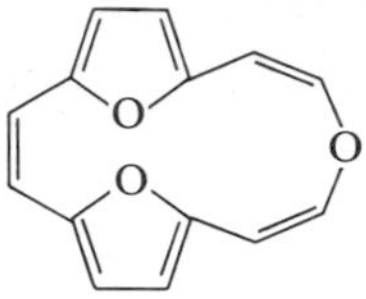

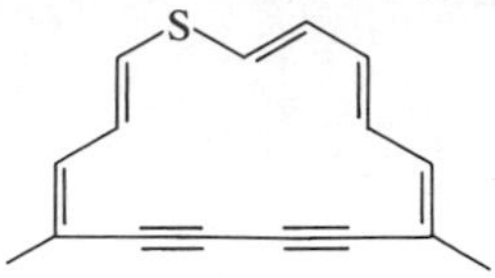

Figure 8.24

The parent oxa (**46**) and aza[17]annulenes (**47a**) have been prepared as mixtures of isomers by the photoirradiation of polycyclic precursors. Aza[17]annulene (**47a**) and its potassium salt **47b** are diatropic with high field (**47a**, R = H, δ 2.5–0.5; **47b**, R = −ve, δ −4.9) inner protons in the ^{1}H NMR spectrum. A variety of dehydro and bridged [17]annulenes have also been prepared with, for example, **48** being diatropic and **49** atropic.

46

47a R = H
47b R = ve

48

49

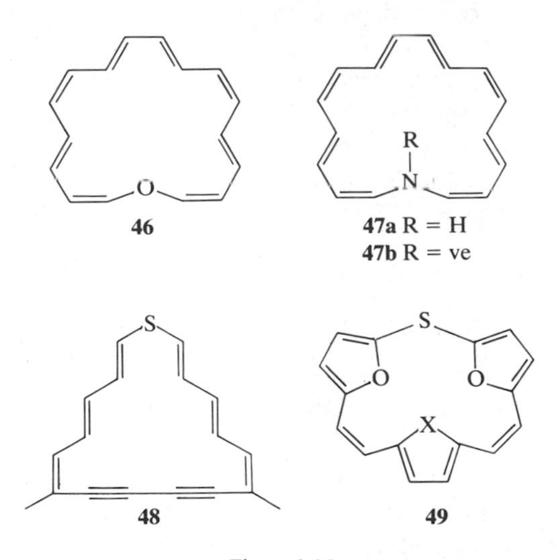

Figure 8.25

A series of bridged hetero[17], [19], and [21]annulenes (**50**) has been prepared by Sondheimer and collaborators, and these compounds exhibit the expected alternation of magnetic properties. In the case of the aza[17] and [21] derivatives, the compounds in which nitrogen is substituted with an electron withdrawing group (COMe, CO_2Et) are also diatropic, which is in contrast to the nine-membered ring systems. The higher rigidity of the planar σ framework allows these systems to be delocalized without incurring an entropic penalty.

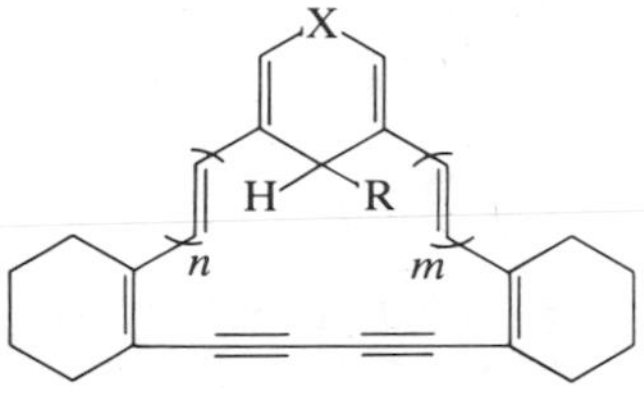

50 X = NR, n = 1, 2; m = 1, 2; R = H, alk
X = O, n = 1, m = 1; R = Et
X = S, n = 1, m = 1; R = H, Et

Figure 8.26

The series of π-excessive systems, although still lacking the larger parent systems, is sufficiently extensive for us to conclude that an alternation of magnetic properties is observed in these compounds. Compared to the annulenes, the degree of diatropicity and paratropicity is less, with a greater tendency for the compounds to be nonplanar, atropic systems. That the nature of the heteroatom is a factor is also apparent, oxygen contributing less readily to the delocalized π system than nitrogen and sulfur. The characteristics of the 6 π heterocycles are thus reflected in the higher members.

8.3. PYRIDINE-TYPE SYSTEMS: π-DEFICIENT COMPOUNDS

The smallest system of this type is azete or azacyclobutadiene (**51**), in which a carbon of cyclobutadiene is replaced by a nitrogen. Azacyclobutadiene is a 4 π-electron system, which like cyclobutadiene should be antiaromatic. The parent system is unknown, but a substituted azacyclobutadiene **53** has been prepared by Gompper and co-workers by flash vacuum pyrolysis of the 1,2,3-triazine **52**. The azacyclobutadiene **53** is a red compound that is only stable in solution at low temperature and decomposes on warming to dimethylcyanamide. "Push–pull" stabilization presumably accounts for the ability to isolate **53**, serving in the same way as for push-pull cyclobutadienes (see Chapter 4).

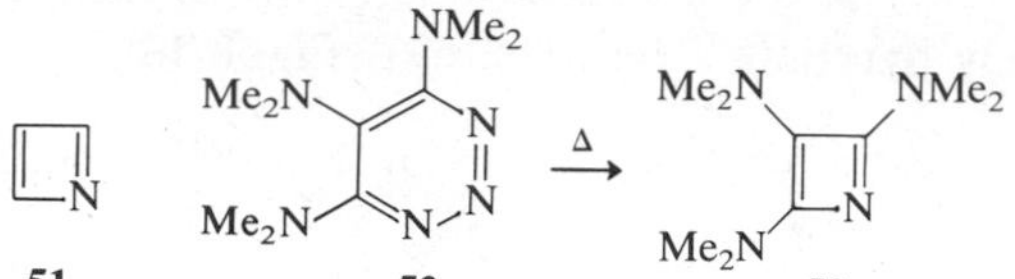

51 52 53

Figure 8.27

Since oxygen is divalent rather than trivalent, it is not possible to replace the nitrogen in pyridine by oxygen to give a neutral system. However, treatment of 4-pyran (**54**) with triphenylmethyl perchlorate removes a hydride ion and gives the corresponding pyrylium salt **55**. The pyrylium salts dissolve in water to give acidic solutions that contain an equilibrium mixture of the cation **55** and the pseudobase **56**. The pyrylium salts are more stable than the corresponding trialkyloxonium salts, but the increase in stability is small. The cationic nature of these compounds tends to mask the aromatic properties, but these salts do not appear to possess the aromatic stabilization observed in pyridine.

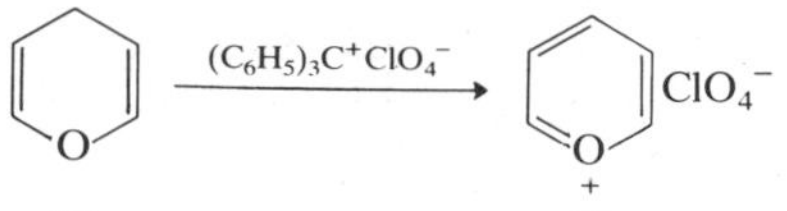

$(C_6H_5)_3C^+ClO_4^-$ ClO_4^-

54 55

Figure 8.28

$X^- + H_2O \rightleftharpoons$ OH + HX

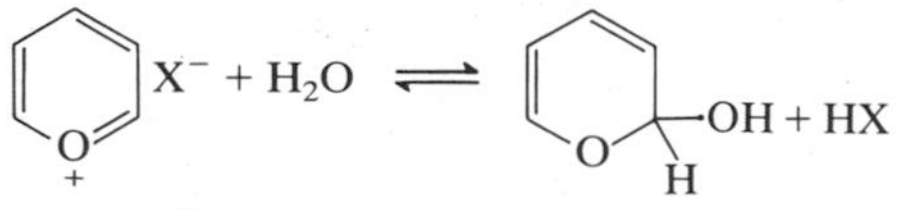

55 56

Figure 8.29

4-Pyran (**54**) is an air-sensitive, unstable liquid with properties resembling a vinyl ether. 4-Pyrone (**57**), a compound that is isoelectronic with tropone, is a much more stable system. 4-Pyrone undergoes electrophilic substitution rather than addition and is stable to acids, two properties unexpected for a vinylogous lactone. Ring opening does, however, occur readily under basic conditions, and the stability of **57** is not as great as was once thought. 2-Pyrone (**58**) has similar properties to **57**, although it is more readily ring opened. The stability of the pyrones to acids arises from the formation of the corresponding pyrylium salts (e.g., **59**) (Fig. 8.30).

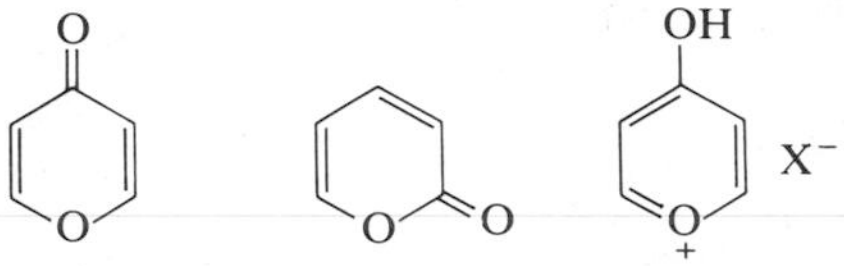

57 **58** **59**

Figure 8.30

Sulfur, like oxygen, is divalent and the thiopyrylium salts are stable, delocalized compounds. Unlike oxygen, however, sulfur has accessible *d*-orbitals that could become involved in bonding. Suld and Price demonstrated that such bonding could occur by treatment of the thiopyrylium salt **60** with phenyllithium to form the S-phenylthiabenzene **61**. This compound was fairly readily oxidized, but it could be isolated as a crystalline compound, melting point 65°C, and its IR, electronic, and ^{1}H NMR spectra were determined. The ^{1}H NMR spectrum shows a single resonance at δ 7.34, indicating that thiabenzene sustains a diamagnetic ring current similar to benzene. The dipole moment (1.88D) suggests that the molecule is not primarily in the ylid form **61b**.

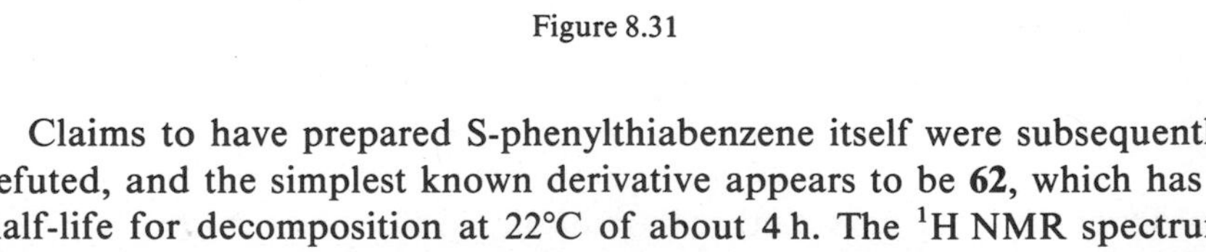

C_6H_5Li

$^-ClO_4$

60 **61a** **61b**

Figure 8.31

Claims to have prepared S-phenylthiabenzene itself were subsequently refuted, and the simplest known derivative appears to be **62**, which has a half-life for decomposition at 22°C of about 4 h. The ^{1}H NMR spectrum shows the H-2,6 protons at high field (δ 4.03), presumably due to the high electron density at these atoms and suggesting a larger contribution of the ylid structure (**62b**) to **62** than of **61b** to **61**.

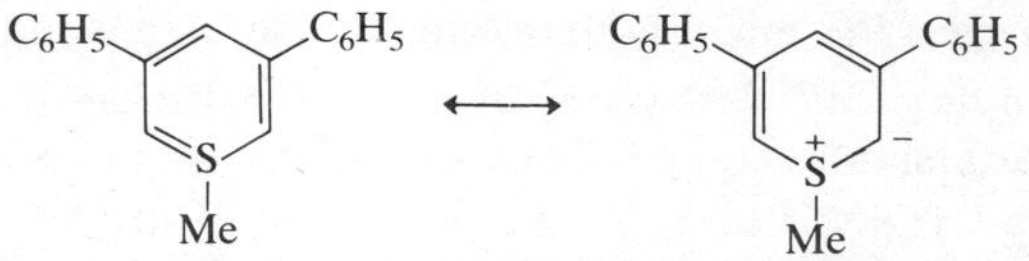

Figure 8.32

The other elements in group V can also be trivalent, and the substitution of these elements for a carbon of benzene to give phosphabenzene (**64**), arsabenzene (**65**), stibabenzene (**66**), and bismabenzene (**67**) was accomplished by Ashe. The stannacyclohexadiene **63** was prepared by reaction of 1,4-heptadiyne with di-*n*-butyl tin hydride, and this could be converted to the desired V-heterobenzenes by reaction with the appropriate reagent (Fig. 8.33).

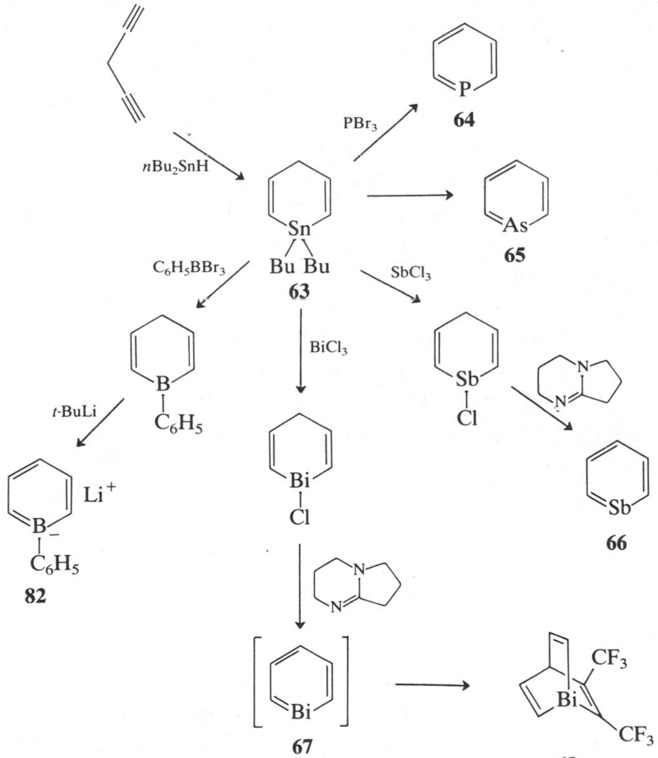

Figure 8.33

Phosphabenzene (**64**) and arsabenzene (**65**) are reasonably stable, air-sensitive compounds; stibabenzene (**66**) is less stable and polymerizes in solution, whereas bismabenzene (**67**) can only be trapped as the Diels–Alder adduct (**68**). The ^{1}H NMR and ^{13}C NMR spectra of the V-heteroannulenes are collected in Table 8.1. The large downfield shift of the C-2,6 and H-2,6

TABLE 8.1. ^{1}H and ^{13}C NMR Chemical Shifts in δ of V-Heterobenzenes

Proton	Carbon	N (ring positions 2–6)	P	As	Sb
H-2,6		8.29	8.61	9.68	10.94
H-3,5		7.38	7.72	7.83	8.24
H-4		7.75	7.38	7.52	7.78
	C-2,6	150.6	154.1	167.7	178.3
	C-3,5	124.5	133.6	133.2	134.4
	C-4	136.4	128.8	128.2	127.4

atoms appears to be derived from diamagnetic ring currents on the heteroatom, but the magnitude of the effect at C-2,6 is too large for it to be described by a simple model for anisotropic magnetic susceptibility.

Microwave and electron diffraction studies on **64** and **65** have been carried out, and the structure shown in Figure 8.33 has been deduced for arsabenzene (**65**). These and the photoelectron spectra suggest that **64** and **65** resemble pyridine and that the heteroatoms participate as one-electron donors to give 6 π-electron systems. However, the stability of these compounds is clearly very different from benzene or pyridine, as exemplified by the ready Diels–Alder addition of alkynes to give the V-1-heterobarrelenes (e.g., **68**).

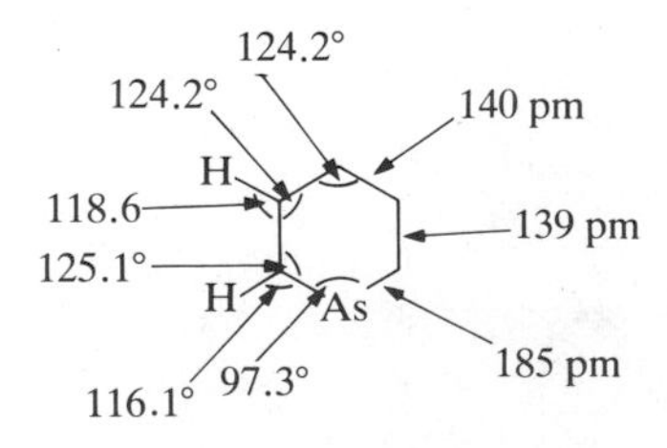

Figure 8.34. Structure of arsabenzene from microwave and electron diffraction data.

A number of substituted phosphabenzenes had earlier been synthesized, mainly by Märkl and co-workers. Thus, 2,4,6-triphenylphosphabenzene (**71**) was prepared by reaction of the pyrylium salt **69** with the phosphine **70**. An x-ray crystallographic study showed that the phosphabenzene ring was a planar distorted hexagon with long (175 pm) C—P bonds and a small CPC bond angle (103°) similar to the structure deduced for arsabenzene in Figure 8.34.

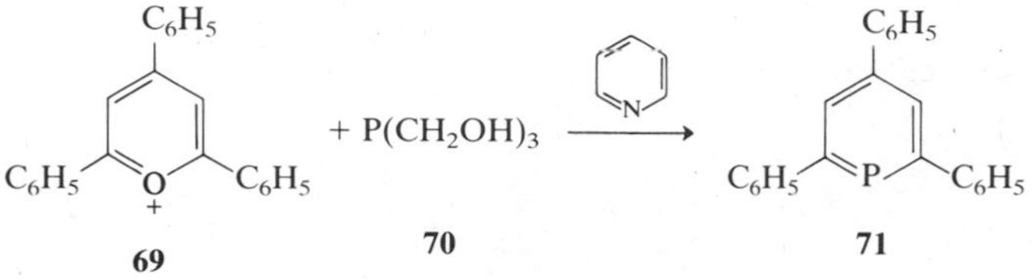

C_6H_5 C_6H_5 C_6H_5 O + **69** + $P(CH_2OH)_3$ **70** → (pyridine) C_6H_5 C_6H_5 P C_6H_5 **71**

Figure 8.35

Phosphorus, like sulfur, has *d*-orbitals available for bonding and a number of compounds had been prepared prior to the preparation of the phosphabenzenes that are analogous to the S-substituted thiabenzenes. 1,1-Diphenylphosphabenzene (**72**) is a yellow, noncrystalline compound that differs considerably in properties from the true phosphabenzenes and appears to have a ylid-type structure. The compound **73**, in which phosphorus is hexacovalent, has also been prepared and has similar properties to **72**.

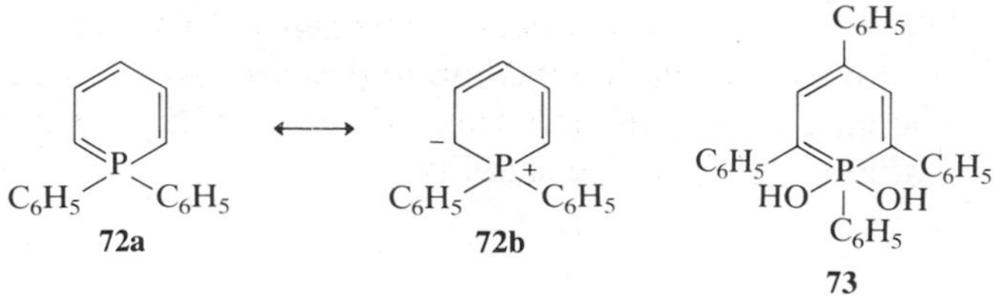

P C_6H_5 C_6H_5 **72a** ⟷ − P+ C_6H_5 C_6H_5 **72b** C_6H_5 C_6H_5 P C_6H_5 HO OH C_6H_5 **73**

Figure 8.36

In Section 8.1, we discussed a simple model for the replacement of carbon by a heteroatom and chose nitrogen as our example because of the long history of the nitrogen heterocycles. We could have taken as a closer analogy to carbon another element from group IV, and the obvious example of such a compound, silabenzene (**74**), has recently been prepared by Maier and co-workers. Silabenzene is a very labile substance that was identified in an argon matrix at 10 K. It was characterized by its IR and electronic spectra but is a very different compound from benzene, undergoing ready phototautomerism to Dewar silabenzene **75** (Fig. 8.37).

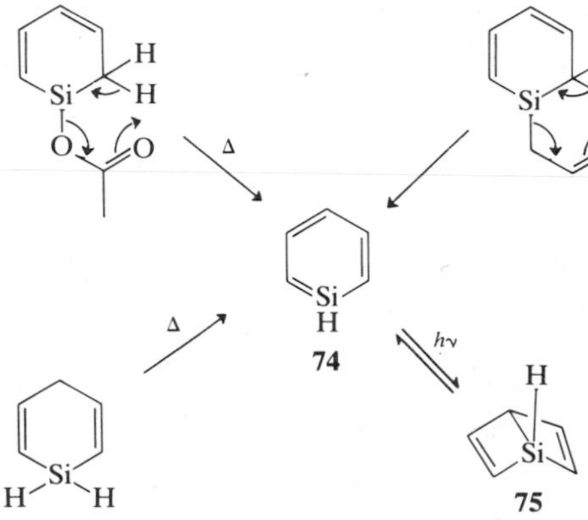

Figure 8.37

The pyrolytic syn elimination method illustrated in Figure 8.37 had previously been used by Barton and Burns to prepare 1-methylsilabenzene (**76**), which was identified from its Diels–Alder adduct with hexafluoro-2-butyne. These authors also prepared **76** by dehydrochlorination of 1-chloro-1-methylsilacyclohexa-2,4-diene, and a similar reaction was used by Märkl and Hofmeister to prepare 1,4-di-*t*-butylsilabenzene. Märkl and co-workers used the pyrolytic syn elimination method to prepare 1,4-dialkylgermabenzenes, for example, by elimination from **77** to give **78**. None of these compounds exhibits any aromatic stability.

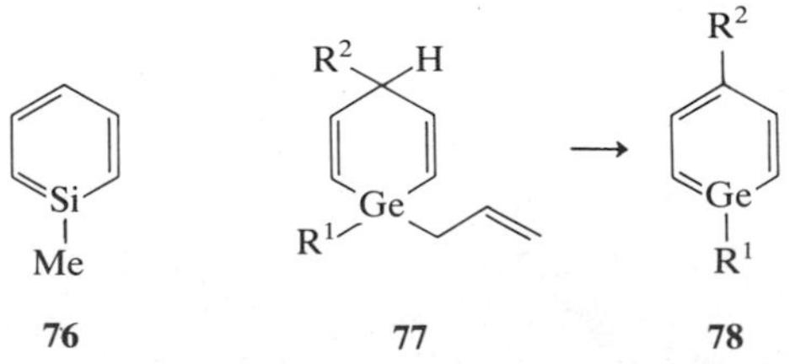

Figure 8.38

In borabenzene (**79**)* the carbon atom has been replaced by an electropositive rather than an electronegative element as in pyridine. The same simple HMO model that we used for pyridine would require a vacant sp^2 orbital rather than the orbital filled by an electron pair. Borabenzene was initially prepared as a ligand in the transition metal complexes **80** and **81** by Herberich and co-workers. The borabenzene anion (**82**) was synthesized shortly thereafter by Ashe, as shown in Figure 8.33. Using either the complexes or the borabenzene anion as a precursor, a wide variety of other borabenzene complexes have been obtained. The borabenzene anion has the properties of a delocalized 6 π-electron system with an increased electron density on the carbon atoms relative to the boron atom.

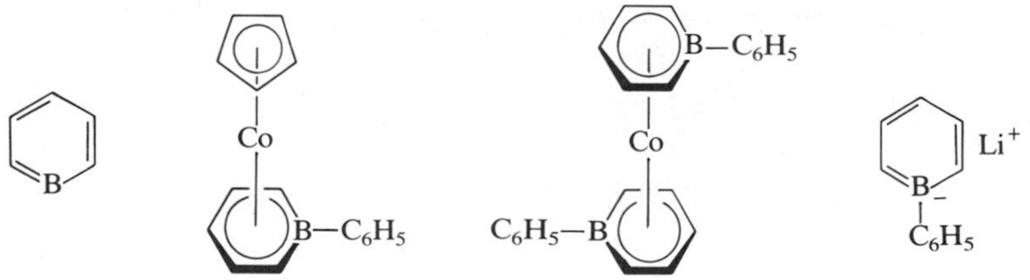

79 **80** **81** **82**

Figure 8.39

Beside these 6 π-electron derivatives, all of the $4n$ and $(4n + 2)$ annulenes can, in principle, give the corresponding heteroannulenes by replacement of one carbon by a heteroatom. A significant number of the higher heteroannulenes has been prepared over the last few years.

Azocine (**83**) was prepared by the flash vacuum pyrolysis of diazabasketane. The ^{1}H NMR spectrum showed signals at δ 7.65, 7.0, and 5.15, each due to one proton, and a multiplet at δ 6.0 due to four protons, and is in accord with the structure shown. Previously, a number of derivatives of **83** had been prepared. 2-Methoxy-1-azocine (**84**) is a yellow liquid, the ^{1}H NMR spectrum of which shows a one-proton doublet at δ 6.54, a four-proton multiplet at δ 6.05–5.75, and a one-proton multiplet at δ 5.12, all due to the ring protons, and a three-proton singlet at δ 3.70 due to the methoxy group. Like azocine itself, compound **84** has a localized structure and both presumably exist in a tub form similar to cyclooctatetraene. The azocines are readily reduced to the corresponding planar dianions either electro-

* The chemical abstract name for **79** is boratabenzene.

chemically or by alkali metals. This reduction differs from that of cyclooctatetraene in that two discrete one-electron additions do not occur.

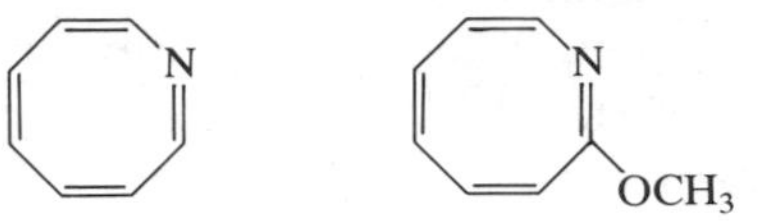

N N OCH_3

83 **84**

Figure 8.40

Azacyclodecapentaene (**85**), the 10 π-electron analogue of pyridine, is unknown, but a number of bridged derivatives have been prepared, all based on the 1,6-methano[10]annulene skeleton. Two isomers are possible, **86** and **87a**, equivalent to homoquinoline and homoisoquinoline. Compound **86** has been prepared, but only derivatives of **87a** have been obtained.

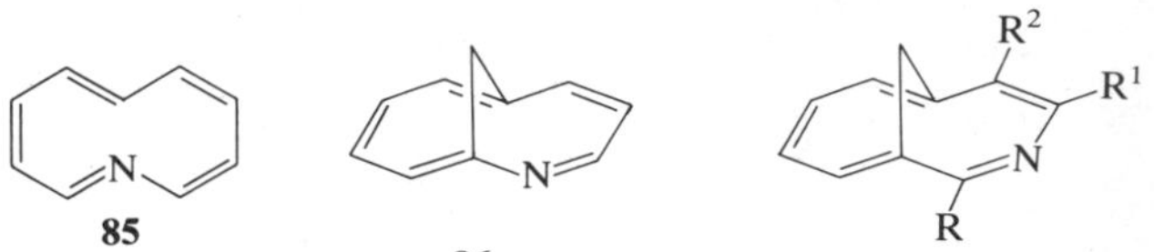

R^2 R^1 N R N N

85 **86**

87a $R = R^1 = R^2 = H$
87b $R = R^2 = H$, $R^1 = OMe, OEt$
87c $R = R^1 = R^2 = CO_2Et$
87d $R = CO_2Et$, $R^1 = R^2 = H$

Figure 8.41

The ^1H NMR spectrum of **86** indicates that it is an aromatic compound, the methylene bridge protons appearing as an AX system at δ −0.4 and 0.65 (J = 8.4 Hz), and the electronic spectrum is similar to that of 1,6-methano[10]annulene. The ^1H and ^{13}C NMR spectra of **87b–d** indicate that these are also aromatic systems. Benzannelated derivatives and more complex systems have been prepared.

Aza[14]annulene (**89**) has been prepared by Röttele and Schröder by photoirradiation of the tricycloazide **88** at low temperature. Aza[14]annulene is a dark violet, crystalline compound. The ^1H NMR spectrum shows 3 inner protons at high field (δ −1.28, −1.12, and −0.58) and 10 outer protons at low field (δ 7.73, 7.90-9.1, 9.60, 9.69). The spectrum is temperature dependent, the signals coalescing at 30°C except for the

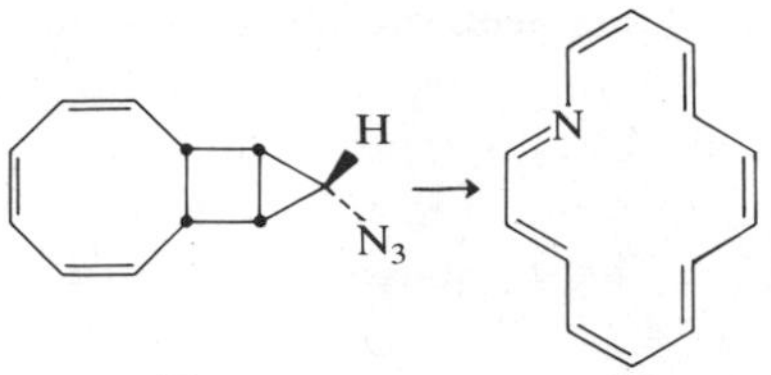

88 **89**

Figure 8.42

signal due to the pseudo *para*-proton H-8, which remains in a high field inner position. Presumably nitrogen and its lone pair always occupy an inner position with protons H-4, 5, 11, and 12 exchanging inner and outer positions and the remainder of the protons except H-8, which is always inside, exchanging but always occupying outer positions (Fig. 8.43). Aza[14]annulene is thus a diatropic molecule that resembles [14]annulene in physical properties.

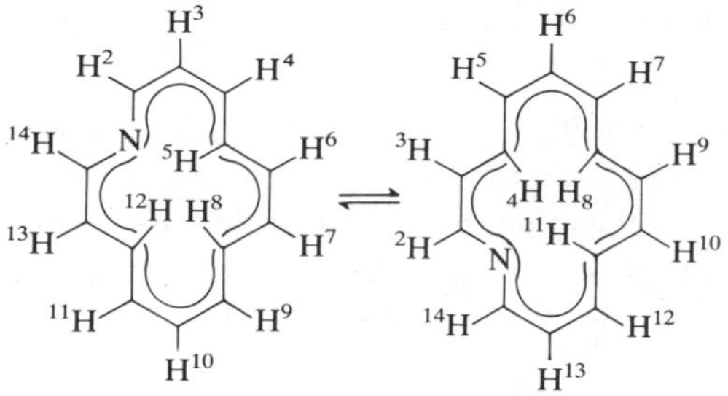

Figure 8.43

A nitrogen-substituted 15,16-dimethyl-15,16-dihydropyrene derivative (**90**) has been prepared and is also a diatropic system.

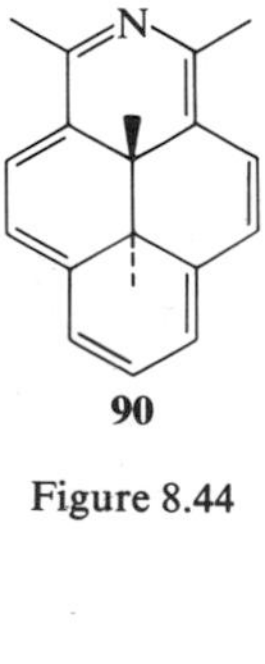

90

Figure 8.44

Aza[18]annulene (**91**) has been prepared by Gilb and Schröder as a dark green, crystalline compound by a route similar to that used for aza[14]annulene. Again the nitrogen occupies an inner position, and the ^{1}H NMR spectrum shows that the molecule is diatropic [δ −1.84 (5H); δ 8.86, 10.05 (12H)]. The spectrum is temperature independent so aza[18]annulene, unlike aza[14]annulene, does not appear to be undergoing conformational exchange. Protonation gives the black-violet, crystalline hydrochloride **92**, which, from the ^{1}H NMR spectrum, exists as a 1:4 mixture of the inner **92a** and outer **92b** isomers. Neutralizing this mixture gives **91**, again solely with the nitrogen having the inner configuration.

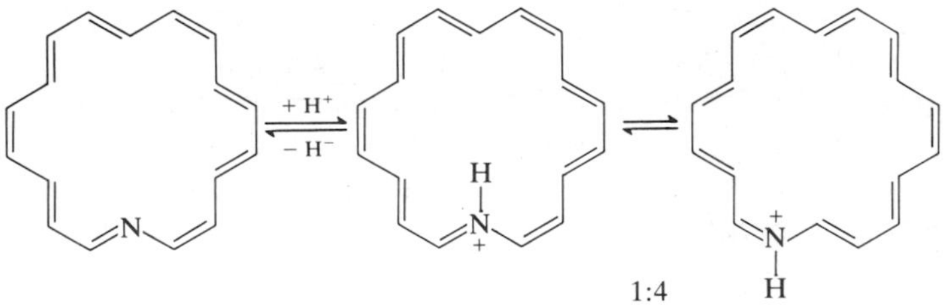

Figure 8.45

Although it is not complete, nevertheless the series of higher pyridine analogues again suggests that the Hückel Rule is being observed. The higher members of the series clearly have physical properties that show a considerable resemblance to the physical properties of the equivalent annulene.

8.4. SYSTEMS WITH TWO HETEROATOMS

Potentially delocalized molecules with two heteroatoms can be classified into three types: (*a*) systems in which both heteroatoms replace carbon-carbon double bond, (*b*) systems in which both heteroatoms replace a carbon atom, and (*c*) systems in which one heteroatom replaces a carbon-carbon double bond and the other a carbon atom. The three types are shown in Figure 8.46.

Figure 8.46

In the first type, both of the heteroatoms contribute two electrons to the delocalized π system, and those compounds with odd numbers of double bonds will have $(4n + 2)$ π electrons, whereas those with even numbers of double bonds will have $4n$ π electrons. For the series shown in Figure 8.47, 3,4-dithiacyclobutene has 6 π electrons, 1,4-dithiin has 8 π electrons, and 1,4-dithiacycloocta-2,5,7-triene has 10 π electrons.

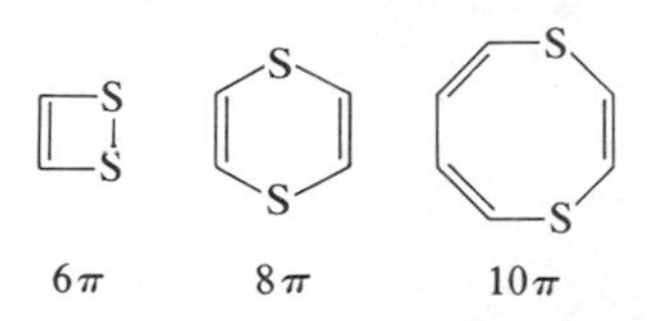

Figure 8.47

1,4-Dithiin (**93**) is an 8 π-electron system if both sulfur atoms contribute 2 π-electrons, whereas if one atom provided two and the other none then a dipolar 6 π-electron system is formed (**94**). 1,4-Dithiin is a thermally stable compound unaffected by strong acids but easily polymerized by Lewis acids. An x-ray crystallographic study showed that the molecule is nonplanar, having a boat conformation with a CSC bond angle of 100° (Fig. 8.49). The bond lengths are within the range expected for normal C—S and C=C bonds. 1,4-Dithiin is thus a nonplanar, nonaromatic molecule.

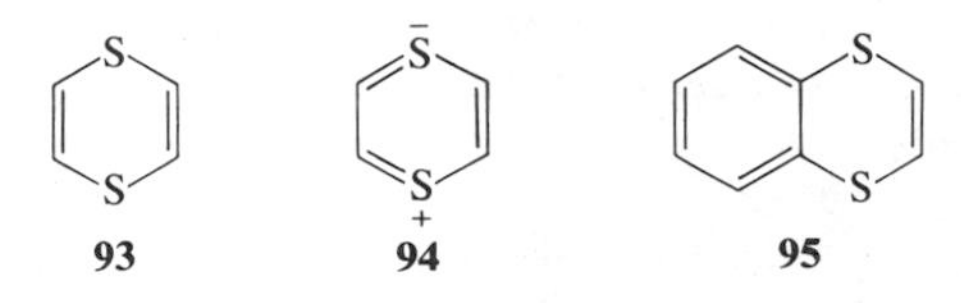

Figure 8.48

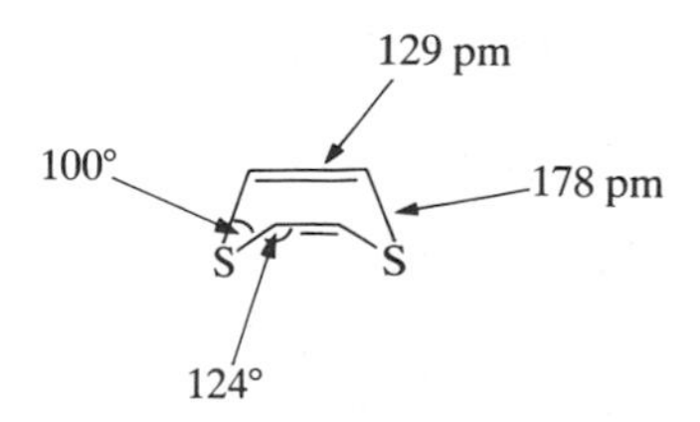

Figure 8.49. X-Ray crystallographic structure for 1,4-dithiin.

By contrast, the properties of benzo-1,4-dithiin (**95**) are more those expected for an aromatic molecule, **95** undergoing a variety of substitution reactions. 1,2-Dithiin has been prepared and is also nonaromatic.

The dihydrodiheteronins are 10 π-electron systems and can exist in two isomeric forms, **96** and **97**. Of these two structures it was suggested by Vol'pin in 1960 that only **96** could be expected to be stabilized since the proximity of the lone pairs of electrons on the 1,2-dihydro-1,2-diheteronins would cause two of them to enter an antibonding molecular orbital. Subsequent calculations suggested that the 1,4-dihydro-1,4-diheteronins were also likely to be destabilized in the planar form.

X X X X

96 **97**

Figure 8.50

A variety of 1,4-dihydro-1,4-diheteronins have now been prepared, including the parent oxygen **98** and nitrogen **99** compounds, but only substituted sulfur derivatives (e.g., **100**) have been obtained. 1,4-Dihydro-1,4-diazocine (**99a**) is a planar, diatropic system, the nitrogen lone pair electrons having delocalized into the ring, and the electronic spectrum resembles that of the cyclooctatetraene dianion. Replacement of the amine hydrogen by an electron-withdrawing group (e.g., **99b**, R = CO_2Me) removes the lone pair electrons, and the molecule adopts a buckled, nonplanar conformation and the electronic spectrum now resembles that of 1,3,6-cyclooctatriene.

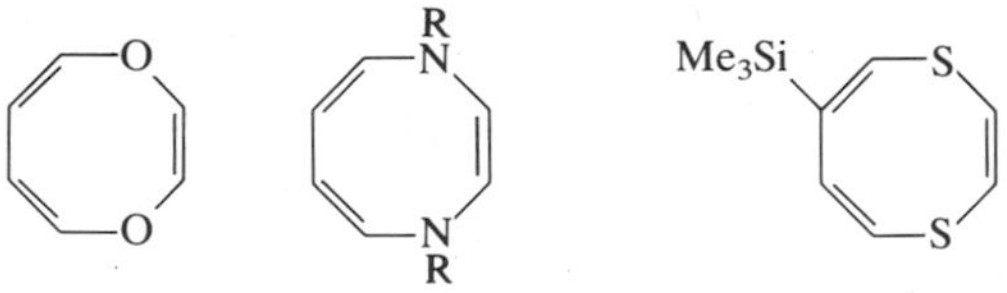

O O R N N R Me_3Si S S

98 **99a** R = H **100**

99b R = CO_2Me

99c R = −ve

Figure 8.51

1,4-Dihydro-1,4-diazocine (**99a**) reacts with potassium amide and the diester **99b** reacts with potassium in liquid ammonia to give the corresponding dianion **99c**, also a 10 π-electron system and an analogue of the cyclooctatetraenyl dianion if the nitrogen lone pair electrons are in the ring plane (Fig. 8.52).

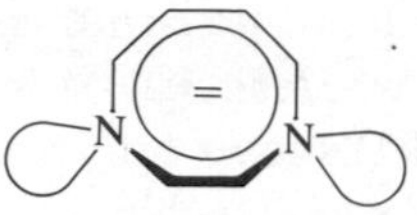

Figure 8.52

Only one 1,2-dihydro-1,2-diheterocine is known, the diazocine **101**. It was suggested that the ^{1}H NMR spectrum was best interpreted if **101** was at least partially planar and highly conjugated, but it is of interest to note that the related system **102** could not be isomerized to the 1,2-diazocine form.

The only larger system of type (*b*) appears to be the 12 π-electron disulfide **103**. Not surprisingly, this is an atropic, buckled system, and the question whether delocalization occurs in dihetero compounds larger than the heterocines must await further experimental work.

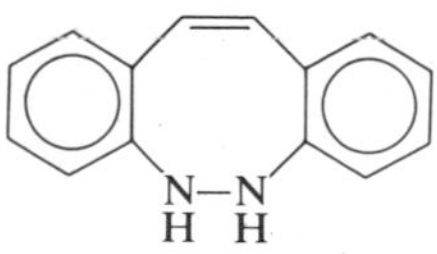

101 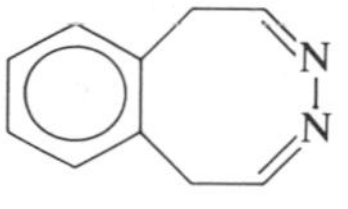 **102** 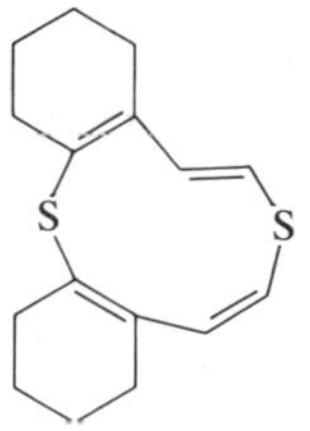 **103**

Figure 8.53

The replacement of a C=C unit in benzene by a B—N unit would give the isoelectronic 6 π-electronic borazarene (**104**), which is a type (*b*) system. In this compound, the nitrogen supplies two electrons to the π system and the boron none, but the boron has an available *p* atomic orbital for combination with the four carbon and one nitrogen atomic orbitals. The nitrogen atom in borazarene is thus electron deficient, like the nitrogen of pyrrole, and the N—B bond is polarized in the opposite manner to that expected from the electronegativities of the two elements. The polar structure **104a** indicates the possible benzenoid structure of borazarene, which might best be represented by **105**. Borazarene is, however, an extremely unstable compound that has not been well characterized. Several derivatives are known, but most of these are also unstable. Dehydrogenation of **106** with palladium on charcoal gave a mixture of **107** and **108**, and **107** could also be prepared by dehydrogenation of **109**. In the ^{1}H NMR spectrum the N-benzyl derivative **108** shows signals for the aromatic protons at δ 7.7–7.2 and a singlet at δ 2.95 for the benzylic protons. Compound **108** is unstable, rapidly decomposing in air, whereas **107** is a relatively stable, crystalline solid. 2-Phenylborazarene (**107**) shows only aromatic protons (δ 7.8–6.3) in the ^{1}H NMR spectrum, and the electronic spectrum resembles that of 2-phenylpyridine. The spectral properties of **107** and **108** thus suggest that

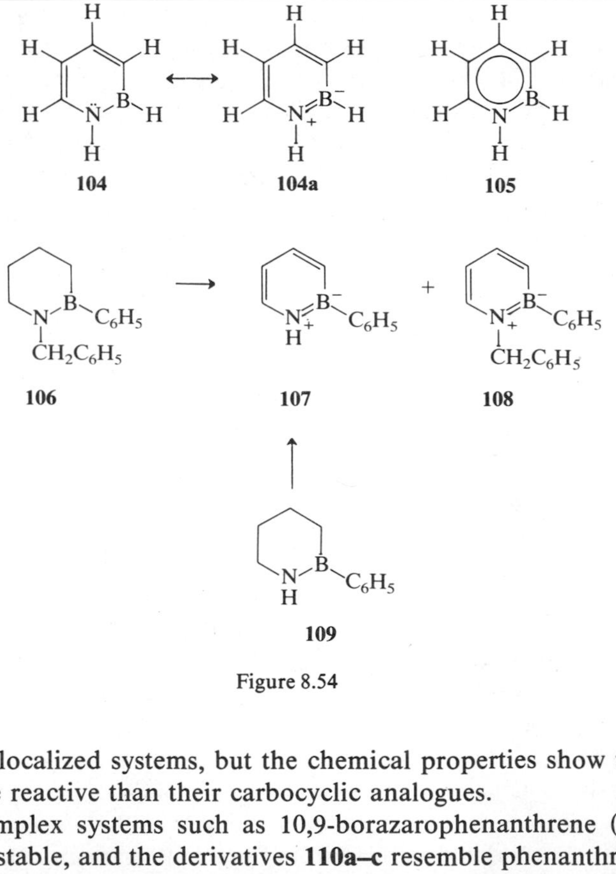

104 **104a** **105**

106 **107** **108**

109

Figure 8.54

these are delocalized systems, but the chemical properties show that they are far more reactive than their carbocyclic analogues.

More complex systems such as 10,9-borazarophenanthrene (**110**) are much more stable, and the derivatives **110a–c** resemble phenanthrene both in physical and chemical properties, the substituted compounds **110b, c** readily undergoing electrophilic substitution. Systems related to naphthalene and anthracene have also been prepared, and those compounds in which the boron and nitrogen atoms are adjacent are much more stable than those in which these atoms are separated (e.g., **111**).

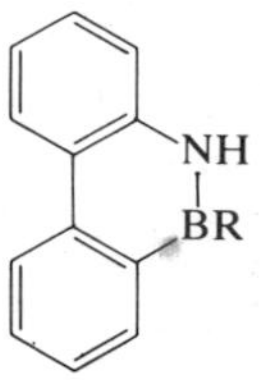

110a R = H
110b R = OH
110c R = CH_3

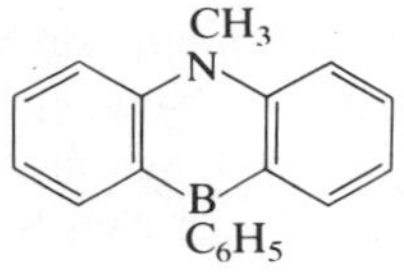

111

Figure 8.55

The majority of 6 π-electron compounds of type (*b*) have two carbon atoms replaced by two nitrogen atoms. The three analogues of benzene are pyridazine (**112**), pyrimidine (**113**), and pyrazine (**114**).

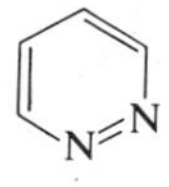

112 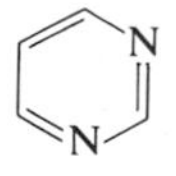 **113** 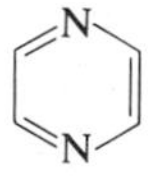 **114**

Figure 8.56

Pyridazine is a liquid, melting point −6°C, whereas pyrimidine and pyrazine are low melting point solids. The dipole moments of these compounds decrease from 3.9D for **112** through 2.4D for **113** to 0D for **114**. All the compounds are less aromatic than pyridine or benzene, the hydroxyl substituted pyrimidines, for example, preferring to exist as the keto rather than the enol tautomer (e.g., cytosine, **115a** ⇋ **115b**). This contrasts with the behavior of resorcinol, which exists primarily in the enolic form, maintaining the 6 π-electron system.

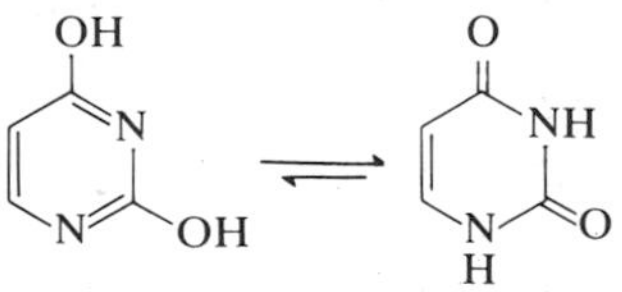

115a **115b**

Figure 8.57

1,2-Diazocine has been prepared from the azoalkane **116** by photoirradiation under a number of specified conditions. The ^1H NMR spectrum shows a sharp singlet at δ 6.03 (4H) and a broad singlet at δ 6.93 (2H). The spectrum is not temperature dependent and is in accord with **117a** rather than **117b**. The ^{13}C NMR spectrum is in accord with a tub-shaped molecule with little, if any, bicyclic valence tautomer present. Thermolysis of **117a** gave a mixture of benzene and pyridine, whereas photoirradiation above 300 nm gave only benzene.

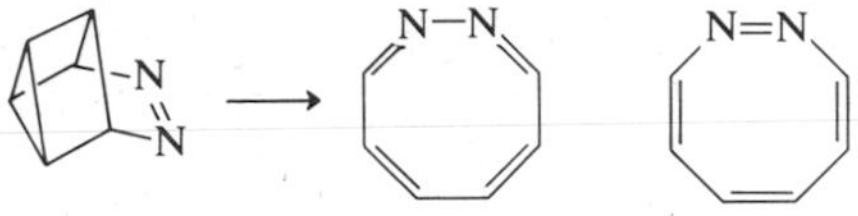

N N N—N N=N

116 **117a** **117b**

Figure 8.58

The derivative of a heterocyclic analogue of oxido[10]annulene **118** has been prepared, and the ^{1}H NMR spectrum suggests that the 10-membered ring is delocalized. The synthesis of macrocyclic analogues of this type would be of interest, particularly since 21H, 23H-porphine (**119**) may be considered to be a 1,10-diaza[18]annulene derivative. The [18]annulene ring is outlined in heavy lines in **119**, and in this ring two of the nitrogens each contribute one electron to the π system, whereas the other two nitrogens and two of the double bonds are not involved. The alternative tautomeric structure in which the nitrogens and pyrrole double bonds reverse roles is also available.

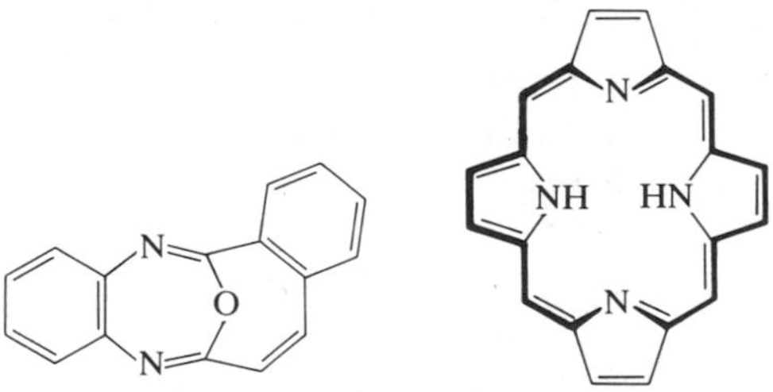

N O N N NH HN N

118 **119**

Figure 8.59

A large number of examples of type (*c*) systems is known in which one heteroatom replaces a C=C bond and the other a carbon atom. Many of them have five-membered rings, and those in which nitrogen replaces carbon are most common. In imidazole (**120**) the two nitrogens are equivalent, the N—H proton readily exchanging its position in the two tautomeric forms (**120a** ⇆ **120b**). Imidazole undergoes electrophilic substitution with the anion as the reactive intermediate.

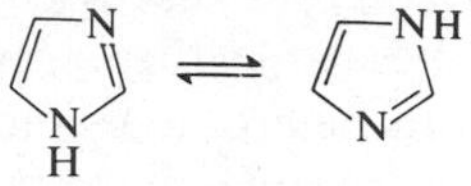

120a **120b**

Figure 8.60

Thiazole (**121**) is also a stable system and is readily quaternized to the corresponding thiazolium salt (e.g., **122**). The thiazolium salts have properties similar to the pyridinium salts, and position C-2 is acidic. Thiazole is not readily reduced, but it does give the 1:2 adduct **123** with dimethyl acetylenedicarboxylate, probably via a dipolar intermediate. A similar adduct can be obtained with pyridine.

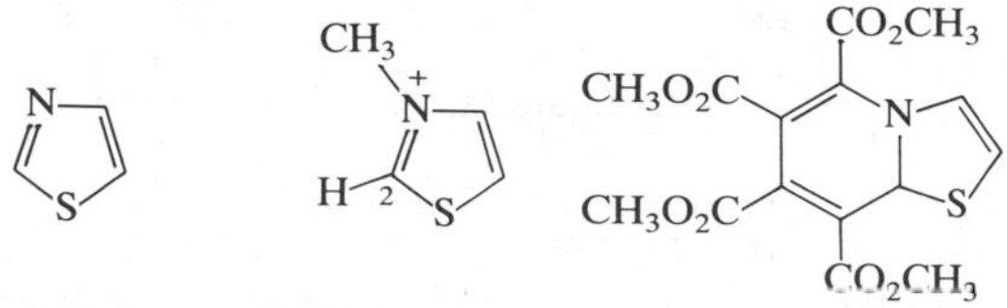

121 **122** **123**

Figure 8.61

These five-membered ring systems appear to be delocalized molecules that exhibit aromatic properties. The only higher homologue appears to be the bridged nine-membered ring system **124** synthesized by Kato and Toda by a route involving a 1,3-dipolar addition to benzocyclopropene.

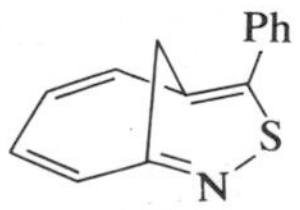

124

Figure 8.62

8.5. SYSTEMS WITH MORE THAN TWO HETEROATOMS

An enormous range of molecules with more than two heteroatoms is possible, and in general the extent of delocalization decreases with the increase in the number of heteroatoms. Numerous examples of compounds analogous to each of the three types discussed for compounds with two heteroatoms are known, and these are either atropic or only weakly delocalized systems. As the number of heteroatoms increases, the stability of the compounds decreases and, for example, the tetra-azine **125** is unstable. One group of compounds that does not have analogies in the systems with one or two heteroatoms is the mesoionic compounds, of which the sydnones were the first examples. The sydnones have the general formula **126**, and it can be seen from this structure that the atoms N-2 and C-4 do not have complete valence shells. A number of polar structures can be written, such as **126a, b**, in which these valencies are satisfied and in which the ring has an aromatic sextet.

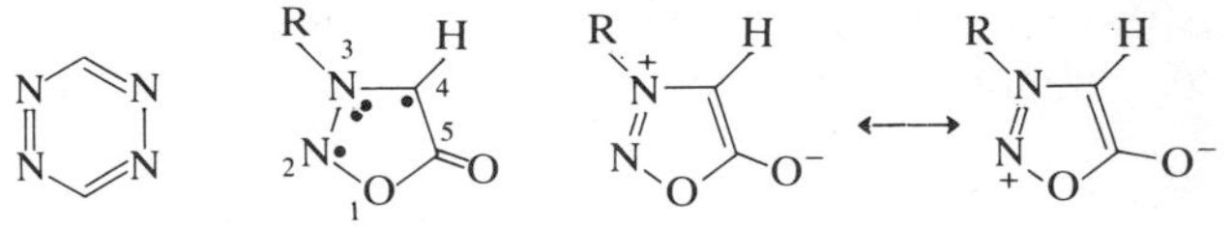

125 **126** **126a** **126b**

Figure 8.63

The sydnones do have large dipole moments in keeping with these dipolar structures, but in the ^{1}H NMR spectrum the proton at C-4 is at much higher field than would be expected for a diatropic system. The sydnones are thermally stable but do react as 1,3-dipolar reagents and add acetylenes at elevated temperatures. The corresponding oxazolanes such as **127**, called münchones like the sydnones from their place of origin, are also mesoionic compounds. These are much more reactive than the sydnones and are readily decomposed by moist air to the corresponding amino acid (e.g., **128**). The münchones are certainly not aromatic in terms of the classic lack of reactivity.

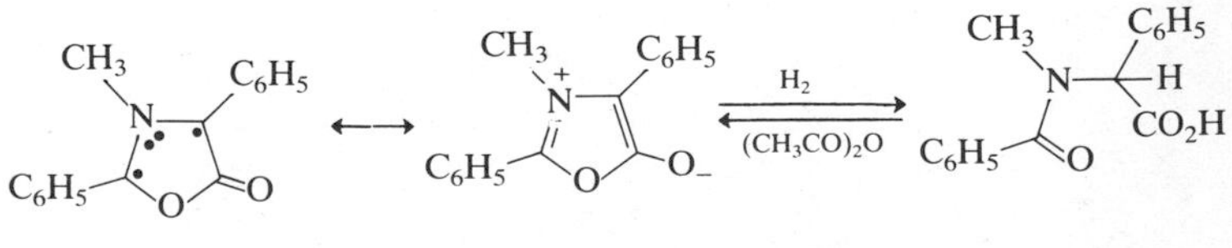

H_2 / $(CH_3CO)_2O$

127a **127b** **128**

Figure 8.64

One group of compounds containing many heteroatoms has been recently discovered that is stable and exhibits aromatic properties. These compounds contain arrays of N—S—N groups, the first neutral examples **129** having been described by Woodward and collaborators in 1981. These eight-ring heterocycles contain 10 π electrons and are planar with nonalternating bonds. Subsequently, Morris and Rees, in a reinvestigation and extension of earlier Japanese work, have prepared **130** and **131** together with a number of derivatives. These seven-ring heterocycles are stable compounds, exhibit long wavelength absorption at about 330 nm characteristic of an aromatic $\pi \rightarrow \pi^*$ transition, and the x-ray determined structures show them to be planar with bond lengths intermediate between those of double and single bonds. A related 14 π-electron bicyclo[5.3.0] derivative has also been prepared. It appears probable that other combinations of group V and VI elements may give stable compounds, and aromatic systems with no carbon atoms are also feasible.

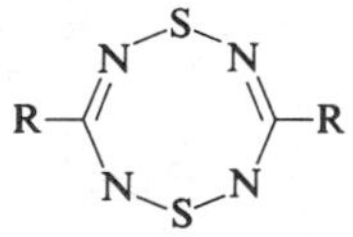

$R_1 = C_6H_5$, $p\text{MeOC}_6H_4$, $p\text{EtO}_2\text{CC}_6H_4$

129 **130** 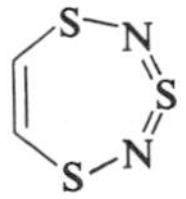 **131**

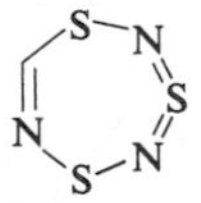

Figure 8.65

FURTHER READING

Extensive accounts of the chemistry of heterocyclic compounds are available including, R. C. Elderfield (Ed.), *Heterocyclic Compounds*, Vols. 1-9, Wiley, New York; C. W. Rees and A. R. Katritzky (Eds.), *Comprehensive Heterocyclic Chemistry*, Pergamon, Elmsford, N.Y., 1984; and the ongoing series, A. Weissberger and E. C. Taylor (series Eds.), *Heterocyclic Compounds*, Interscience, New York.

For a discussion on the aromaticity of heterocycles, see M. J. Cook, A. R. Katritzky, and P. Lunda, *Advances in Heterocyclic Chemistry*, 1974, **17**, 255.

For general texts on heterocyclic aromatic compounds, see R. M. Acheson, *An Introduction to the Chemistry of Heterocyclic Compounds*, Ed. 3, Wiley, New York, 1976; A. Albert, *Heterocyclic Chemistry*, Ed. 2, Athlone Press, London,

1968; L. A. Paquette, *Principles of Modern Heterocyclic Chemistry*, Benjamin, New York, 1968; J. A. Joule and G. F. Smith, *Heterocyclic Chemistry*, Ed. 2, Van Nostrand, London, 1978.

For the chemistry of the azepins and related systems see L. A. Paquette in J. P. Snyder (Ed.), *Nonbenzenoid Aromatics*, Vol. 1, Academic Press, New York, 1969.

For a review of π-excessive compounds containing more than eight carbon atoms, see A. G. Anastassiou and H. S. Kasmai, *Advances in Heterocyclic Chemistry*, 1978, **23**, 55.

For a discussion of the heteronins, see A. G. Anastassiou, in T. Nozoe, R. Breslow, K. Hafner, Shô Itô, and I. Murata (Eds.), *Topics in Nonbenzenoid Chemistry*, Vol. 1, Hirokawa, Tokyo, 1973, p. 153.

For a review of the azocines, see H. D. Perlmutter and R. B. Trattner, *Advances in Heterocyclic Chemistry*, 1982, **31**, 115.

For a review of thiepin, see I. Murata and K. Nakasuji, *Top. Curr. Chem.*, 1981, **97**, 33.

For a review on phospha and arsa-aromatic compounds, see C. Jongsma and F. Bickelhaupt, in T. Nozoe, R. Breslow, K. Hafner, Shô Itô, and I. Murata (Eds.), *Topics in Nonbenzenoid Chemistry*, Vol. 2, Hirokawa, Tokyo, 1977, p. 139.

For group V heterobenzenes, see A. J. Ashe, *Accounts Chem. Res.*, 1978, **11**, 153.

For silabenzene, see G. Maier, *Pure Appl. Chem.*, 1986, **58**, 197.

For borabenzene, see C. W. Allen and D. E. Palmer, *J. Chem. Educ.*, 1978, **55**, 497.

For 1,2-diazepines, see V. Snieckus and J. Streith, *Accounts Chem. Res.*, 1981, **14**, 348.

For mesoionic compounds see C. G. Newton and C. A. Ramsden, *Tetrahedron*, 1982, **38**, 2965; K. T. Potts, in A. Padwa (Ed.), 1,3-*Dipolar Cycloaddition Chemistry*, Vol. 2, Wiley, New York, 1984, p. 1; W. D. Ollis, S. P. Stanforth, and C. A. Ramsden, *Tetrahedron*, 1985, **41**, 2239.

For polyheterocyclic systems with NSN groups, see I. Ernst, W. Holick, G. Rihs, D. Schomburg, G. Shoham, D. Wenkert, and R. B. Woodwàrd, *J. Amer. Chem. Soc.*, 1981, **103**, 1540 and J. L. Morris and C. W. Rees, *Pure Appl. Chem.*, 1986, **58**, 197.

9

POLYCYCLIC SYSTEMS

9.1. INTRODUCTION

Polycyclic systems present a number of problems that do not arise in monocyclic systems. Thus, in monocyclic compounds if conjugation occurs, it does so over the complete cycle, whereas polycyclic compounds may behave as if composed of individual discrete cycles rather than as a single system. Counting the number of π electrons and applying the Hückel Rule cannot, in general, be expected to supply very meaningful predictions of the properties of polycyclic systems. Other simple methods, such as counting the number of π electrons on the periphery, are also unlikely to be reliable for predictive purposes.

Another problem, partly arising out of the first, is how to classify polycyclic systems. Various classification schemes are possible, but only two will be considered in this chapter. The first scheme is based on the π-electron properties of the component monocyclic rings. There are three types: (*a*) molecules formed by the fusion of two $(4n + 2)$ π-electron units, (*b*) molecules formed by the fusion of two $4n$ π-electron units, and (*c*) molecules formed by the fusion of a $4n$ and a $(4n + 2)$ π-electron unit. The molecules that result from (*a*) and (*b*) have a total of $(4n + 2)$ π electrons, whereas those resulting from (*c*) have a total of $4n$ π electrons. This classification system is tabulated in Table 9.1 and exemplified by the molecules shown in Figure 9.1.

TABLE 9.1

Type	Unit A	Unit B	Total No. of π-Electrons
(*a*)	$4n + 2$	$4n + 2$	$4n + 2$
(*b*)	$4n$	$4n$	$4n + 2$
(*c*)	$4n + 2$	$4n$	$4n$

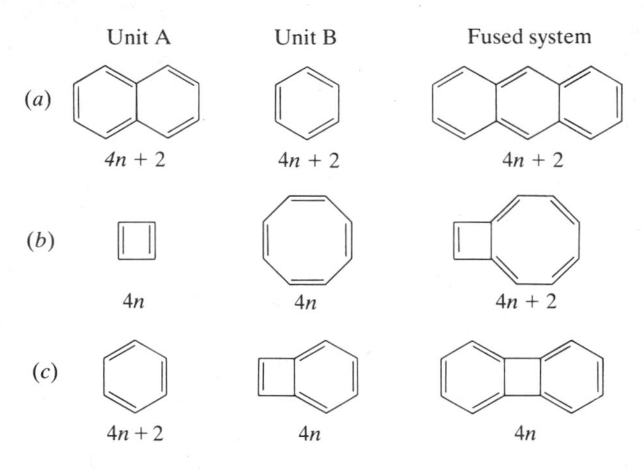

Figure 9.1

Systems of type (*a*), resulting from the fusion of two $(4n + 2)$ π-electron rings are well known, whereas systems belonging to types (*b*) and (*c*) have only recently been investigated.

The other classification that will be used is to group together those compounds containing benzene rings as the benzenoid hydrocarbons and those without such rings as the nonbenzenoid hydrocarbons. Difficulties arise with this classification in those cases in which the benzene ring is merely annelated onto the nonbenzenoid system and in those cases in which the benzene ring, although an integral part of the system, has had its properties greatly modified. Arbitrarily, in the subsequent discussion the annelated systems will be grouped with the nonbenzenoid systems and the modified molecules with the benzenoid. The benzenoid compounds will be considered first and then the nonbenzenoid, and both groups will be further classified into type (*a*), (*b*), or (*c*) systems wherever possible.

9.2. BENZENOID SYSTEMS

Naphthalene (**1**) is conceptually formed by the fusion of two benzene molecules and is consequently a type (*a*) system with 10 π electrons. Naphthalene is a white, crystalline solid, melting point 80°C, with a characteristic benzenoid electronic spectrum (see Fig. 9.7). It is diatropic and undergoes electrophilic substitution; thus bromine, for example, gives 1-bromo- and 1,4- and 1,5-dibromonaphthalenes. It is unreactive toward dieneophiles and gives only a low yield of the adduct **2** after prolonged reaction with maleic anhydride.

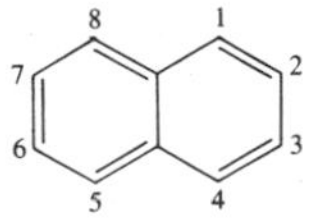

1 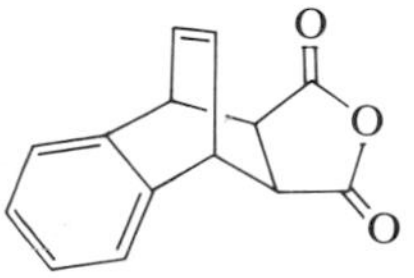 **2**

Figure 9.2

The physical and chemical properties of naphthalene are thus those expected for a classic aromatic system. In the Hückel model, the σ framework is fixed and each of the 10 carbon atoms supplies one $2p$ atomic orbital to the π system. These are combined to give 10 molecular orbitals, the energies of which can be determined by the HMO or more sophisticated methods. In the HMO method, the determinant is no longer cyclic since the two carbon atoms at the ring junction are joined to three, rather than two, other carbon atoms. This determinant can be evaluated, but a simpler and more valuable method has been introduced for these types of system by Coulson, Longuet-Higgins, and Dewar. This is the perturbational molecular orbital (PMO) method and it is based on the properties of the nonbonding orbital in odd alternant hydrocarbons. *Alternant hydrocarbons* are defined as those systems in which the carbon atoms can be divided into two sets such that the atoms of one set are *only* joined to atoms of the other set. This property is true for naphthalene but not azulene as illustrated in Figure 9.3, one set of carbons being starred.

Figure 9.3

Odd alternant hydrocarbons, that is, those having an odd number of conjugated atoms, have an NBMO that has the property that there is zero electron density on one set of atoms and the *sum* of the coefficients on the adjoining atoms is always zero. An estimate of the relative delocalization energies of naphthalene, [10]annulene, and decapentaene can be determined by considering these molecules to be formed by the addition of one carbon atom to the remaining nine carbon fragment. Both the one carbon and the nine carbon fragment are odd, alternant systems and consequently have nonequal sets of alternant atoms. The larger set is starred, and the smaller set has zero electron density. The nine-carbon fragment is shown in Figure 9.4 (a) and the coefficients of the NBMO in Figure 9.4 (b). The addition

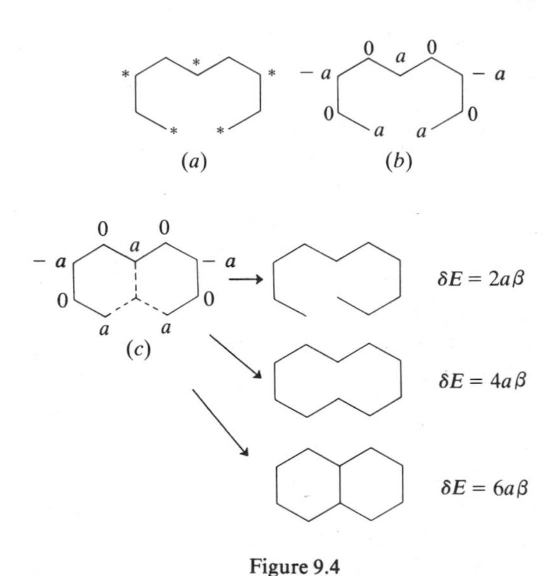

Figure 9.4

of the single carbon atom, which is the simplest alternant hydrocarbon, then introduces the perturbation shown in Figure 9.4 (c). The change in energy on the union of an odd alternant hydrocarbon is given by equation (9.1):

$$E = 2\sum a\beta \tag{9.1}$$

where a is the value of the coefficient at the carbon atom(s) to which the single carbon is joined and β is the resonance integral.

The change in energy, δE, for decapentaene is $2a\beta$, for [10]annulene is $4a\beta$, and for naphthalene is $6a\beta$. The difference in energy between decapentaene and [10]annulene is thus $2a\beta$ and between decapentaene and naphthalene it is $4a\beta$. These energy terms can then be equated to the delocalization energies of [10]annulene and naphthalene and suggests that the transannular bond in naphthalene, besides removing the nonbonded interactions, also leads to a system of lower energy.

The PMO method is very simple to apply and provides a useful indication of the probable gain in energy to be expected for a cyclic or polycyclic system when compared to the acyclic analogue. The theoretical justification for the PMO method has been fully discussed by Dewar (see Chapters 1 and 9, Further Reading), and the method will be used throughout the remainder of this chapter.

Anthracene (**3**) and phenanthrene (**4**) are conceptually formed by fusion of a third benzene ring to naphthalene, and both are thus type (a) systems. Both are white, crystalline solids with benzenoid electronic spectra (see Fig. 9.7). Besides typical aromatic behavior both molecules, however, undergo reactions that are more typical of unsaturated rather than aromatic systems. Thus, anthracene smoothly reacts with maleic anhydride to give the Diels–Alder adduct **5** and phenanthrene adds bromine across the 9,10-double bond to give **6**.

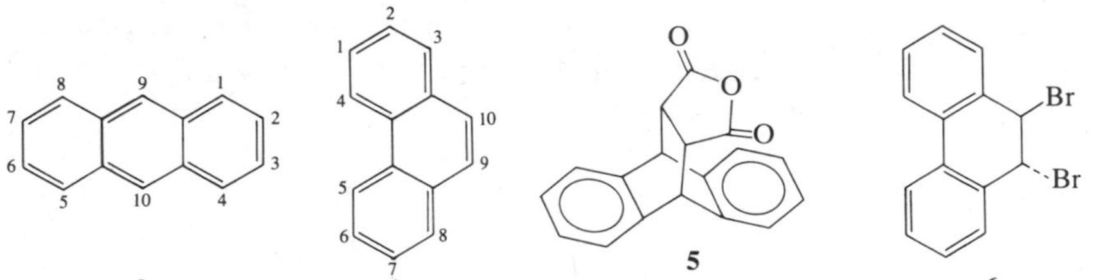

Figure 9.5

The decrease in aromatic character observed on the addition of a further ring to naphthalene has been developed into a general theory by Clar. This theory suggests that the effect of *annelation** on the physical and chemical properties of a system will depend on the number of aromatic sextets that can be formed. The aromatic sextets are depicted by the circle notation that is used as in its original formulation by Armit and Robinson rather than in its more common modern usage merely to depict delocalization. Any π electrons that do not participate in aromatic sextets are depicted as double bonds (Fig. 9.6).

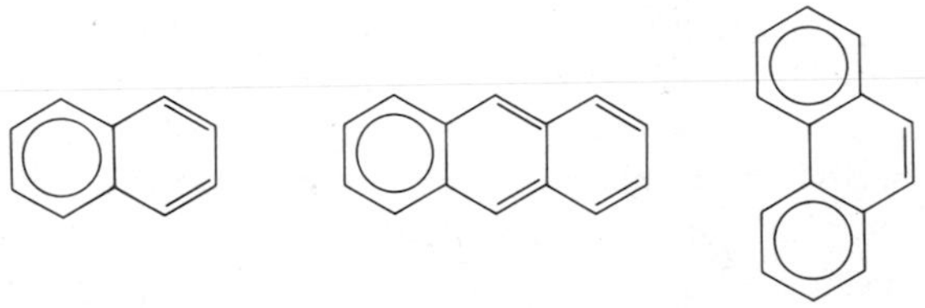

Figure 9.6

The formulas in Figure 9.6 depict only one possible "sextet" structure for these systems, and the sextet can be drawn in any of the six-membered rings. In the case of phenanthrene, the formula with the aromatic sextet in the central ring has both the other rings polyenic and in the theory this structure is therefore less favorable. The theory thus accounts for the reactivity of the 9,10-double bond in anthracene. Similarly, converting anthracene to the Diels–Alder adduct **5** produces a system with two aromatic sextets, whereas anthracene had only one. The changes in the electronic spectra of these systems were correlated by Clar, who established that these absorption bands arose from the same transitions as those in benzene. These electronic spectra are shown in Figure 9.7, and it can be seen that each compound has three main absorption bands and that the effect of annelation in any series is to cause a bathochromic shift to longer wavelengths. The large polycyclic systems are colored, the colors changing from yellow through red to blue with increased annelation.

Although Clar's theory gives good correlations among the properties of various compounds, it should be used with caution. Thus, the ^{1}H NMR spectrum of anthracene shows that the 9,10-protons are not deshielded, appearing at δ 8.31, and the 9,10-protons of phenanthrene, although appearing at highest field in the ^{1}H NMR spectrum, are still at lower field (δ 7.71) than those of benzene.

* It has been suggested that annulation rather than annelation would be more "correct" to describe the conceptual addition of another ring, but this appears contrary to standard English usage (e.g., annelides); if a distinction on ring size is to be made, annulation could be applied to the addition of large rings (i.e., 20 or more carbon atoms).

Biphenylene (**7**) is a molecule in which two benzene rings have been joined in the *ortho*-positions rather than fused as in naphthalene. The system contains 12 π-electrons and is a type (c) system formally derived from

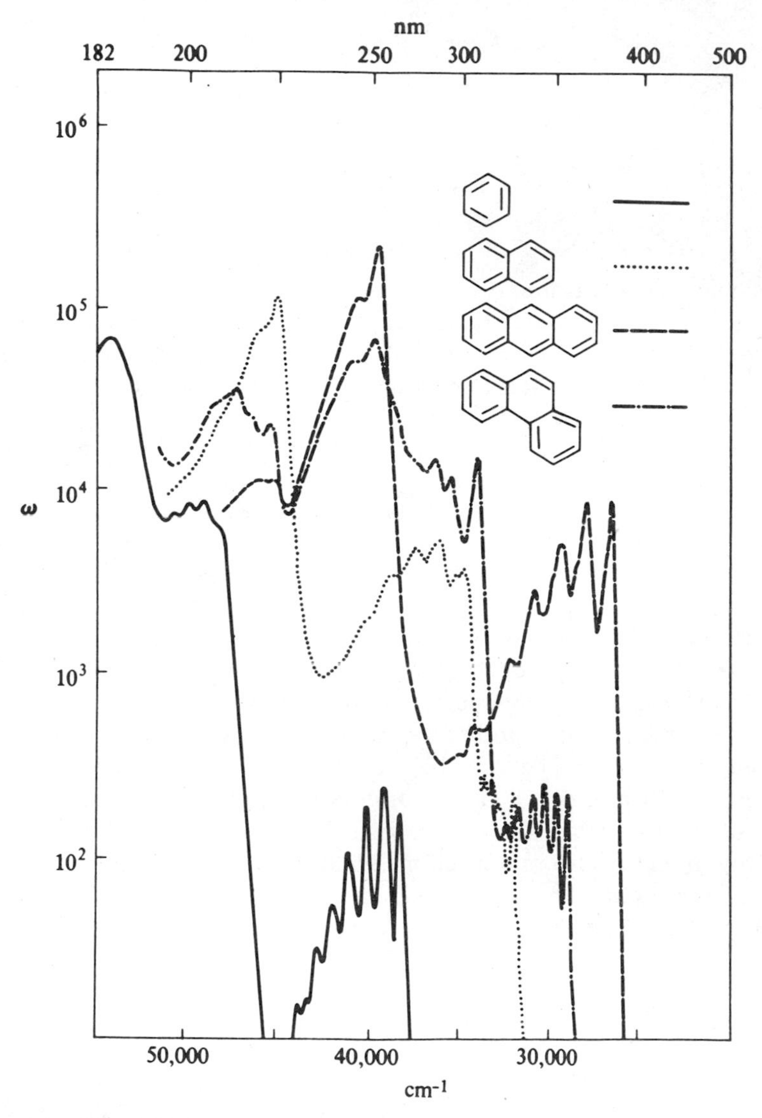

Figure 9.7. Electronic spectra of benzene, naphthalene, anthracene, and phenanthrene in hexane (adapted from spectra due to Perkampus and Kassebeer, and Roe in the *DMS U.V. Atlas*, Butterworths, Woburn, Mass., 1966).

benzocyclobutadiene, a $4n$ π-electron unit, with benzene. Application of the PMO method suggests that biphenylene will have *less* delocalization energy than diphenyl (**8**), the extra bond *destabilizing* **7** compared to **8** due to the antiaromatic character of the four-membered ring.

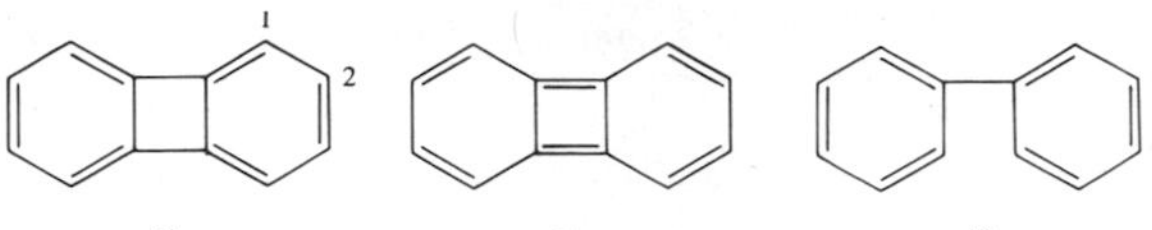

Figure 9.8

Biphenylene is a white, crystalline solid with a characteristic electronic spectrum, showing two main bands at 235–260 and 330–370 nm. It undergoes electrophilic substitution preferentially at position 2, although position 1 is more acidic. The four-membered ring can be ruptured, but this normally requires fairly vigorous conditions.

An x-ray crystallographic study of biphenylene revealed that the bonds connecting the phenyl rings are long (151.4 pm), and there thus appears to be little contribution from cyclobutadiene structures such as **7a**. The ^{1}H NMR spectrum shows the protons are at higher field than those of benzene (H-1, δ 6.7; H-2, δ 6.6), and this has been attributed to a paramagnetic contribution from the four-membered ring.

Both the heterocyclic analogues **9a** and **10** and the homologue **11** have been prepared. Both **9a** and **11** show similar effects in the ^{1}H NMR spectra to biphenylene, and the chemical properties of the thiophene ring in **9a** are greatly modified, readily adding bromine and oxidizing to the sulfone **9b** without dimerization. The ^{13}C NMR spectrum of biphenylene shows a large downfield shift of C-3, which was originally ascribed to the paramagnetic effect of the four-membered ring. However, the same effect was observed not only in **9a** but also in the sulfone **9b** and is thus seen to arise from the strain in the system. The isomer **10** is much less stable than **9a**, presumably because the bismethylene structure is now not available. The two methylene

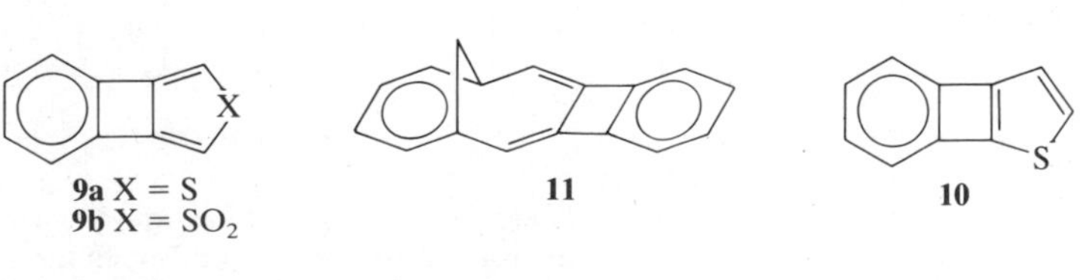

Figure 9.9

protons in **11** exhibit very different chemical shifts, and **11** is best represented by the homobenzene structure shown.

Phenalene (**12**) is a member of an interesting group of compounds, which on oxidation give potentially delocalized systems for which no Kekulé structure can be written. The chemical properties of phenalene indicate that it is difficult to designate any one of its rings as nonaromatic since rapid interconversion occurs between structures in which different rings are aromatic. Phenalene is readily converted into the anion, cation, or free radical, all of these systems being best represented as having a 12 π-electron periphery with the charge or odd electron on the central carbon atom (e.g., **13**). Another example of this type of system is triangulene (**14**), which is predicted by MO calculations to be a triplet diradical but is presently unknown.

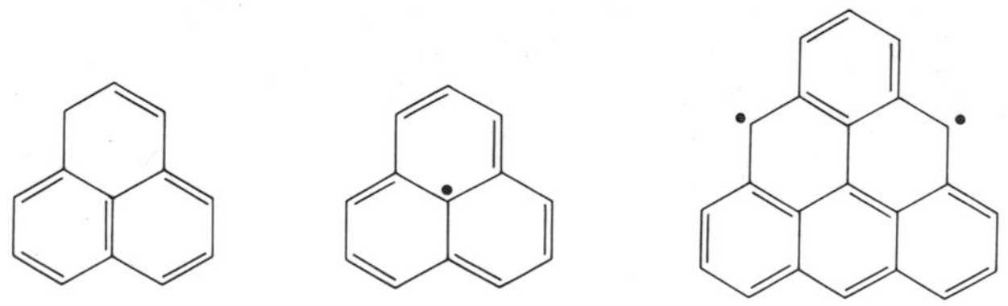

12 **13** **14**

Figure 9.10

The phenalene-type systems are not readily classified by the criteria discussed in Section 9.1, and another molecule that does not fit easily into this classification is corannulene (**15**). This molecule consists of five fused benzene rings, and the molecule has a total of 20 π electrons. However, the canonical structure **15b** suggests that the molecule might behave as a fulvalene with a 14 π-electron periphery. Corannulene readily forms the corannulene radical anion **16** with the central system as cyclopentadienyl anion and the periphery as a 15 π-electron radical.

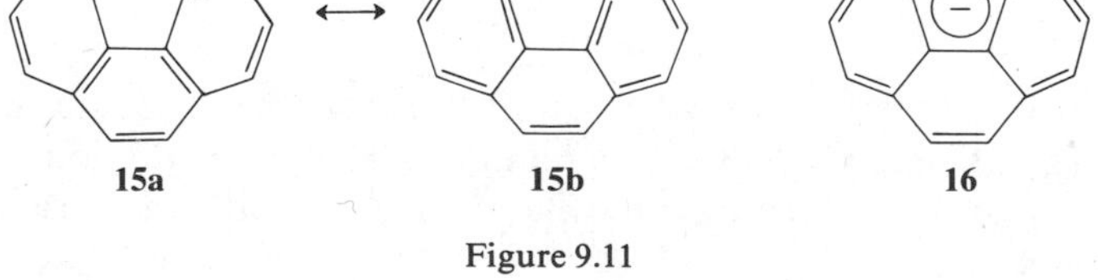

15a $\longleftrightarrow$ **15b** **16**

Figure 9.11

A number of polycyclic systems have been prepared that might be considered as annelated annulenes. An example is the molecule **17**, which appears as a hexaphenyl[18]annulene. However, for conjugation to occur in the macrocyclic ring, the benzenoid conjugation must be disrupted and **17** does not have any properties to suggest such extended conjugation. Diederich and Staab have prepared [12]kekulene (**18**), a 48 carbon polycyclic benzenoid system with an [18]annulene cavity. An x-ray crystallographic structural analysis showed that the molecule exhibited considerable variations in bond length, and **18** is best represented by the structure shown,

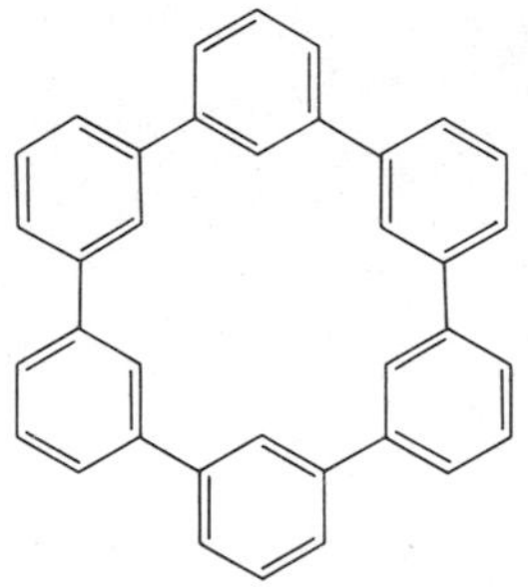

17

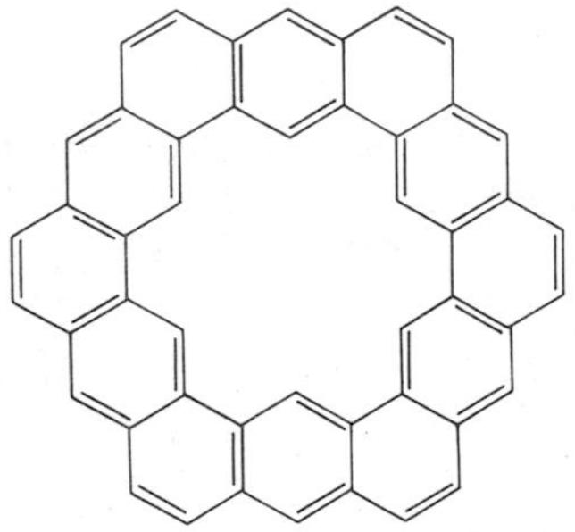

18 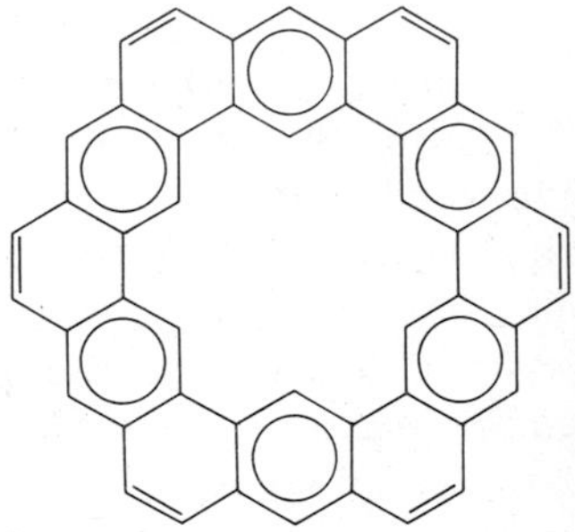 **18a**

Figure 9.12

consisting of three phenanthrene units rather than as an annelated [18]annulene. Compound **18** is thus related to **17** by the addition of double bonds *ortho* to the interconnecting *meta* linkages and can be represented in Clar's notation as **18a**.

9.3. NONBENZENOID SYSTEMS

Interest in nonbenzenoid polycyclic systems was stimulated by some early HMO calculations of Brown who predicted that pentalene (**19**) and heptalene (**20**) should have a considerable delocalization energy.

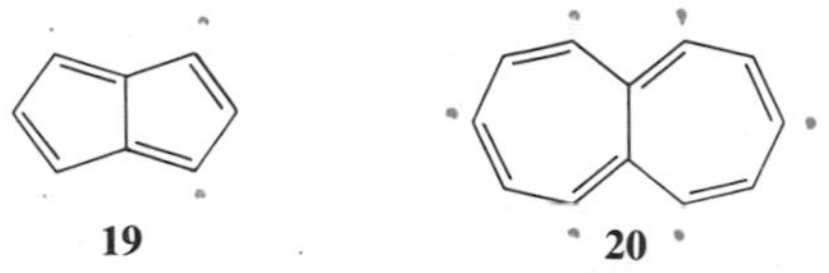

19 **20**

Figure 9.13

Subsequent valence bond and more sophisticated MO calculations suggested that the structures with localized bonds would be more stable, and the PMO method agrees with these results. As shown in Figure 9.14, the PMO method predicts that both pentalene and heptalene are *antiaromatic* systems, the transannular bond not contributing to the stabilization of these compounds, which can be considered as perturbed cyclooctatetraene and [12]annulene, respectively.

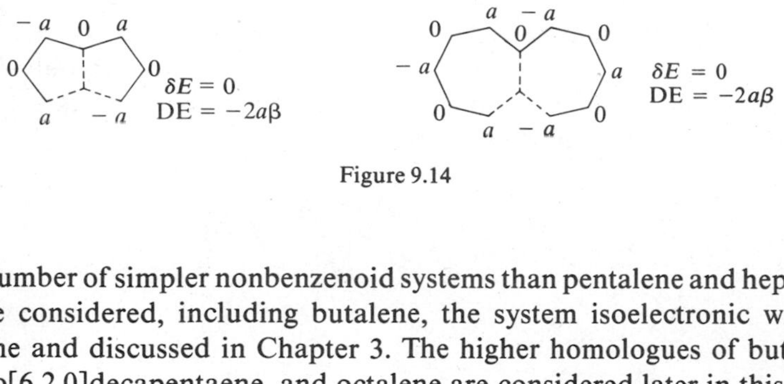

$\delta E = 0$
$\mathrm{DE} = -2a\beta$

$\delta E = 0$
$\mathrm{DE} = -2a\beta$

Figure 9.14

A number of simpler nonbenzenoid systems than pentalene and heptalene can be considered, including butalene, the system isoelectronic with *m*-benzyne and discussed in Chapter 3. The higher homologues of butalene, bicyclo[6.2.0]decapentaene, and octalene are considered later in this chapter. Others, such as bicyclo[1.1.0]butadiene (**21**) are highly strained systems, but some of the charged molecules, such as the bicyclo[2.1.0]pentadienyl anion (**22**) may be attainable, even if only as fleeting intermediates.

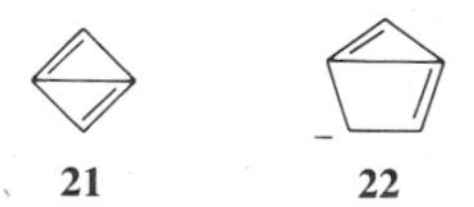

Figure 9.15

Pentalene (**19a**), 1-methylpentalene (**19b**), and 1,3-dimethylpentalene (**19c**) have all been identified as fugitive species at low temperature (about −196°C) by their IR or electronic spectra. On warming to −100°C the IR spectrum of **19b** disappears and a new spectrum appears due to the dimers **23b** and **24b**. Both **19a** and **19c** give similar mixtures of dimers on warming, and these dimers may be photochemically cleaved to restore the monomers.

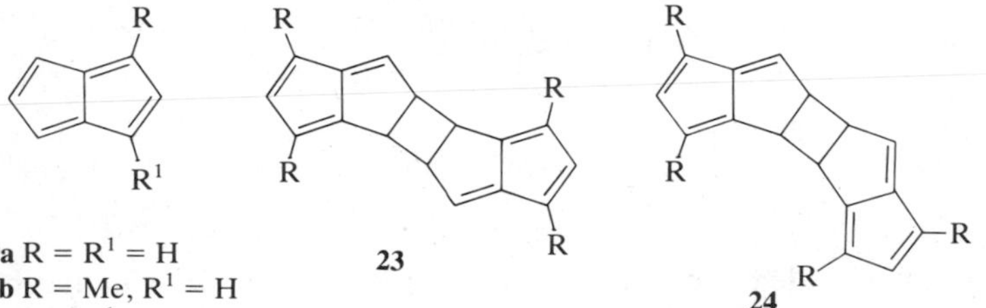

19a R = R^1 = H
19b R = Me, R^1 = H
19c R = R^1 = Me

Figure 9.16

The simplest pentalene derivative to have been isolated as a thermally stable compound is 1,3,5-tri-*t*-butylpentalene (**25**) prepared by Hafner and Süss by the route outlined in Figure 9.17. The ^{1}H NMR spectrum of **25** shows three singlets at δ 0.98 [$(CH_3)_3C$], 4.72 (H-2) and 5.07 (H-4,6) and the ^{13}C NMR spectrum has nine lines, the 5-*t*-butyl group resonating at a slightly different position to the 1,3-*t*-butyl groups. An x-ray crystallographic analysis indicates the molecule has the structure shown in Figure 9.18(a) with alternate single and double bonds.

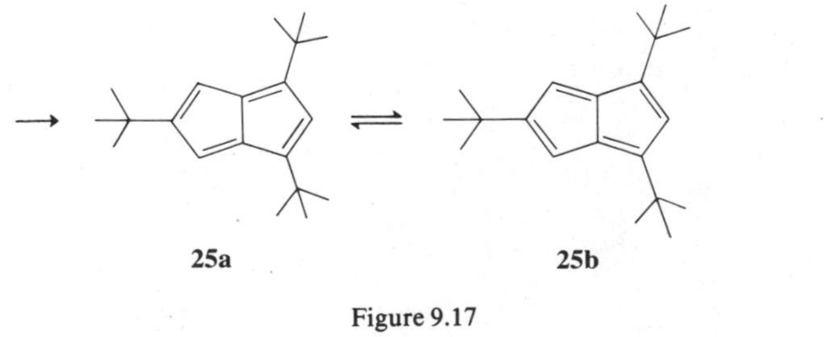

Figure 9.17

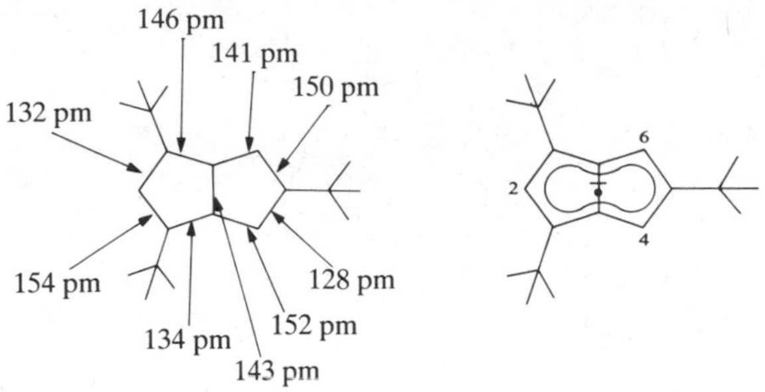

Figure 9.18 (a) X-Ray crystallographic structures of 1,3,5-tri-*t*-butylpentalene. (b) Radical anion of **25**.

Related derivatives have also been prepared in which the 5-*t*-butyl group has been replaced by other functional groups, and these also show bond alternation. The ^{13}C NMR spectrum of **25** indicates that in solution the tautomers **25a** and **25b** are interconverting at 93 K with an activation energy of 16 kJ mol^{-1}. 1,3,5-Tri-*t*-butylpentalene can be readily reduced or oxidized to give the corresponding radical anion or cation, as expected for a molecule with a low-lying LUMO and a moderate ionization potential. The hyperfine coupling in the ESR spectrum of the radical anion indicated a large charge density at C-4,6 and a lower density at C-2, in keeping with the predictions of simple MO theory. The ESR spectra give no indication of bond shift down to low temperatures, and it would appear that bond alternation is reduced in the radical anion (Fig. 9.18(b)).

Earlier, a number of more complex pentalene derivatives had been prepared including the 1,3-bis(dimethylamino)pentalene (**26**), to which the dipolar structure **26b** makes an important contribution, and the crystalline hexaphenylpentalene (**27**).

Heptalene (**28**) was prepared by Dauben and Bertelli by the route shown in Figure 9.20. The ditosylate **29** was ring expanded by solvolysis in a

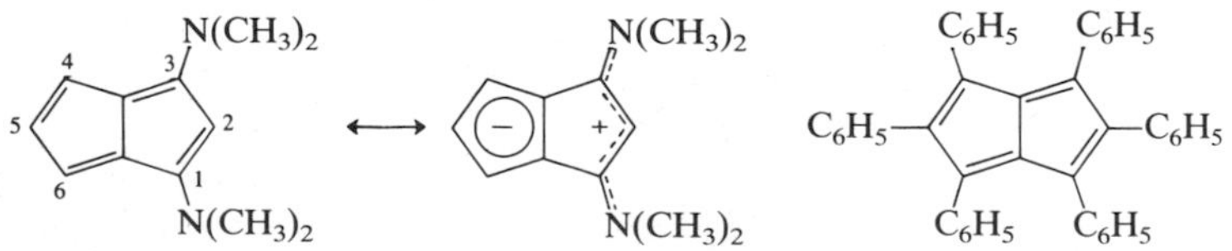

Figure 9.19

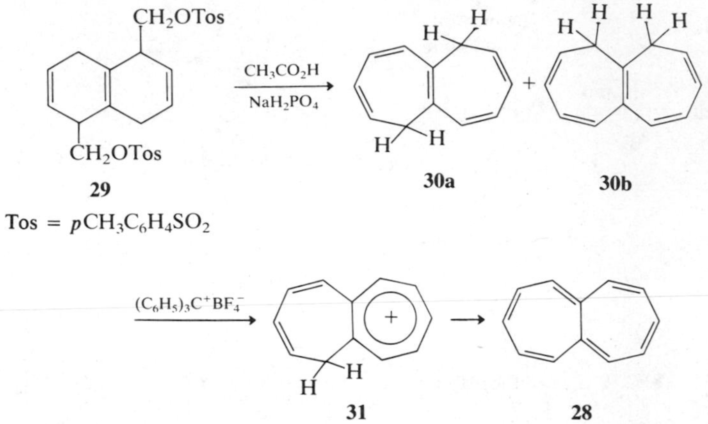

CH_2OTos

CH_3CO_2H

NaH_2PO_4

H H H H H H H H

CH_2OTos

29

Tos = $p CH_3C_6H_4SO_2$

30a **30b**

$(C_6H_5)_3C^+BF_4^-$

+

H H

31 **28**

Figure 9.20

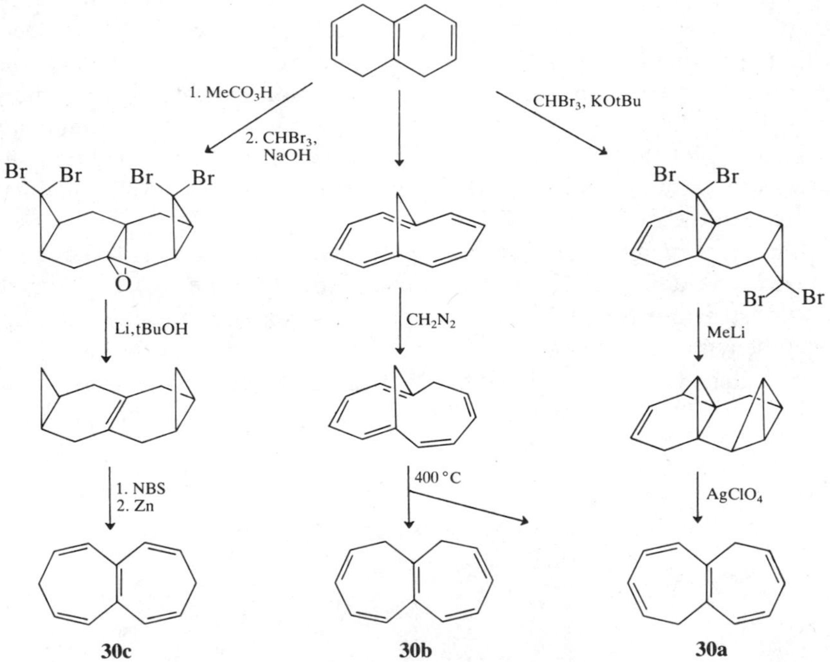

1. $MeCO_3H$

2. $CHBr_3$, NaOH

$CHBr_3$, KOtBu

Br Br Br Br

O

Br Br

Br Br

Li,tBuOH

CH_2N_2

MeLi

1. NBS

2. Zn

400 °C

$AgClO_4$

30c **30b** **30a**

Figure 9.21

mixture of acetic acid, NaH_2PO_4, to give the dihydroheptalene **30a**. Hydride abstraction with trityl fluoroborate gave the heptalenium cation **31**, which on deprotonation gave heptalene. Heptalene is a reddish-brown liquid that is rapidly polymerized by oxygen or by warming to 50°C. The electronic spectrum has bands at 265 and 352 nm, and the ^{1}H NMR spectrum shows resonance signals centered at δ 5.8 and 5.1. The spectral properties and lack of diatropicity indicated that heptalene is not an aromatic system.

After a considerable period of inactivity, interest in heptalene chemistry was restored by new dihydroheptalene syntheses from a number of groups from the mid-1970s onwards. The synthetic routes leading to unsubstituted dihydropentalenes are outlined in Figure 9.21.

The more ready availability of heptalene led to a reexamination of its physical properties using more recently available methods. Bertelli had shown by simulation of the ^{1}H NMR spectrum that heptalene must be nonplanar, and the ^{13}C NMR spectrum at -100°C now allowed the observation of the interconversion of the equivalent structures by bond shift and ring inversion.

A variety of substituted heptalenes has also become available, and some of these have proved valuable precursors to other nonbenzenoid systems.

Although pentalene and heptalene thus appear to be nonaromatic, even antiaromatic compounds, they could, in principle, be converted to aromatic systems by the addition or removal of two electrons. Of the four possible charged ions the two with 10 π-electrons are type (a) systems and the other two are type (b) systems. The four ions are shown in Figure 9.22.

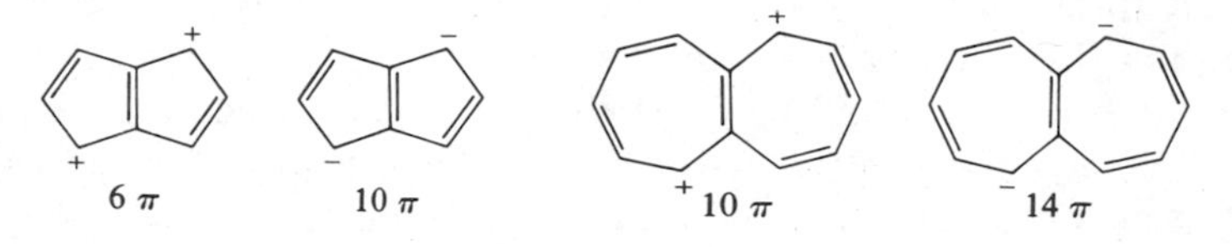

Figure 9.22

At the present time, only the dianions **32** and **33** have been prepared, neither of the dications having been observed despite considerable efforts to prepare the heptalenium dication **34**.

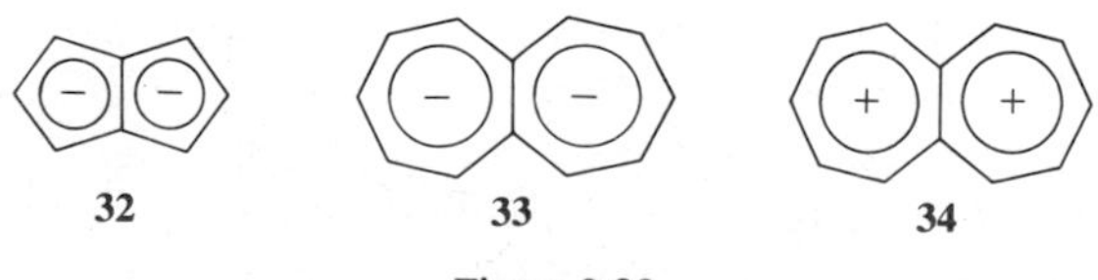

Figure 9.23

The pentalenyl dianion **32**, a type (*a*) system derived conceptually from the fusion of two 6 π-electron cyclopentadienyl anions, was prepared by Katz and Rosenberger by the reaction sequence shown in Figure 9.24.

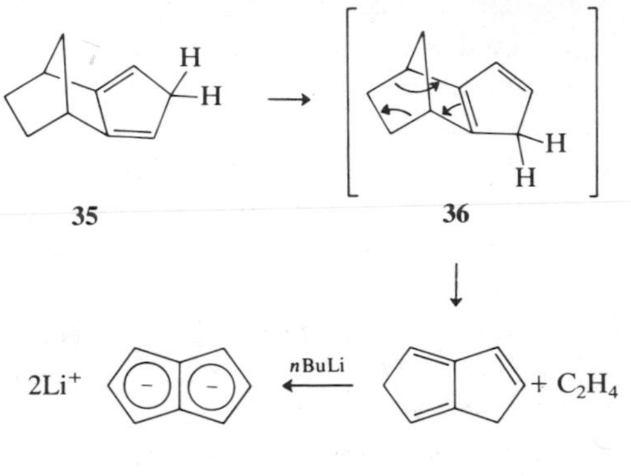

35 → [36] ↓ 37 + C_2H_4 —nBuLi→ $2Li^+$ 32

Figure 9.24

Isobicyclopentadiene (**35**) was pyrolyzed at 575°C under nitrogen to dihydropentalene (**37**) and ethylene, presumably via an initial rearrangement of **35** to **36**, which then undergoes a retro-Diels–Alder reaction to **37**. Deprotonation of **37** with *n*-butyl lithium gave a solution of the pentadienyl dianion **32** from which it could be precipitated as a crystalline solid. The ^{1}H NMR spectrum showed a two-proton triplet at δ 5.73 for H-2, 5 and a four-proton doublet at δ 4.98 for the remaining protons. The high-field position of these protons reflects the high negative charge density that offsets the diatropicity. The pentalene dianion is related to pentalene in exactly the same way that the cyclooctatetraenyl dianion is related to cyclooctatetraene (Chapter 5).

Reaction of heptalene with lithium in THF-d_8 at −80°C gave the heptalenium dianion (**33**). The ^{1}H NMR spectrum shows signals at δ 7.41 (H-1), 6.13 (H-3), and 5.64 (H-2), indicating that **33** is diatropic. The differences in chemical shift reflect the differences in charge density, C-2 having the highest coefficient in the HOMO. Comparison of the chemical shifts of **32** and **33** reveals the smaller electron densities in **33** in which two electrons are delocalized over 12 rather than 8 atoms. The heptalene dianion is stable at 100°C, and it thus appears to be a diatropic, 14 π-electron aromatic system and is best considered a perturbed [12]annulenyl dianion in the same way that **32** is a perturbed [8]annulenyl dianion. In the case of the heptalene dianion, however, charge delocalization has prevailed over the antiaromatic nature of the contributing monocyclic systems since **33** is a type (*b*) system.

The heptalenium dication (**34**), having the advantage of two contributing $4n + 2$ rings, has not yet been prepared, but recent theoretical calculations favor its formation and should encourage further experimental work. A benzannelated pentalene dication has been prepared, and the possibility of forming the pentalene dication thus remains.

Two systems that are closely related to pentalene are the *s*- and *as*-indacenes. Of these two systems *s*-indacene (**38**) has been prepared but *as*-indacene (**39**) is at present unknown. Both compounds have a 12 π-electron periphery and can be considered to be perturbed [12]annulenes. *s*-Indacene is a red oil that is thermally labile and susceptible to oxidation. Mild hydrogenation converts **38** into *s*-hydrindacene (**40**), and bromination gives hexabromo-*s*-hydrindacene (**41**), emphasizing the tendency of the central ring to become benzenoid.

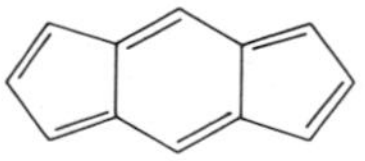

38 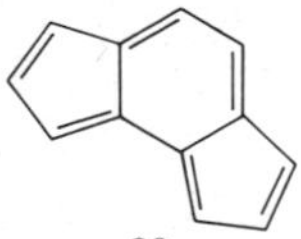 **39**

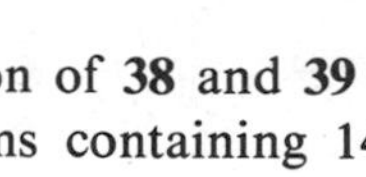

Br H H Br H Br H Br H Br Br H

40 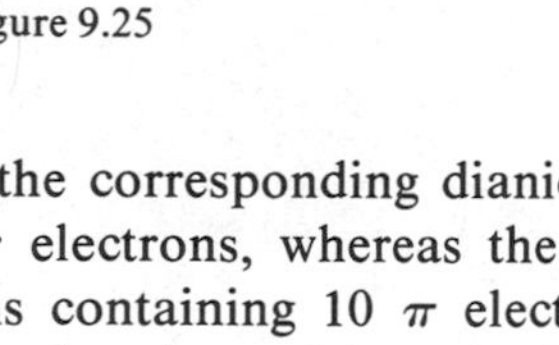 **41**

Figure 9.25

The reduction of **38** and **39** to the corresponding dianions would give type (*a*) systems containing 14 π electrons, whereas the corresponding cations would be type (*b*) systems containing 10 π electrons. Both the dianions **42** and **43** have been prepared as air-sensitive salts that on treatment

42 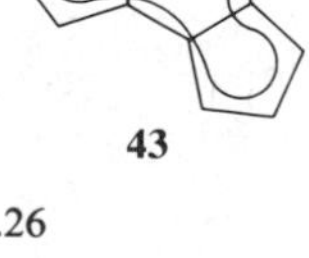 **43**

Figure 9.26

with water give the corresponding dihydroindacenes. The ^{1}H NMR spectrum of **42** is consistent with this dianion being a planar, delocalized system.

The 18 π-electron anion **44** has been prepared, and the low-temperature x-ray crystallographic analysis indicates it has a sandwich structure with a lithium atom between the terminal six-membered rings at each end of the dimer. The anion is derived from a molecule for which, like phenalene, a completely delocalized structure cannot be written.

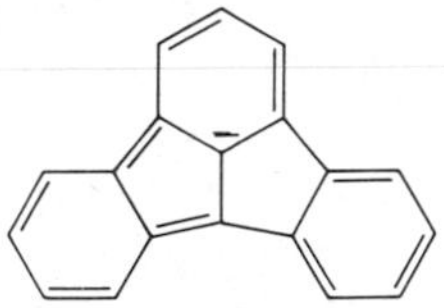

44

Figure 9.27

Whereas the combination of two five-membered or two seven-membered rings gives rise to nonaromatic (or antiaromatic) systems, the fusion of a five- with a seven-membered ring gives the aromatic system azulene (**45**). The dipolar form of azulene **45b** is a type (a) system, conceptually formed by the fusion of a cyclopentadienyl anion with a tropylium cation. The PMO method indicates that azulene will have the same delocalization energy as [10]annulene, the transannular bond acting as a minor perturbation (Fig. 9.29). The dipolar structure **45b** might be expected to make a significant contribution to the ground state of azulene but the dipole moment (1.08D) suggests that the contribution is not large.

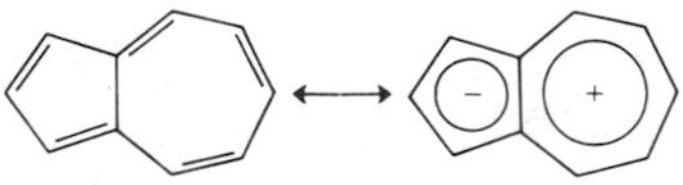

45a ⟷ **45b**

Figure 9.28

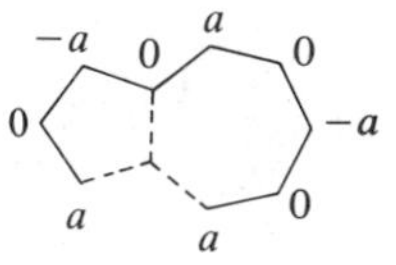

$-a$ 0 a 0 0 $-a$ 0 a a

$\delta E = 4n\beta$
$\text{DE} = 2a\beta$

Figure 9.29

Azulene is a deep blue compound that was originally prepared by Plattner and Pfau by the palladium on charcoal dehydrogenation of the alcohol **46**. Several other dehydrogenation methods were subsequently reported, but a much superior synthesis that does not involve dehydrogenation is outlined in Figure 9.31. The salt **47**, which is readily prepared by treatment of N-(2,4-dinitrophenyl)pyridine with N-methylaniline, reacts with base to give the aldehyde **48**. Condensation with cyclopentadiene gives the fulvene **49**, which on pyrolysis gives azulene. The method can be modified to prepare substituted azulenes.

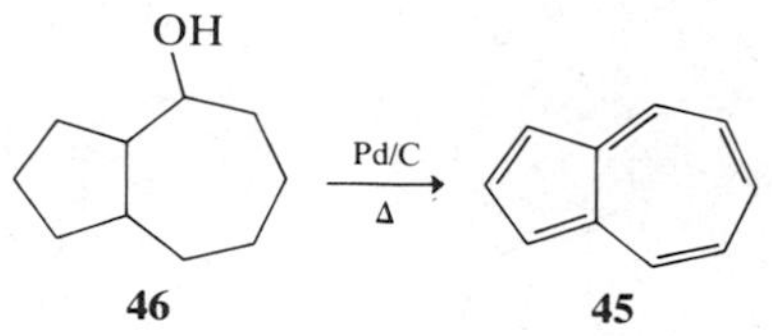

Figure 9.30

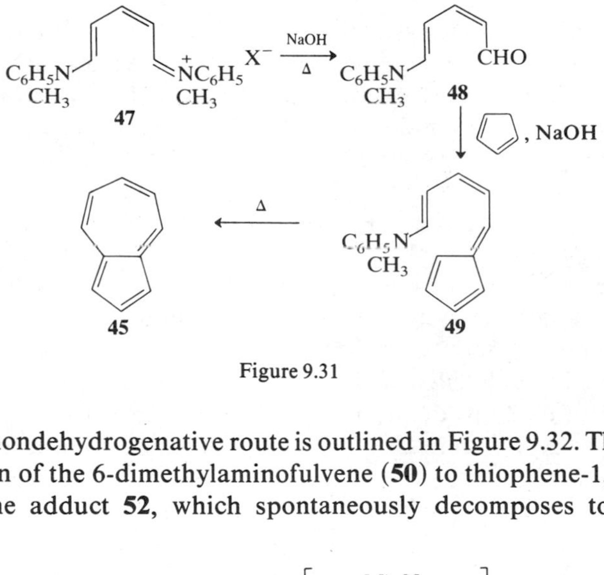

Figure 9.31

Another nondehydrogenative route is outlined in Figure 9.32. The (6 + 4) cycloaddition of the 6-dimethylaminofulvene (**50**) to thiophene-1,1-dioxide (**51**) gave the adduct **52**, which spontaneously decomposes to azulene.

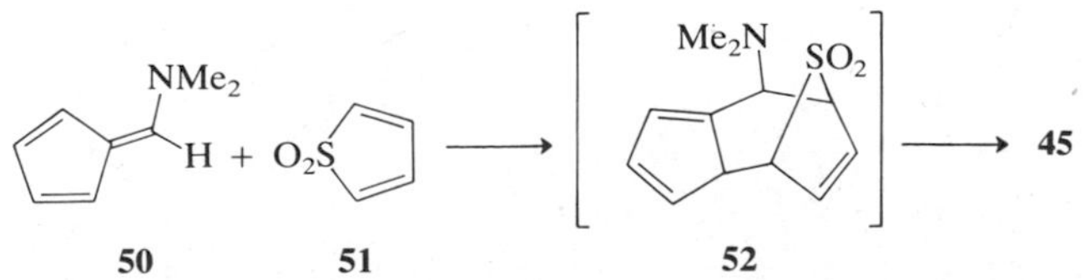

Figure 9.32

Substituted azulenes can be prepared by this method in higher yield than the parent compound.

Azulene is readily protonated, the proton adding in the 1 position of the cyclopentadiene ring to form the azulenium cation **53**. Electrophilic substitution occurs at this position, whereas nucleophilic substitution takes place in the seven-membered ring, presumably via the intermediate **54**. The reactions of azulene are thus controlled by the formation of new 6 π-electron systems during the course of the reaction.

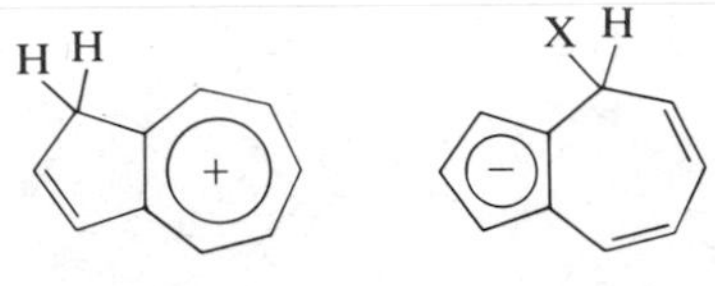

H H X H

53 54

Figure 9.33

A variety of polycyclic systems have been prepared containing a number of five- and seven-membered rings. The tricyclic system **55** can be considered to be either a derivative of pentalene or azulene. The ^{1}H NMR spectrum and chemical properties indicate that **55** is best considered to be an azulene with an exocyclic double bond. The tetracyclic compound **56** may be considered to be a derivative of pentalene, heptalene, or azulene. The spectral and chemical properties are best interpreted on the basis that **56** has a 14 π-electron periphery and a central double bond. The tricyclic system **57** is again best considered as an azulene derivative with the addition of an exocyclic diene system, but this compound shows no diamagnetic exaltation unlike a derivative of **55**, which has $\Lambda = 30$ (see Chapter 2).

Azulene and naphthalene are isomeric, isoelectronic compounds, but a comparison of the PMO delocalization energies (Fig. 9.4 and 9.29) indicates that whereas the transannular bond in naphthalene stabilizes the system, in

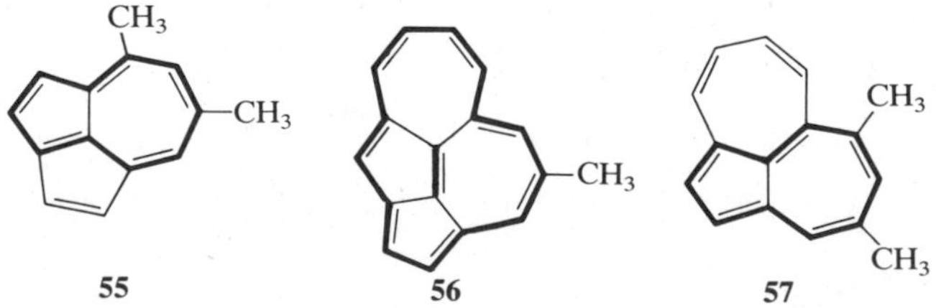

CH_3 CH_3 CH_3 CH_3 CH_3

55 56 57

Figure 9.34

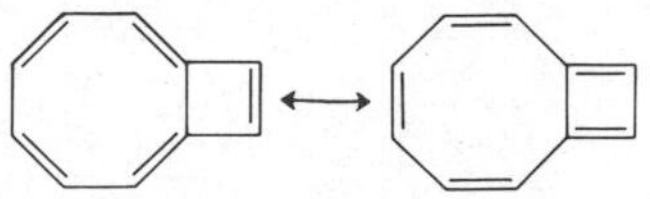

58a **58b**

Figure 9.35

azulene it acts as only a minor perturbation. This finding is supported by the x-ray crystallographic analysis of azulene, which indicates that the central bond is long. The fact that azulene is aromatic whereas [10]annulene is not appears to be due to the removal of nonbonded interactions in the former compound. A third system that is isomeric and isoelectronic with azulene and naphthalene is bicyclo[6.2.0]decapentaene (**58**), which is a type (*b*) system formed by fusion of cyclooctatetraene and cyclobutadiene. The PMO method indicates that **58** has the same energy as the acyclic decapentaene, that is, the delocalization energy is zero and the transannular bond *destabilizes* the system (Fig. 9.36).

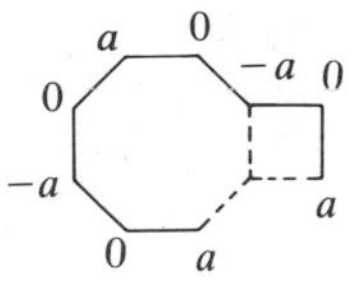

$\delta E = 2a\beta$
DE = 0

Figure 9.36

The parent compound **58** has recently been prepared by Oda and Oikawa by the route shown in Figure 9.37. 2 + 2 Addition of the enedione **59** with *cis*-dichloroethene (**60**) gave **61**, which on reduction gave the diol-dichloride **62**. Treatment of **62** with sodium in liquid ammonia gave the dienediol **63** as a mixture of *cis* and *trans* isomers. The *cis* isomer was mesylated and pyrolyzed to give the tricyclic tetraene **65** that valence tautomerized at 100°C to **58**.

The 9,10-diphenyl derivative **66a** was prepared in a similar fashion, and earlier Schröder and co-workers had prepared the 9,10-dimethyl and a number of other derivatives (**66b–e**). The x-ray crystallographic analysis of **66a** indicates that this is a fairly planar molecule with alternate long and short bonds (Fig. 9.39). Most striking is the long transannular bond (153.5 pm), which, taken together with the planarity, suggests that **66a** is a peripheral 10 π-electron system. The alternations in bond length, however, preclude this being an aromatic system.

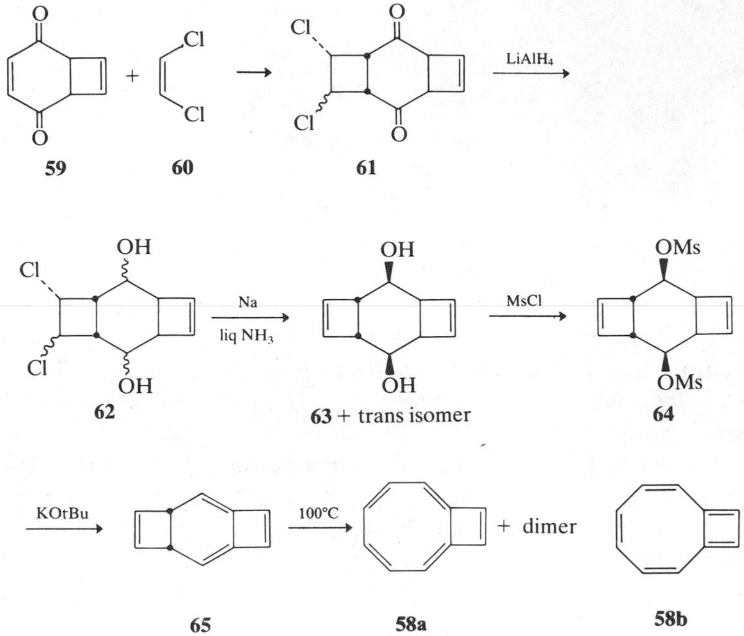

59 60 61

LiAlH4

62 Na liq NH_3 63 + trans isomer MsCl 64

KOtBu 65 100°C 58a + dimer 58b

Figure 9.37

Allinger and Yuh, using a molecular mechanics method, calculated that the 1,4-transannular interaction was an antiaromatic perturbation but found that the planar conformation and bond localization indicated in **58a** is an energy minimum although a nonplanar conformation of **58b** surprisingly has a similar energy.

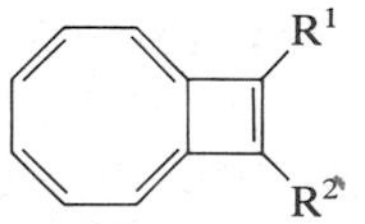

66a $R^1 = R^2 = C_6H_5$
66b $R^1 = R^2 = CH_3$
66c $R^1 = H$, $R_2 = OtBu$
66d $R^1 = R^2 = tBu$
66e $R^1 = Cl$, $R^2 = OtBu$

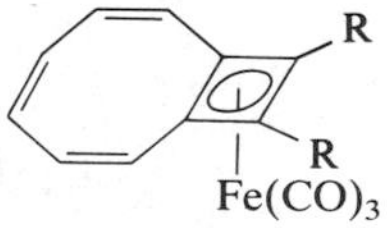

$Fe(CO)_3$

68a R = H
68b R = Me

Figure 9.38

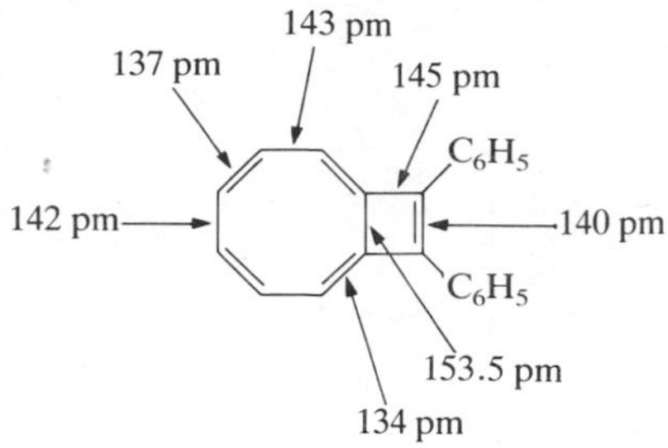

Figure 9.39. X-Ray crystallographic structure of 9,10-diphenylbicyclo[6.2.0]deca-1,3,5,7,9-pentaene.

Compounds **66c–e** react with tetracyanoethylene to give Diels–Alder adducts by addition across the C-1, 4 bonds rather than 2 + 2 adducts as was originally thought. Thermolysis of **66b** at 680°C gave a mixture of 1,2-, 2,3-dimethylnaphthalenes and 1,2-dimethylazulene, possibly via the bis-allene **67** (Fig. 9.40). At low temperature, **66b** forms a mixture of complexes with iron tricarbonyl, which on warming to 60°C gives the single isomer **68b**.

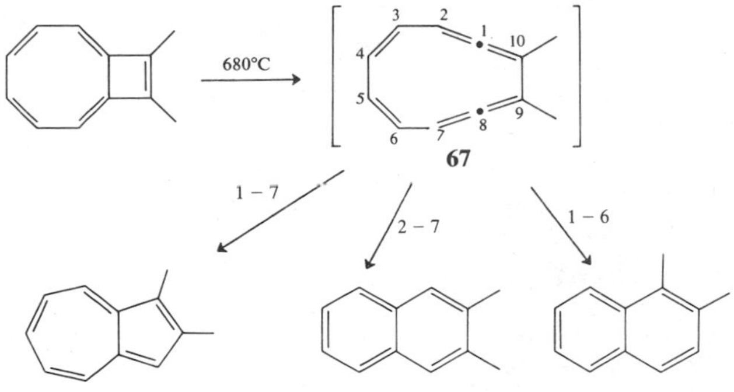

Figure 9.40

The benzannelated derivative **69** is a crystalline solid and the spectral and chemical properties suggest that the eight-membered ring is nonplanar. The polycyclic system **70** has been synthesized and in this compound the eight-membered ring is a planar paratropic system. The benzannelated iron tricarbonyl complex **71** has also been prepared.

Benzocyclobutadiene (**73**) has been isolated at 20 K, and its IR, electron, and photoelectron spectra have been reported. The preparations involve the dehalogenation of di-iodobenzocyclobutene **72** (X = I) with zinc at

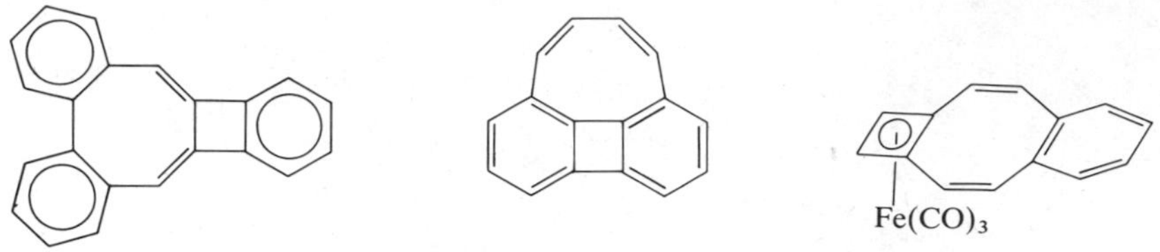

69 70 71

$Fe(CO)_3$

Figure 9.41

230°C or α,α'-dibromo-*o*-xylene with magnesium at 440°C (Fig. 9.42). The electronic spectrum showed bands at 243, 246.5, 256, 264, 270, 281.5, and 289 nm, and the photoelectron spectrum was interpreted in favor of structure **73a** with little contribution from **73b**. The PMO method indicates that the

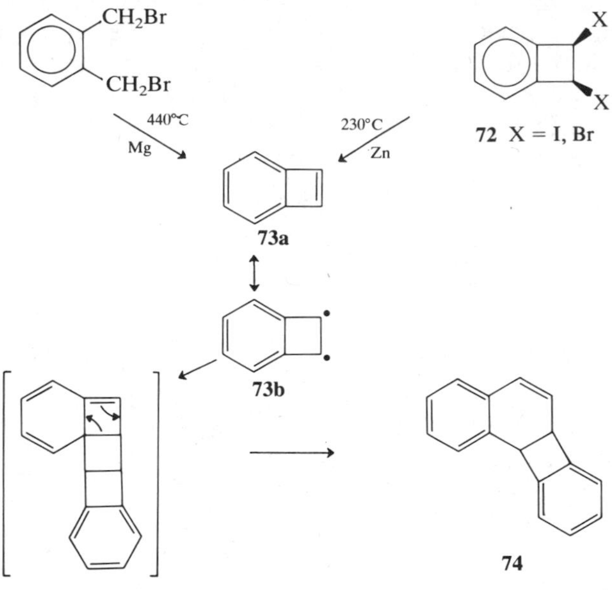

CH_2Br CH_2Br 440°C Mg 230°C Zn X X **72** X = I, Br **73a** **73b** **74**

Figure 9.42

transannular bond stabilizes benzocyclobutadiene as compared to planar cyclooctatetraene, benzocyclobutadiene being nonaromatic rather than antiaromatic (Fig. 9.43)

Benzocyclobutadiene had previously been recognized as a transient intermediate in a number of reactions by the isolation of the dimer **74**. Thus,

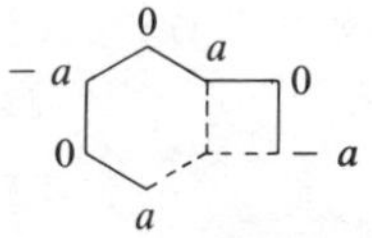

$\delta E = 2a\beta$
DE = 0

Figure 9.43

zinc debromination of the dibromide **72** (X = Br) gave **74**, and the reaction was presumed to proceed by Diels–Alder dimerization of cyclobutadiene followed by tautomerism of the resulting cyclobutene to **74**.

A number of substituted benzocyclobutadienes have been prepared as stable compounds and their structures investigated. The least substituted derivative is the 1,2-bis(trimethylsilyl) system **75** prepared by Vollhardt and Lee by thermal cyclization of 1,8-bis(trimethylsilyl)octa-3,5-diene-1,7-diyne. In the ^{1}H NMR spectrum the ring protons are at δ 6.20 and 5.63, indicating a considerable paratropic shift. The x-ray crystallographic analysis of **76**, the first benzocyclobutadiene to be isolated, shows some alternation of bond length, particularly in the four-membered ring (C-1,2, 135.7 pm, C-2,2a, 153 pm) but also in the six-membered ring (C-2a,3, 137 pm, C-3,4, 145 pm, C-4,5, 137 pm, C-6a,2a, 140 pm). This bond alternation may again be attributed to the 8 π-electron nature of **76**.

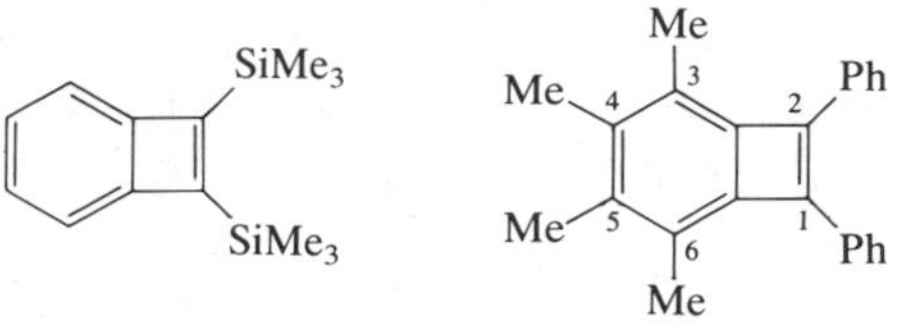

75 **76**

Figure 9.44

Stable naphthocyclobutadiene and anthrocyclobutadiene derivatives have been prepared. 1,2-Diphenylnaphthocyclobutadiene (**78**) was obtained by debromination of the dibromide **77**. The ^{1}H NMR spectrum shows, besides the aromatic protons, a two proton singlet at δ 6.5 due to the H^a protons. Compound **78** reacts with 1,3-diphenylisobenzofuran (**79**) to give the adduct **80**. The canonical structure **78b** is presumably predominant, resulting in the loss of only one aromatic sextet and the slightly smaller paratropic shift of the H^a protons compared to those in **75**.

Compounds in which the benzene ring of benzocyclobutadiene has been replaced by a five-membered heterocyclic ring **81** (X = O, S; R = H, Ph)

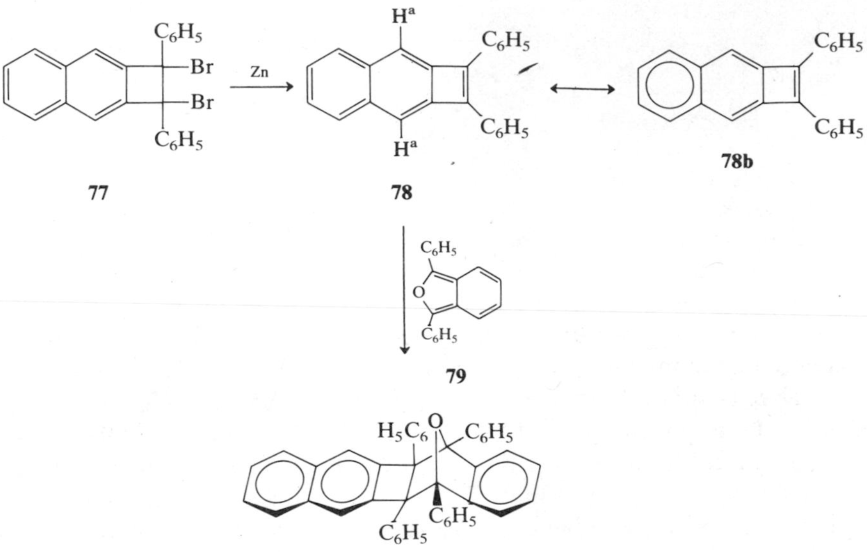

77 78 78b

79

80

Figure 9.45

have been prepared, and the heterocyclic ring protons show considerable upfield shifts (about 0.7 ppm) in the ^{1}H NMR spectrum compared to the corresponding protons in the cyclobutaheterocycles. These compounds are again behaving as 8 π-electron paratropic systems.

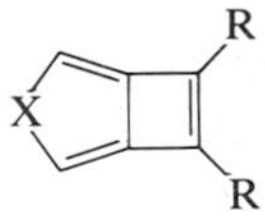

81 X = O, S; R = H, Ph

Figure 9.46

That the cyclobutadiene ring actually destabilizes these types of systems has been emphasized by Breslow and co-workers. The hydroquinone **82** is an unstable compound that slowly dimerizes. On electrochemical oxidation **82** gives the quinone **83**, which is even less stable than **82**. The oxidation of **82** to **83** was shown to occur less readily than the oxidation of naphthohydroquinone to naphthoquinone, whereas the dihydro compound **84** was more readily oxidized than naphthohydroquinone. An estimate of 50 kJ mol^{-1} was made for the *destabilizing* effect of the double bond.

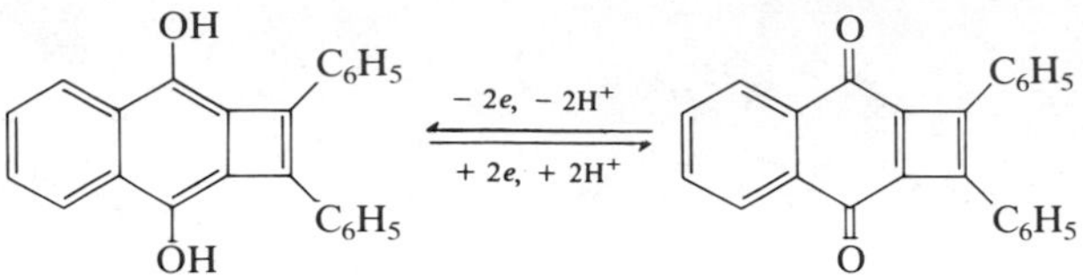

Figure 9.47

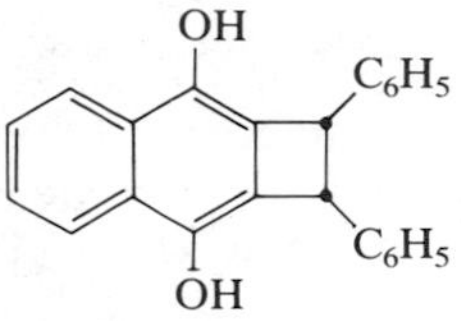

In a related series it has been observed that the bicyclo[3.2.0]heptatrienyl anion (**85**) and the norbiphenylene anion (**86**) are both much less stable than the cyclopentadienyl anion. The anion **85** can be considered to be a perturbed cycloheptatrienyl anion with **86** as a benzannelated derivative of that system. It has been estimated that **85** is about 60 kJ mol^{-1} *less* thermodynamically stable than the cyclopentadienyl anion. The anion **86** is presumably a little more stable in the same way that biphenylene is more stable than benzocyclobutadiene. The rather artificial nature of our classification is apparent here as biphenylene, which we discussed in Section 9.2, might more readily be considered to be a nonbenzenoid compound.

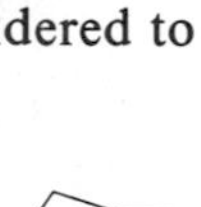

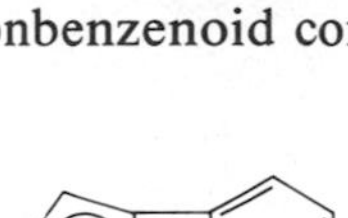

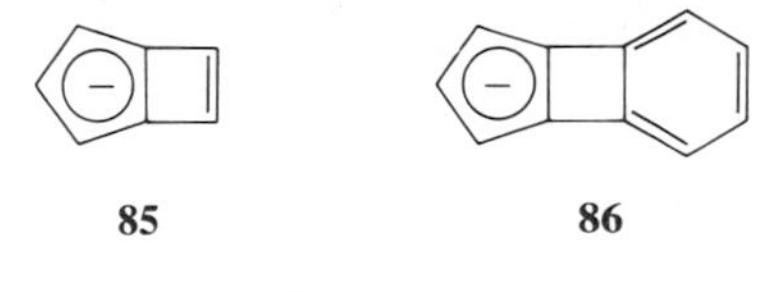

Figure 9.48

The 1,2-dimethylbenzocyclobutadiene dication (**87**) has been prepared by treatment of the corresponding benzocyclobutene-1,2-diol with SbF_5, SO_2ClF at −78°C. In the ^{13}C NMR spectrum, C-1,2 resonate at δ 186 and the cation is clearly an 8C-6π-electron system. The corresponding 1,2-diphenylbenzocyclobutadienyl dianion (**88**) has also been observed by deprotonation of the corresponding benzocyclobutene.

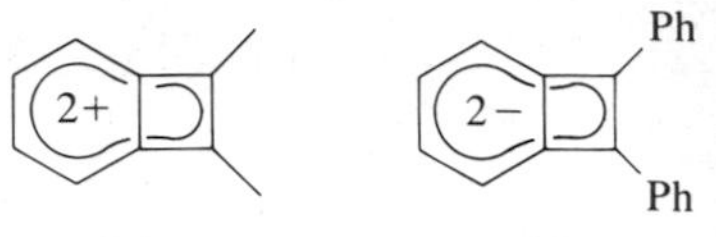

87 **88**

Figure 9.49

Like cyclobutadiene, benzocyclobutadiene is stabilized by complexing with iron tricarbonyl (Fig. 9.50). The dibromide **72** on treatment with iron enneacarbonyl gives the complex **89**, which in the ^{1}H NMR spectrum shows a four-proton multiplet at δ 6.95 and a two-proton singlet at δ 4.02 for the aromatic and cyclobutadienyl protons, respectively. Oxidation of the iron group leads to dimerization, the nature of the product depending on the oxidizing agent used.

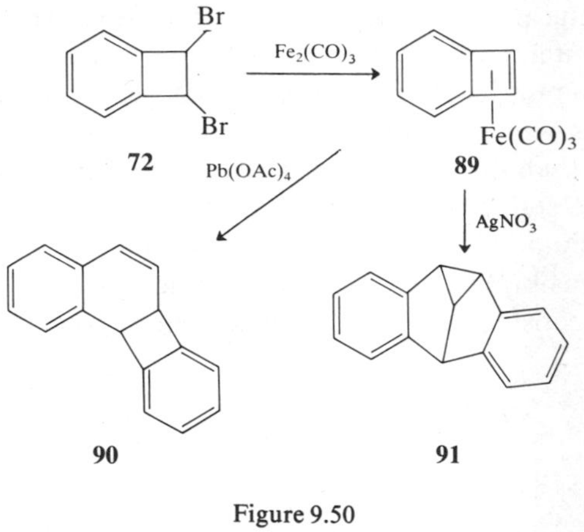

Br Br $Fe_2(CO)_3$ $Fe(CO)_3$ **72** **89** $Pb(OAc)_4$ $AgNO_3$ **90** **91**

Figure 9.50

Iron complexes of the anion **85** have also been prepared, and although these are air sensitive and decompose in the solid state they are much more stable than the anion itself.

Toda and Ohi have prepared the 10 π-electron tricyclic system **92** as a blue, crystalline compound that rapidly oxidizes. This compound appears to behave as an annelated benzocyclobutadiene rather than a 10 π system, and it can be readily converted to benzocyclobutadiene derivatives.

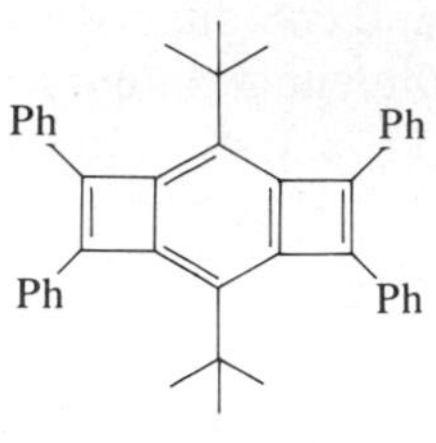

92

Figure 9.51

The cation **93**, a type (*b*) system formed by the fusion of cyclooctatetraene with the cyclopentadienium cation, has 10 π electrons and might be aromatic. Although this cation has not been prepared, the ketone **94** has but this shows no aromatic properties.

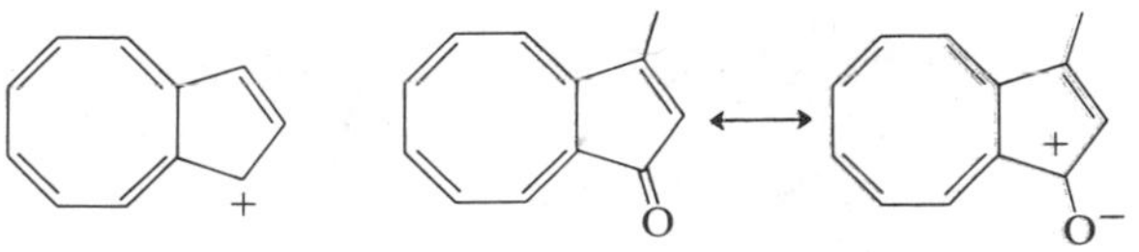

93 94a 94b

Figure 9.52

The interesting polycyclic bicalicene system **95** has recently been prepared. The physical properties, including the x-ray crystallographic structure, suggest that this is a delocalized system with a major contribution from the dipolar structure **95b**. Not unexpectedly, it does not appear

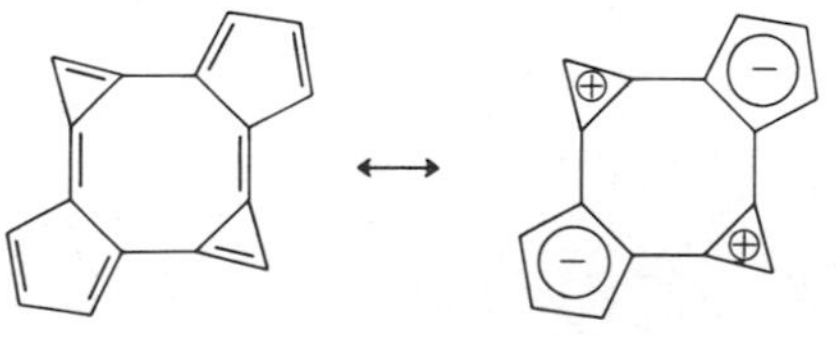

95a 95b

Figure 9.53

important that **95** possesses a 16 π-electron periphery. The PMO theory does not make satisfactory predictions for the structure of **95**.

Octalene (**96**) is a type (*b*) system formed conceptually by the fusion of two cyclooctatetraene rings and contains 14 π electrons. It is thus a member of the series of $4n + 4n$ annulenoannulenes of which butalene (6 π) and bicyclo[6.2.0]decapentaene (10 π) are other members we have discussed.

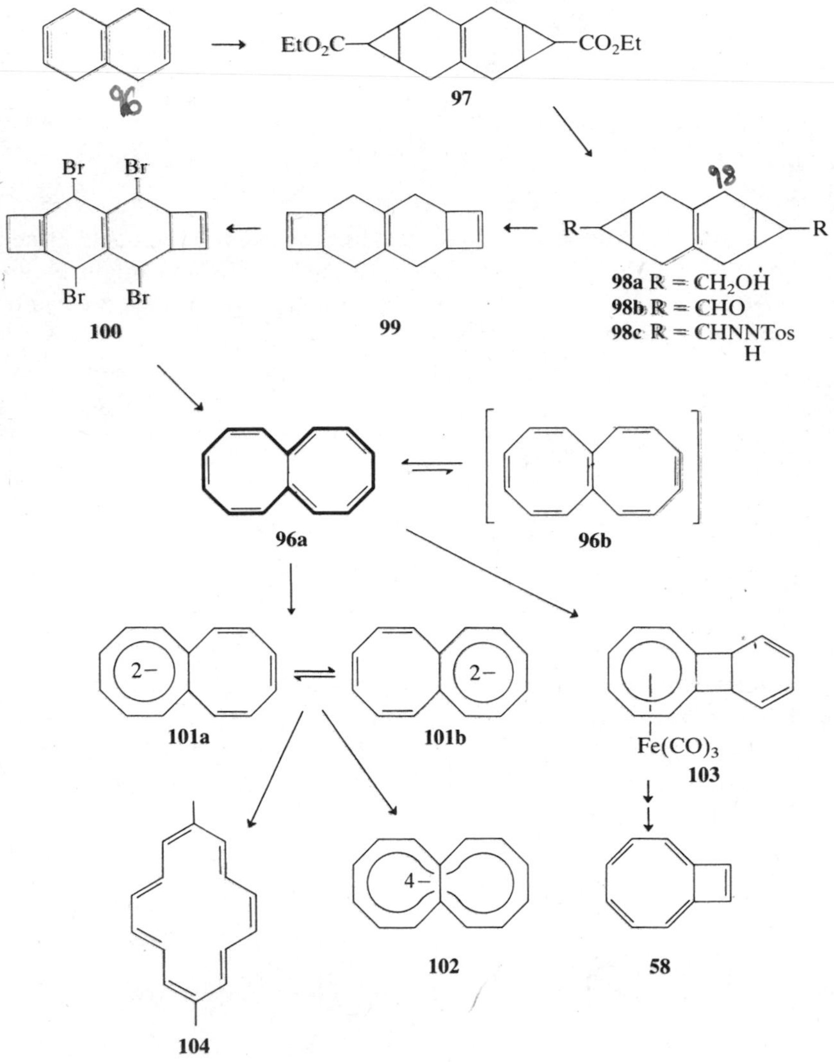

96 → EtO_2C—97—CO_2Et → 98 (R—…—R) → 99 → 100 (Br, Br, Br, Br) → 96a ⇌ [96b]

98a R = CH_2OH
98b R = CHO
98c R = CHNNTos
H

101a (2−) ⇌ **101b** (2−)

102 (4−)

103 $Fe(CO)_3$ → **58**

104

Figure 9.54

This series complements the type (*a*) annulenoannulenes of which naphthalene (10 π) and benzo[10]annulene (14 π) are the corresponding members. The PMO method predicts that octalene will have zero delocalization energy and will be less stabilized than [14]annulene.

Octalene has been prepared by Vogel and co-workers by the route outlined in Figure 9.54. The bisadduct **97**, formed by the reaction of ethyl diazocarboxylate with isotetralin, was reduced to the diol **98a**, which was oxidized to the dialdehyde **98b** and converted to the bistosylhydrazone **98c**. Ring expansion of **98c** via the sodium salt gave the tetracyclic **99**, which on bromination gave the tetrabromide **100**. Debromination of **100** gave octalene (**96**) as a lemon yellow liquid, melting point −5 to −4°C. The ^{1}H NMR spectrum showed olefinic absorptions at δ 5.65 and 6.30, and the ^{13}C NMR spectrum has seven lines, indicating that octalene has the structure **96a** rather than the symmetric structure **96b**. The ^{13}C NMR spectrum is temperature dependent, the signals broadening at −100°C and a 14-line spectrum appearing at −150°C, indicating that a single conformation of **96** is chiral and that the room temperature spectrum arises from the equilibration of the two enantiomers. At temperatures above 80°C line broadening occurs as π-bond shift becomes sufficiently fast to be observed on the NMR time scale, but decomposition occurs before a temperature is attained at which the four-line spectrum can be observed.

Octalene can be reduced to the 16 π-electron dianion **101**, which behaves as a 10 π-electron cyclooctatetraenyl dianion annelated by a hexatriene but in which the two rings equilibrate (**101a** $\leftrightarrows$ **101b**). Further reduction gives the 18 π-electron tetra-anion **102**, which now represents a type (*a*) system.

Octalene forms the iron tricarbonyl complex **103** in the presence of light in which one ring has valence tautomerized to the bicyclo[4.2.0] form. The complex **103** can be converted through a series of reactions to bicyclo[6.2.0]decapantaene (**58**). Alkylation of the dianion **101** gives 1,8-dimethyl[14]annulene (**104**).

A number of octalene derivatives had been prepared prior to the synthesis of the parent system, including benzooctalene (**105**) and furooctalene (**106**). Both **105** and **106** are nonplanar systems with the eight-membered rings probably in tub conformations. Hexabenzooctalene (**107**) has also been prepared and, not surprisingly, is again a nonplanar system.

A variety of other annulenoannulenes have been prepared for the most part with one of the rings benzenoid. Benzobisdehydro[8]annulene (**108**) was synthesized by Sondheimer and Wong as an unstable yellow oil that showed in the ^{1}H NMR spectrum the double bond protons at δ indicating that **108** is a paratropic planar system. More highly annelated dehydro[8]annulenes have been prepared, as have other more highly annelated planar [8]annulenes (e.g., **70**).

105 106 107

Figure 9.55

Staab and co-workers have prepared benzo[14]annulene (**109**) and benzo[18]annulene (**110**), and Mitchell and co-workers have prepared a number of benzannelated 15,16-dimethyldihydropyrans (e.g., **111**, **112**). These are all diatropic systems in which the diatropicity of the macrocyclic ring has been diminished in comparison to the parent annulene.

In compounds **109** and **110** the benzene ring dominates in its requirements. Thus, in the ^{1}H NMR spectrum of **109** the inner protons appear at δ 4.3–5.1, whereas those of [14]annulene are at δ −0.61, and in **110** the inner protons are at δ 4.75–4.95, whereas in [18]annulene they are at δ

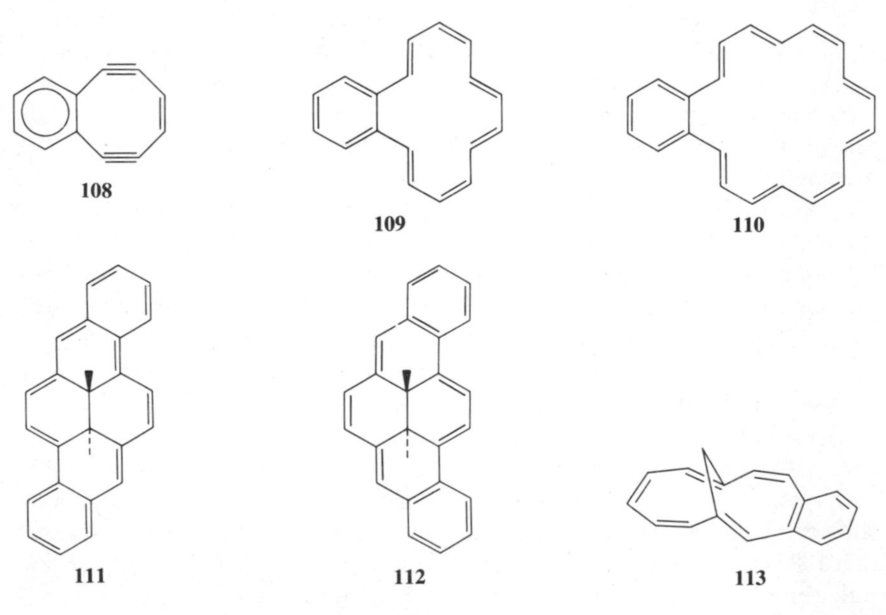

Figure 9.56

−2.99. For polyannelated systems such as **111** and **112**, the relative orientation of the aromatic rings is important; **111** resembles anthracene and the diatropicity of the central 14 π-electron ring is not greatly diminished, whereas **112** resembles phenanthrene and the diatropicity of the central ring is greatly decreased. Many other more highly benzannelated compounds have been prepared, a large number of which contain triple bonds in the annulene ring.

Bridged systems, such as 4,5-benzo-1,7-methano[12]annulene (**113**), otherwise benzo[b]homoheptalene, have been prepared and show similar changes in magnetic properties. Thus, in the ^1H NMR spectrum **113** shows a decreased paratropicity (methylene protons δ 4.43, 4.52; nonbenzenoid ring protons δ 5.61–6.22).

Fascinating compounds in which both rings are macrocyclic have been prepared, the first examples having been independently announced in 1975 in the same issue of the *Journal of the American Chemical Society*. The compounds are of two types; in that synthesized by Cresp and Sondheimer the rings have one common double bond, whereas in the compound synthesized by Nakagawa and co-workers the rings have five cumulated common bonds. The compounds synthesized by both groups are dehydroannulenes and are substituted. The synthetic routes are outlined in Figures 9.57 and 9.58.*

Compound **114** was obtained as dark red-brown prisms that decomposed above 200°C. The electronic spectrum was complex with absorption out to 607 nm with a main maximum at 387 nm (ε 45,400). Compound **115** was obtained as dark green crystals and also has an extensive electronic spectrum with absorption out to 932 nm and a main maximum at 470 nm (ε 75,200). Both groups modified their synthetic routes to prepare series of related compounds, including a [14]annuleno[16]annulene derivative by Cresp and Sondheimer.

The ^1H NMR spectra of the $(4n + 2)$ $(4n + 2)$ systems show a decrease in diatropicity of the component rings. The effect in **114** is not as large as that observed in the corresponding benzo[14]annulene derivative **116**, which is indicative of a greater demand by the phenyl ring to maintain its aromatic nature and also the equivalence of the two Kekulé structures in **114**. Compound **115** also shows a decreased diatropicity of the 18-membered

* These new annulenoannulenes pose some nomenclature problems since the Nakagawa compounds are the first to have more than one common bond. Trivial nomenclature systems were suggested by both groups, which can be combined and used for all compounds. The ring size is given as for the annulenes and the ring junctions indicated by the bicycloalkane nomenclature. Further dehydrogenation or substitution is indicated in the normal way. Thus, **115** is 5,10,18,23-tetra-*t*-butyl-6,8,19,21,27,29-hexakisdehydro[12.12.4][18]annuleno[18]annulene (this is a *trivial* name).

ring compared to [18]annulene but a *greater* diatropicity than the corresponding [30]annulene derivative **117**, which has an equivalent periphery. The same applies to the [10.10.2][14]annuleno[14]annulene derivative **118** and its peripheral equivalent the [22]annulene derivative **119**.

The ^{13}C NMR spectrum of **118** shows only one type of *sp*-hybridized carbon, and it would appear that each ene-yne-ene: butatriene group is equivalent and that **118** is best represented by **118a**.

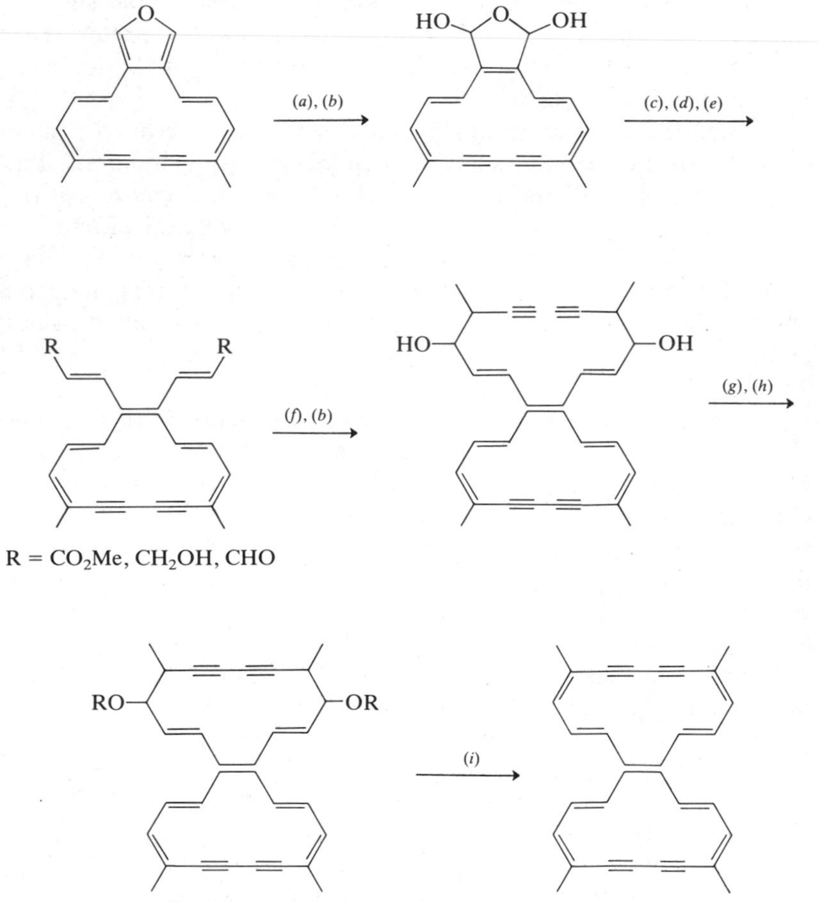

HO O OH

(*a*), (*b*)

(*c*), (*d*), (*e*)

R R

(*f*), (*b*)

HO OH

(*g*), (*h*)

R = CO_2Me, CH_2OH, CHO

RO OR

(*i*)

R = $MeSO_2$

114

(*a*) $Pb(OAc)_4$, HOAc; (*b*) $(CO_2H)_2$; (*c*) $Ph_3\overset{+}{P}\overset{-}{C}HCO_2Me$;
(*d*) iBu_2AlH; (*e*) MnO_2; (*f*) $BrMgCH(Me)C \equiv CH$;
(*g*) $Cu(OAc)_2$, DMF; (*h*) MsCl; (*i*) DBN

Figure 9.57

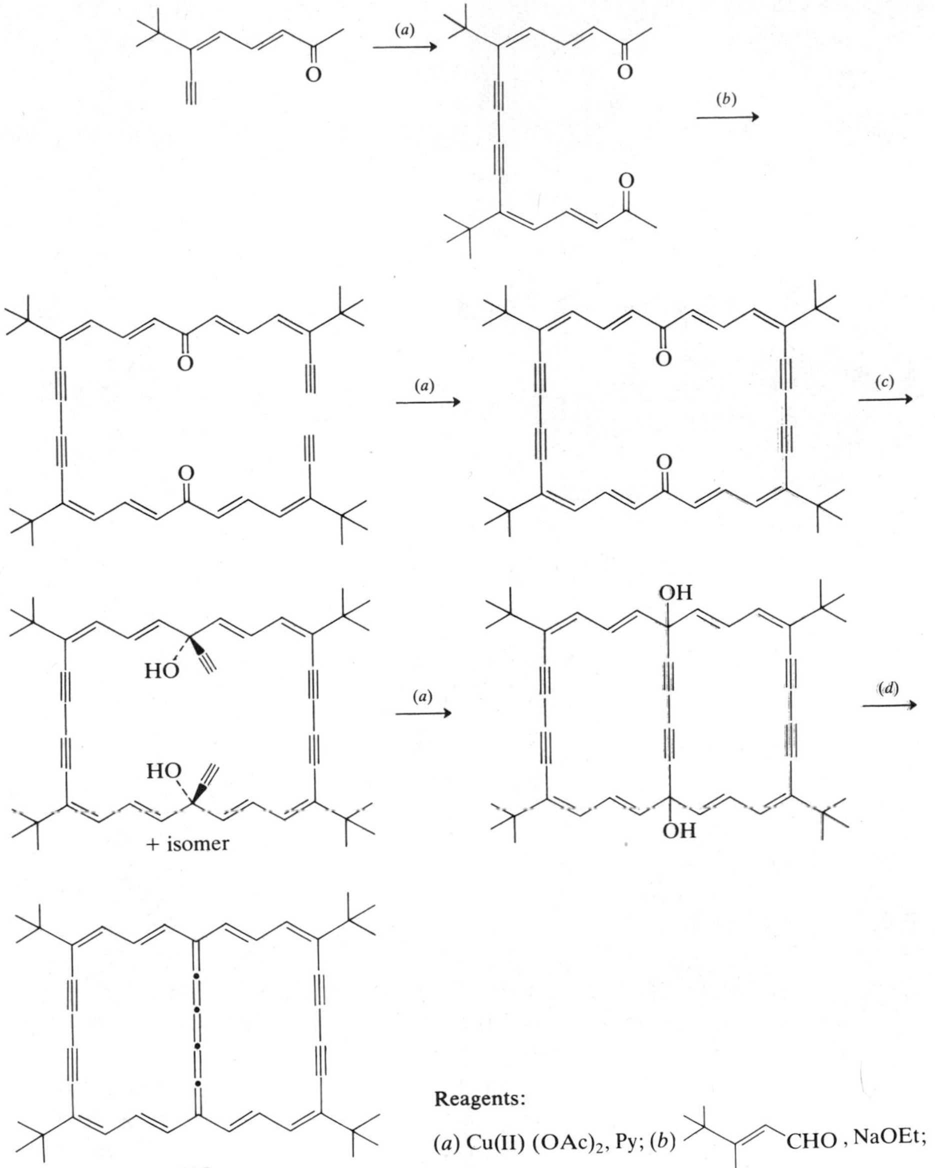

(a)

(b)

(a)

(c)

HO

HO

+ isomer

(a)

OH

OH

(d)

115

Reagents:

(a) Cu(II) $(OAc)_2$, Py; (b) CHO, NaOEt;

(c) ≡Li (H_2N, NH_2); (d) Sn(II)Cl_2, Et_2O, HCl

Figure 9.58

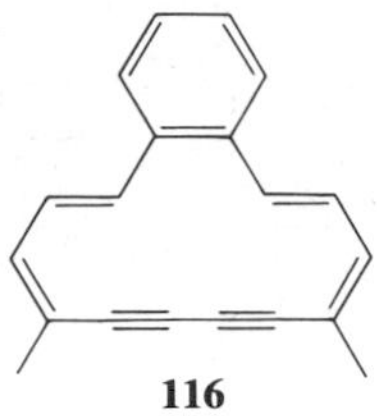

116

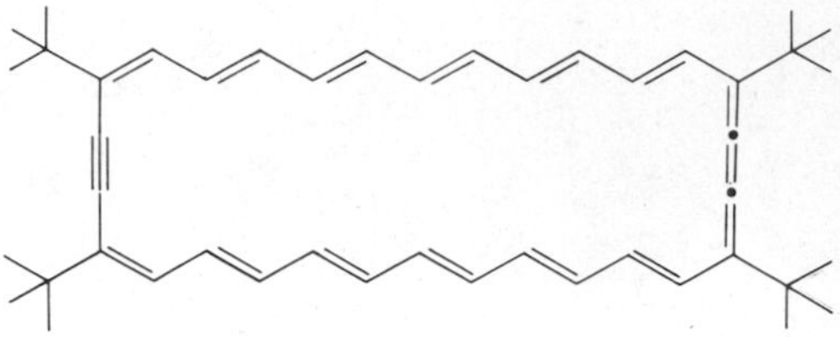

117

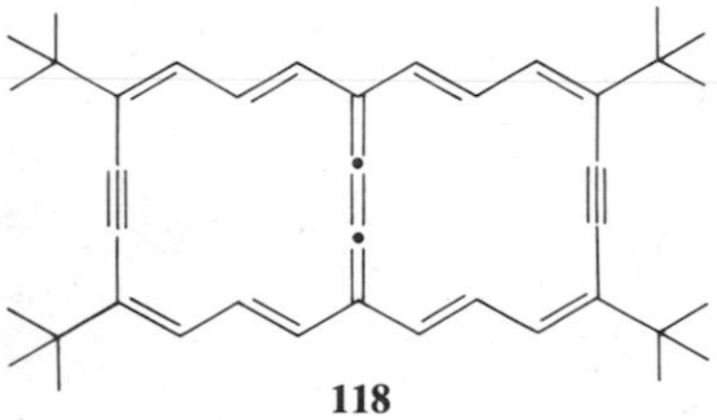

118

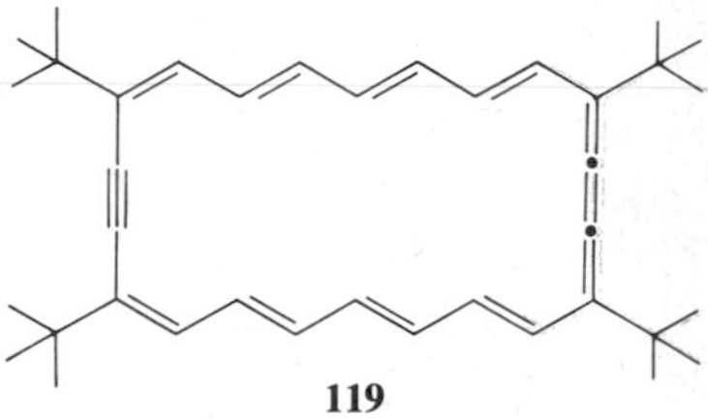

119

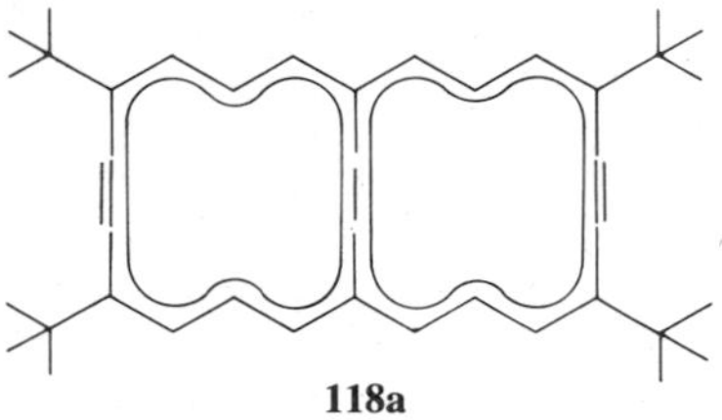

118a

Figure 9.59

9.4. CONCLUSIONS

Only in the type (*a*) systems are the properties of the individual rings consistent with the theoretical properties of the overall number of π electrons. In both type (*b*) and type (*c*) systems, the properties of the individual rings are different from those expected from the total number of π electrons, and it appears that in these molecules it is the properties of the individual rings that are normally important. The only systems of these types in which the overall number of π electrons has become dominant are, as expected, charged systems where charge delocalization favors extended conjugation. The PMO method is particularly useful in its application to polycyclic systems and is preferable to making simple HMO calculations for such compounds.

FURTHER READING

For a comprehensive discussion of the PMO method, see M. J. S. Dewar and R. C. Dougherty, *PMO Theory of Organic Chemistry*, Plenum, New York, 1975.

For a discussion of benzenoid hydrocarbons, see E. Clar, *Polycyclic Hydrocarbons*, Vols. 1 and 2, Academic Press, New York, 1964; and for nonbenzenoid polycyclic systems, see P. J. Garratt and M. V. Sargent, in E. C. Taylor and H. Wynberg (Eds.), *Advances in Organic Chemistry*, Vol. 6, Wiley-Interscience, New York, 1969, p. 1, and D. Lloyd, *Non-benzenoid Conjugated Carbocyclic Compounds*, Elsevier, Amsterdam, 1984.

For the Robinson–Clar theory of the aromatic sextet, see E. Clar, *The Aromatic Sextet*, Wiley, New York, 1972.

For a discussion of biphenylene and similar systems, see M. P. Cava and M. J. Mitchell, *Cyclobutadiene and Related Compounds*, Academic Press, New York, 1967, and J. W. Barton in J. P. Snyder (Ed.), *Nonbenzenoid Aromatics*, Vol. 1, Academic Press, New York, 1969; For the ^{1}H NMR spectrum of biphenylene and annelated biphenylenes, see J. W. Barton and D. J. Rowe, *Tetrahedron*, 1985, **41**, 1323.

For a review of the recent chemistry of heptalene, see L. A. Paquette, *Israel J. Chem.*, 1980, **20**, 233.

For review on benzannulenes, see R. H. Mitchell, *Israel J. Chem.*, 1980, **20**, 294.

For a review on annulenoannulenes, see M. Nakagawa, *Angew. Chem. Int. Ed. Engl.*, 1979, **18**, 202.

For annelated [8]annulenes, see N. Z. Huang and F. Sondheimer, *Accounts Chem. Res.*, 1982, **15**, 96.

For a review of some complex polycyclic systems, see K. Hafner, *Angew. Chem. Int. Ed. Engl.*, 1964, **3**, 165; *Pure Appl. Chem.*, 1982, **54**, 939.

For polycyclic compounds containing four-membered rings, see P. J. Garratt, *Pure Appl. Chem.*, 1975, **44**, 783.

For charged polycyclic systems, see M. Rabinovitz, I. Willner, and A. Minsky, *Accounts Chem. Res.*, 1983, **16**, 298.

10

HOMOAROMATIC AND TOPOLOGICALLY RELATED SYSTEMS

In the preceding chapters, a wide variety of compounds have been discussed, the main link between them being that the possibility existed that the π electrons could be delocalized over all the annular atoms. In the present chapter, a group of compounds will be examined in which such delocalization is interrupted by one or more saturated atoms. The general question to be answered will, however, be the same: Can the energy of the system be lowered by cyclic delocalization of the π electrons over the remaining atoms?

The first suggestion that delocalization might occur in such a system was advanced by Thiele to explain the decreased acidity of the methylene protons of cycloheptatriene compared to those in cyclopentadiene. Thiele postulated a 1,6-interaction in cycloheptatriene (**1**) would give a benzenelike system and that this "aromatic character" would be lost on deprotonation, which therefore does not occur readily.

Although this is not the currently accepted explanation for the difference in acidity between cycloheptatriene and cyclopentadiene, nevertheless the occurrence of such an interaction in cycloheptatriene has found supporters. Of particular interest is the observation by Dauben and co-workers that the

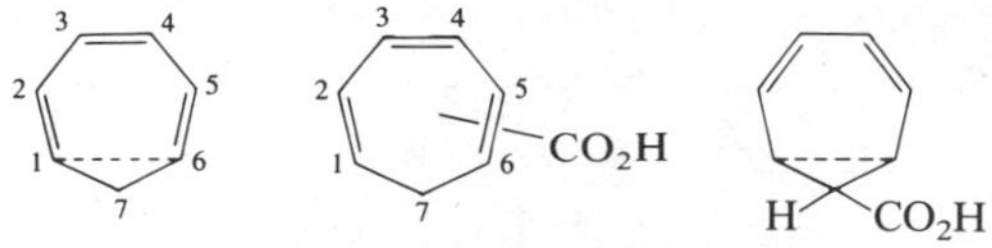

Figure 10.1

diamagnetic exaltation (Λ) for cycloheptatriene is 8.1, which they suggest is only consistent with cycloheptatriene possessing a partially delocalized structure. In their earlier studies of the Buchner acids, von Doering and collaborators had suggested a similar interaction in these compounds could best explain the observed properties. These investigators showed that there were only four Buchner acids (**2**) corresponding to the four possible isomers with the carboxylic function at positions 1, 2, 3, and 7, and they were suggested to have planar aromatic structures through 1,6 interaction, as illustrated in **3**. Although the cycloheptatrienes are now known to be nonplanar, the results of Dauben and others suggest that the 1,6-interaction is appreciable.

Winstein and co-workers attempted to observe this type of interaction in 1,4,7-cyclononatriene (**4**), which might be considered to be benzene with three interrupting methylene groups as illustrated in **5**.

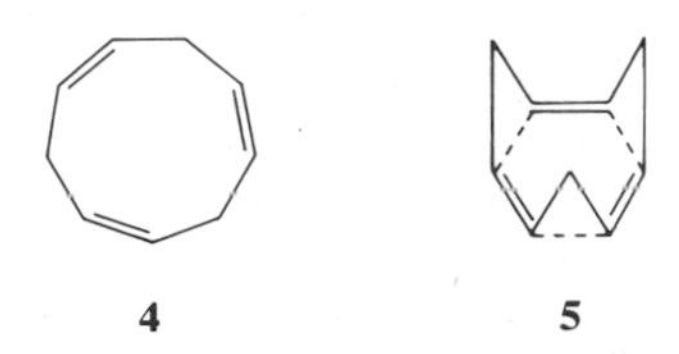

Figure 10.2

The term *homoaromatic* was introduced to describe this phenomenon of interrupted delocalization, and the structure **5** then represents trishomobenzene. The properties of **4** do not suggest that **5** makes any contribution to the structure. The first convincing cases of homoaromatic participation arose from a study of potentially homoaromatic cations. The homotropylium cation (**7**) was first prepared by von Rosenberg, Mahler, and Petit in 1962 by treatment of cyclooctatetraene (**6**) with concentrated sulfuric acid or with $SbCl_5$, HCl in nitromethane. The hexachloroantimonate of **7** has a complex ^1H NMR spectrum, showing a multiplet at δ 8.5 (H-2,3,4,5,6), a multiplet at δ 6.6 (H-1,7), and multiplets at δ 5.2 and −0.6 due to H^b and H^a, respectively.

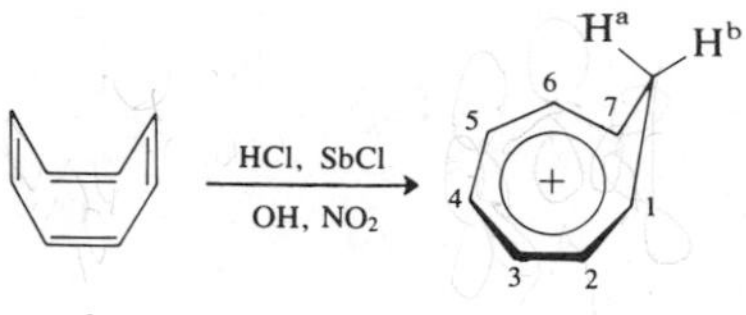

HCl, SbCl
OH, NO2

6 7

Figure 10.3

This NMR spectrum is best accommodated by the open structure **7** in which one of the methylene protons is above and inside the ring (H^a) and the other is above and outside (H^b). The classic structure **8** would be expected to have H-1,7 at higher field (about δ 3.0 rather than δ 6.6) with larger coupling constant between the *cis* protons H^b and H-1 than the *trans* protons H^a and H-1. In fact, the coupling constant between H^a and H-1 (10 Hz) is greater than that between H^b and H-1 (7.5 Hz). Structure **7** was substantiated by a high field ^1H NMR study in which the coupling constants for all coupled protons were determined and were found to have the expected values for **7**. The difference in chemical shift between H^a and H^b can be attributed to the diatropicity of **7**, which shields the inner and deshields the outer proton.

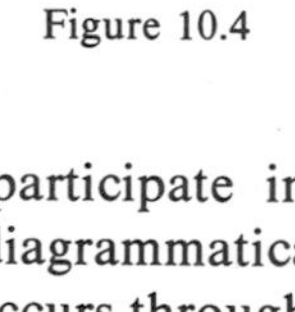

8

Figure 10.4

The p orbitals that must participate in the formation of the seven molecular orbitals are shown diagrammatically in Figure 10.5. The orbital overlap between C-1 and C-7 occurs through only one lobe of the p-atomic orbital and thus partially resembles a σ-type bond. The structure **7** nevertheless appears to be a better representation of the homotropylium cation than does **8**.

Simple HMO calculations can be made for the homotropylium cation by treating it as a tropylium cation and adjusting the values of the resonance integral across the 1,7-atoms. Using a value for this integral $\beta_{1,7} = 0.5\beta$, Winstein calculated that the delocalization energy for homotropylium was 2.423β compared to 2.988β for the tropylium ion itself.

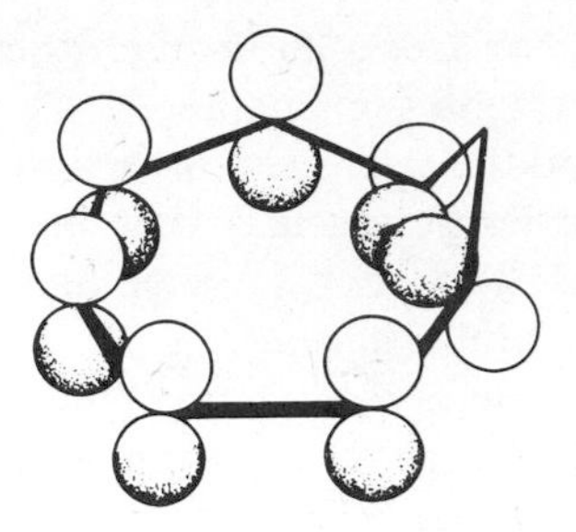

Figure 10.5

The electronic spectrum of **7** is quite similar to that of the tropylium ion, having absorption bands at 232.5 nm (ε 33,000) and 313 nm (ε 3000). These bands show a bathochromic shift compared to those of the tropylium cation, but the spectrum is very different from that of the heptatrienyl cation. A resonance integral of 0.73β must be assumed to predict the position of the long-wavelength band by an HMO calculation, but such calculations are not very reliable. A measurement of the diamagnetic exaltation of **7** gave a value of Λ of 18, of the same order as that of benzene and the tropylium cation and again in accord with **7** being a delocalized system.

When the homotropylium cation is prepared by treatment of cyclooctatetraene with D_2SO_4 below $-15°C$, it is found that 80% of the deuterium is incorporated in the endo position (Fig. 10.6). The deuterium thus adds to cyclooctatetraene from a position inside the ring. On warming the solution equilibration occurs between the endo and exo positions, presumably via

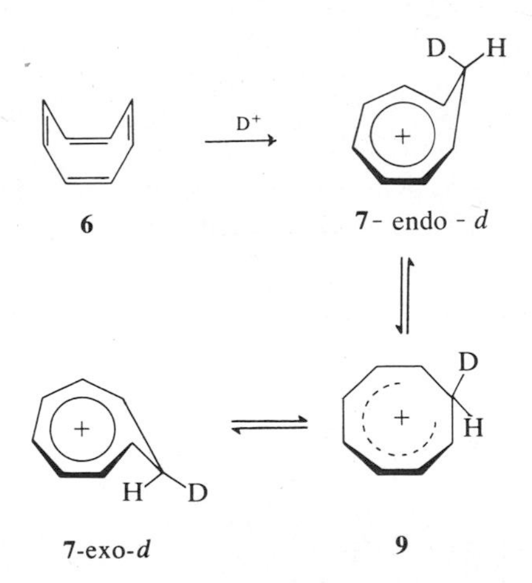

Figure 10.6

ring inversion (7-endo-*d* ⇆ 7-exo-*d*) through the planar cation **9** (for the inversion of cyclooctatetraene see Chapter 2). This equilibration occurs on the NMR time scale, and the barrier to inversion (ΔG) was determined to be 94 kJ mol^{-1}. This barrier height may then be assumed to be equivalent to the *gain* in delocalization energy on going from the cyclooctatrienyl cation **9** to the homotropylium cation **7**.

A number of transition metal complexes of this cation have been prepared that nicely illustrate the balance between the electronic requirements of the metal and those of the ligand. The molybdenum tricarbonyl (**10**) and tungsten tricarbonyl (**11**) complexes have the ligand in the open, delocalized form **7**, whereas the iron tricarbonyl complex **12** has the ligand in the partially delocalized form **8**. Whereas molybdenum and tungsten prefer to coordinate to a 6 π system, which would be provided by either of the delocalized cation **7** or **9**, the iron atom prefers to coordinate to a 4 π system, which is provided by the cation **8**. The ^{1}H NMR spectrum of **12** has signals for the H-1,7 protons at δ 2.52, the ring protons are at higher field than those in **7**, and H^a and H^b both resonate near δ 1.5. The spectra of **10** and **11**, in contrast, show a large chemical shift difference between H^a and H^b, and the chemical shifts of the ring protons and of H-1,7 are intermediate in position between those of **7** and **12**.

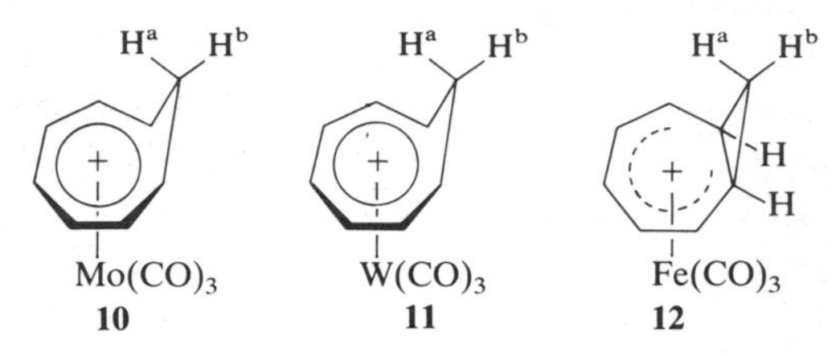

Figure 10.7

When substituted cyclooctatetraenes are protonated, the homotropylium ions that are formed are derived by protonation at a position next to the substituent. Thus methyl (**13a**) and phenylcyclooctatetraene (**13b**) give the corresponding 1-methyl (**14a**) and 1-phenylhomotropylium (**14b**) cations.

H_2SO_4

13a R = CH_3
13b R = C_6H_5

14a R = CH_3
14b R = C_6H_5

Figure 10.8

Chlorination of cyclooctatetraene has been shown to proceed through the intermediate endochlorohomotropylium cation (**15**), which then reacts further to give *cis*-7,8-dichlorobicyclo[4.2.0]octa-2,4-diene (**16**) (see Chapter 2).

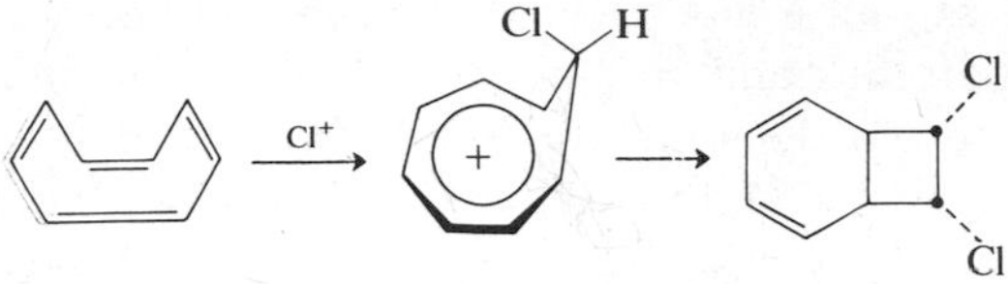

Cl H
Cl^+
Cl
Cl

6 15 16

Figure 10.9

In general it therefore appears that cyclooctatetraene is more readily attacked by electrophiles from a position inside the ring and that the resulting cation rearranges to the corresponding delocalized, 6 π-electron homotropylium cation. Protonation appears to occur preferentially next to a substituent, but in all cases the homotropylium structure is more stable than the corresponding cyclooctatrienyl cation.

By analogy with the relationship between the tropylium ion and tropone, some stabilization might be expected for 2,3-homotropone (**18**). This compound was prepared by Holmes and Petit by the sequence of reactions shown in Figure 10.10. Addition of sodium hydroxide to the salt **12** gave the alcohol **17**, which on oxidation with chromium trioxide followed by removal of the iron tricarbonyl unit with cerium(IV) ammonium nitrate gave homotropone **18**.

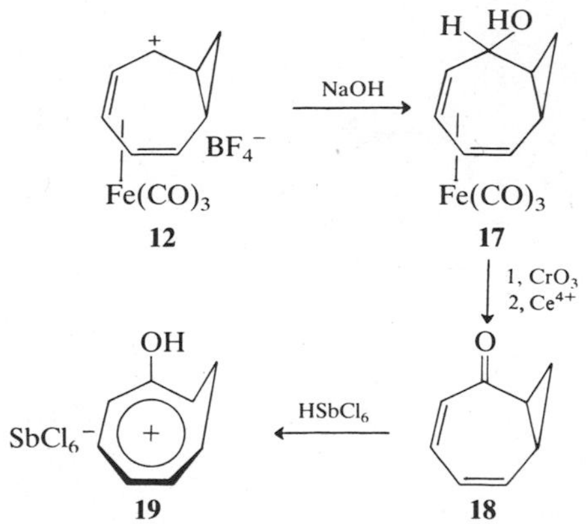

BF_4^- NaOH H HO
$Fe(CO)_3$ $Fe(CO)_3$
12 17
1, CrO_3
2, Ce^{4+}
OH O
$SbCl_6^-$ $HSbCl_6$
19 18

Figure 10.10

The ^1H NMR spectrum suggests that there is little delocalization in this system, although an enhanced basicity of the carbonyl group was observed. Treatment of **18** with concentrated sulfuric acid or with $HSbCl_6$ gave the corresponding 2-hydroxyhomotropylium cation (**19**), obtained by the second method as a yellow, crystalline hexachloroantimonate. The ^1H NMR spectrum of **19** showed a downfield shift of the ring protons and three of the cyclopropyl protons and an *upfield* shift of the remaining cyclopropyl proton. Correlation of the structure in solution with that in the solid was made by a comparison of the solution and "magic angle" solid NMR spectra, and an x-ray crystallographic analysis of **19** gave the structure shown in Figure 10.11, C-8 being out of the ring plane. Compound **19** is clearly a homotropylium cation.

118.2 pm
137.0 pm OH 142.2 pm
137.8 pm 162.6 pm
148.9 pm

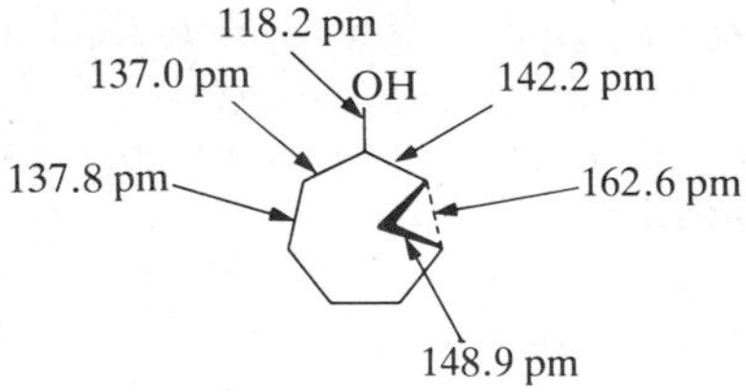

Figure 10.11. X-Ray crystallographic structure of the 2-hydroxyhomotropylium cation.

4,5-Homotropone (**20**) has also been prepared, and this also exhibits little delocalization through the cyclopropyl ring. Protonation of **20** gives the 4-hydroxyhomotropylium cation (**21**), which has the characteristic ^1H NMR spectrum of a homotropylium cation, the methylene cyclopropane

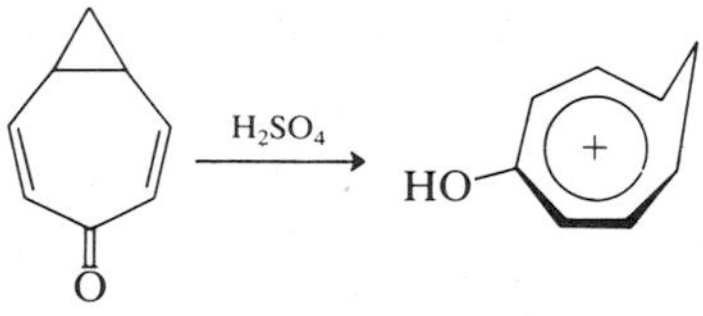

H_2SO_4
O HO +
20 **21**

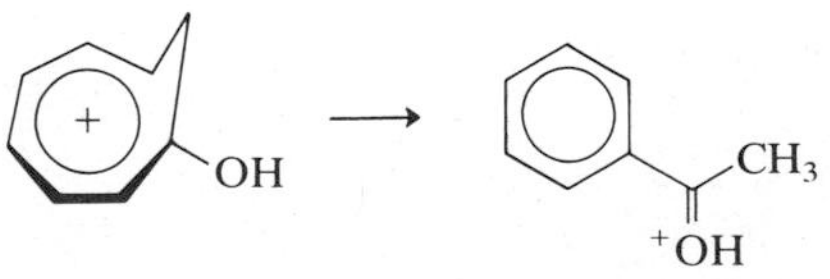

+ OH CH$_3$ $^+$OH
22 **23**

Figure 10.12

hydrogens showing a chemical shift difference of 5.4 ppm. The corresponding 1-hydroxytropylium cation (**22**) has also been prepared, and this is again a delocalized species that is stable at low temperatures but that rearranges to protonated acetophenone (**23**) at room temperature. Interestingly, photoirradiation of **19** at low temperatures gives **22**, a process involving a circumambulation of C-8 around the ring.

The homotropylium ions that have so far been described have all had their cyclic delocalization interrupted by one methylene group. A number of bishomotropylium ions has also been prepared in which the delocalized system is interrupted by two methylene groups. The simplest 1,4-bishomotropylium cation **26** was prepared by protonation of *cis*-bicyclo[6.1.0]nonatriene (**24**), possibly via the intermediate *trans*-cation **25**. The ^{13}C NMR spectrum of **26** shows signals at δ 31.5 (C-1,7), 117.9 (C-3,5), 137.0 (C-2,6), 141.4 (C-8,9), and 183.4 (C-4) and the diatropicity is less than for **7**.

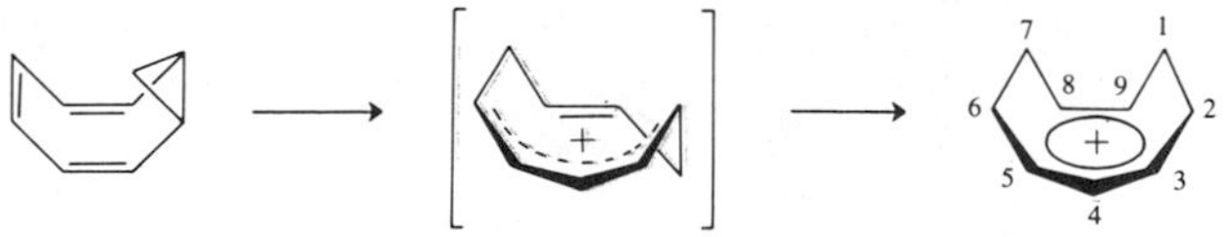

Figure 10.13

The bicyclo[4.3.0]nonatrienium cation (**29**) has been prepared from the alcohol **27** by treatment with a mixture of fluorosulfonic acid and SO_2ClF at low temperature. The initial product is the cation **28**, which rearranges

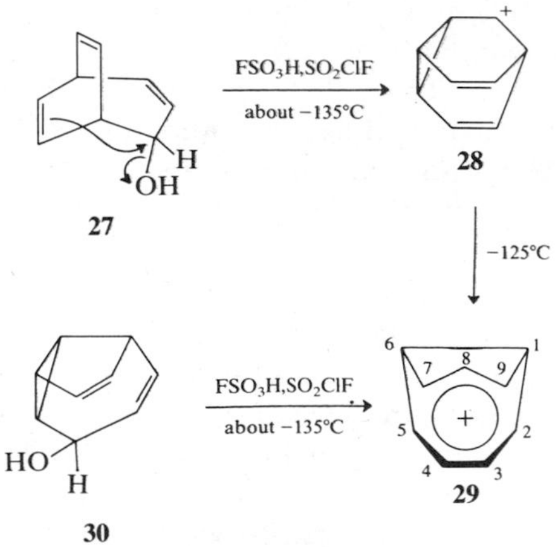

Figure 10.14

above $-125°C$ to give **29**. The cation **29** can also be obtained by the similar treatment of the alcohol **30**. The ^1H NMR spectrum of **29** shows resonance signals at δ 8.23 (H-7,9), 7.52 (H-3,4), 7.38 (H-8), 6.40 (H-2,5), and 3.63 (H-1,6), consistent with a delocalized structure for the ion. The 1-methyl derivative of **29** has also been prepared and has a delocalized structure.

The bicyclo[4.3.1]decatrienium cation (**32**) was prepared from the hydrocarbon **31** by protonation with fluorosulfonic acid in a mixture of SO_2 and deuteromethylene chloride at $-75°C$. The ^1H NMR spectrum of **32** shows the H-3,4 protons at δ 8.01, the bridgehead protons H-1,6 at δ 4.44, and the remaining ring protons at about δ 6.8. The proton H^b appears as a multiplet at δ 1.01 and the proton H^a as a doublet at δ -0.03. The rearrangement of **31** to the cation **32** indicates the enhanced stability of this ion, and the ^1H NMR spectrum is fully in accord with the delocalized structure.

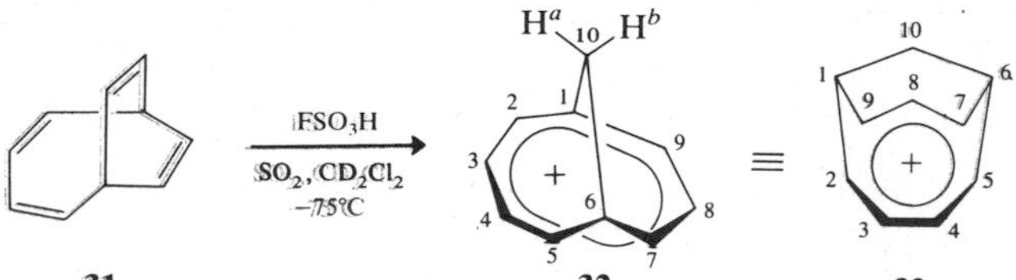

Figure 10.15

The relationship among the three cations **26**, **29**, and **32** can be readily seen by inspection of the structures illustrated.

The properties of the homotropylium cations serve to indicate that there might be other homoaromatic ions isoelectronic with the monocyclic ions discussed in Chapter 5. Thus, the corresponding homocyclopentadienyl anion (**33**) might be envisaged as the counterpart to the cyclopentadienyl anion. The anion **33** has not so far been prepared, and it may be less stable than the cyclohexadienyl anion **34**, which can maintain a planar geometry without much bond angle distortion unlike the cation **9**. A bridged bishomocyclopentadienyl anion **36** analogous to **32** has, however, been prepared and appears to be delocalized. Brown and Occolowitz first postulated the anion **36** as being involved in the deprotonation of **35** which occurs $10^{4.5}$ times faster than deprotonation of the corresponding monoene **38**. The anion **36** was subsequently prepared by Brown by treatment of the methyl ether **37** with sodium-potassium alloy, and a similar preparation was carried out by Winstein and collaborators. Brown found that the exoisomer **37a** reacts rapidly with the alloy in THF, whereas the endoisomer **37b** reacts much more slowly. It appears likely that in the case of **37b** the proton is abstracted rather than the ether being cleaved. When solutions of **36** are quenched with water or methanol, the hydrocarbon **35** is isolated. The ^{1}H NMR spectrum of **36** in THF-d_8 shows signals at δ 5.41 (H-3), 3.65 (H-6,7), 2.87 (H-2,4), 2.55 (H-1,5), 0.87 (H^a), and 0.42 (H^b). The symmetry of the spectrum supports the formation of at least a partially delocalized molecule, and the *upfield* chemical shift of H-6,7 and the *downfield* shift of H-3 in **36** compared to these protons in **35** are indicative of the delocalization of the negative charge over the C-6,7 atoms. All the proton chemical shifts are at high field for an aromatic system, which suggests that the diamagnetic ring current in **36** must be small. This view is supported by the finding of Bergman and Rajadhyaksha that the 2-bromobishomocyclopentadienyl anion **40** is not obtained by treatment of the bromide **39** with base, the product being **41**. These workers suggest that **40** is formed but subsequently loses Br^- and rearranges to **41**.

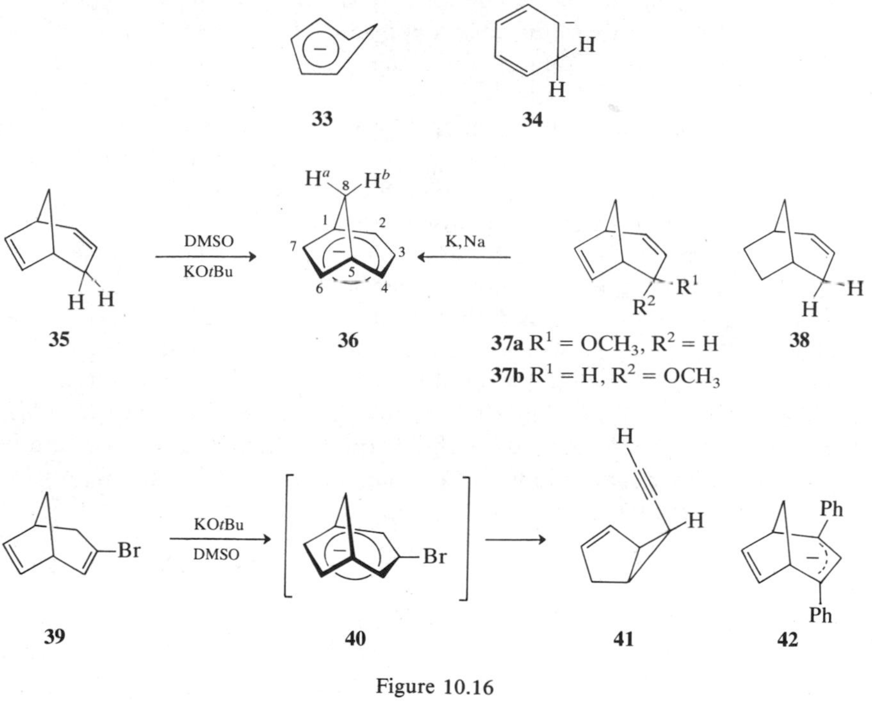

33 **34**

35 **36** **37a** $R^1 = OCH_3$, $R^2 = H$ **38**

37b $R^1 = H$, $R^2 = OCH_3$

39 **40** **41** **42**

Figure 10.16

The diphenyl-substituted anion **42** has been prepared, and this shows considerable shielding of the H-6,7 protons in the ^{1}H NMR spectrum but very *little* stabilization when the deprotonation rate was compared with model compounds. This caused doubt to be cast on whether **36** was a homoaromatic species, and semiempirical calculations on **36** concluded that there was no stabilization of the anion by participation of the 6,7-double bond. However, more recent calculations considering geometry and using a sphere-charge technique to determine the charge distribution concluded that charge was transferred to the 6,7-double bond and that **36** is homoconjugatively stabilized. This latter view has received further experimental support from an investigation of the ^{13}C NMR spectra of **36** and deuterium-substituted derivatives. The C-6,7 atoms show a large upfield shift on going from the parent hydrocarbon to the anion (39.0, 48.5 ppm), and these shifts are not significantly changed on deuterium substitution, indicating that **36** is a delocalized system and not a set of equilibrating ions.

Examples of analogues of both the 6 π-electron species, the cyclopentadienyl anion and the tropylium cation, have thus been prepared in which conjugation is interrupted by one or more methylene groups but that nevertheless appear to be delocalized aromatic systems. The enhanced stability of the homotropylium cation appears to be greater than that of the bishomocyclopentadienyl anion, and it may be anticipated that an increase in the number of interrupting groups will decrease the importance of homoaromatic stabilization. The geometry of the system will also be expected to be important. In the 6 π-electron species it is clear that, although some interaction may occur in neutral cycloheptatriene, the effect is dramatically emphasized in the case of charged compounds. It is for ionic species that other examples of homoaromatic systems containing 2 or 10 π electrons might be expected, and these have indeed been found.

Katz and co-workers found that treatment of 1,2,3,4-tetramethyl-3,4-dichlorocyclobutene (**43**) with silver hexafluoroantimonate at −70°C in SO_2 gave the cation **44**. The cation **44** could also be prepared as the chloroaluminate by treatment of **43** with $AlCl_3$ in methylene chloride. The ^{1}H NMR spectrum of **44** showed there were three types of methyl resonance, and in the electronic spectrum the absorption maximum was at 253 nm, a position intermediate between that of the cyclopropenium cation (about 185 nm) and the allyl cation (300 nm). The position of the electronic absorption

CH3 CH3 Cl Cl CH3 CH3 → SO2 CH3 CH3 + Cl CH CH3 SbF_6^-

43 **44**

Figure 10.17

maximum suggests that a large 1,3-interaction occurs in **44** and a simple HMO calculation required a value of $\beta_{1,3} = 0.33\beta$ to account for the band position. A later x-ray crystallographic analysis of the crystalline chloroaluminate gave the structure illustrated in Figure 10.18. The distance between C-2,4 is only 177.5 pm, which is indicative of a strong transannular interaction.

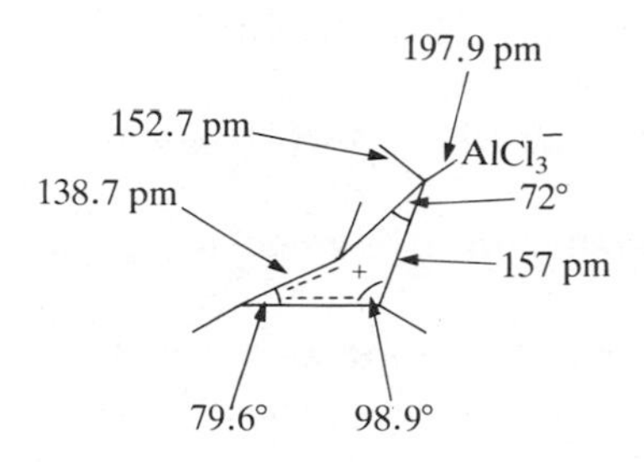

Figure 10.18. X-Ray crystallographic structure of 1,2,3,4-tetramethylcyclobutenium trichloroaluminate.

The parent homocyclopropenium cation **45** and a number of simple derivatives were subsequently synthesized by Olah and co-workers. In the ^{1}H NMR spectrum the parent cation **45** shows signals at δ 9.72 (H-2), 7.95 (H-1,3), and 4.53 (H-4), and in the ^{13}C NMR spectrum it shows signals at δ 178.7 (C-2), 141.6 (C-1,3), and 63.7 (C-4). Both the positions of the H-2 and C-2 resonances are deshielded compared to the corresponding 1,3-atoms, in complete contrast to the cyclopentenium and higher cyclic allyl cations in which the 1,3-atoms are most deshielded. This implies a rehybridization of C-1,3 in the cyclobutenium cation to provide the necessary overlap between the 1,3-atoms. A gradation of behavior was observed with the substituted cations. The 1,3-diphenyl derivative **46** showed no transannular interaction, and the effect increased through to the parent system.

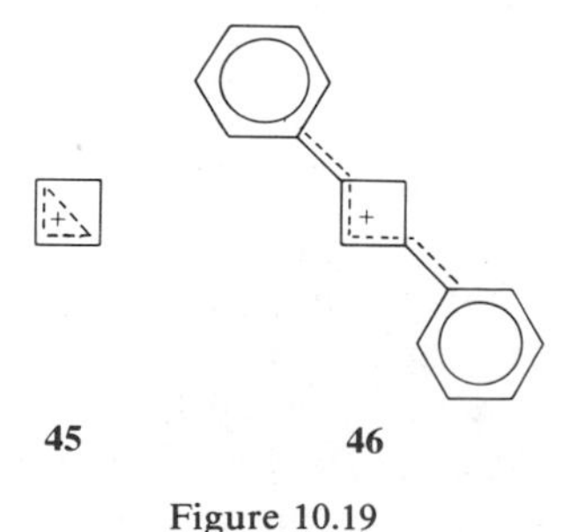

Figure 10.19

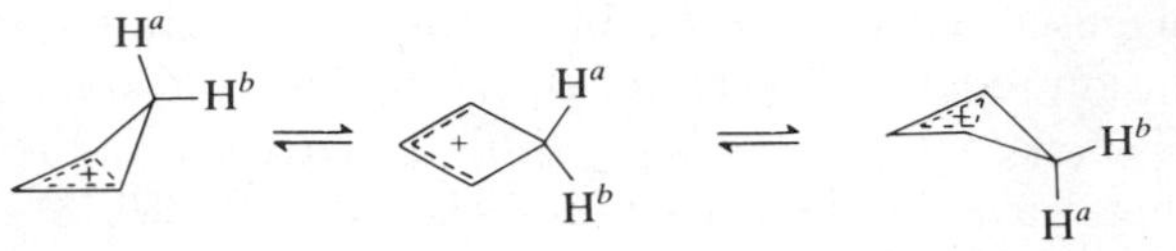

Figure 10.20

Paralleling this effect, the cation **45** exists in a nonplanar conformation with a barrier between equivalent forms of 35 kJ mol^{-1} representing the planar transition state (Fig. 10.20), whereas **46** has a planar ground state. This barrier is much lower than that observed for the homotropylium cations, which explains why the latter do not change their ground-state conformation on substitution.

The fascinating dihomocyclopropenium dication **48** has been prepared by oxidative cleavage of the tricyclic diene **47**.

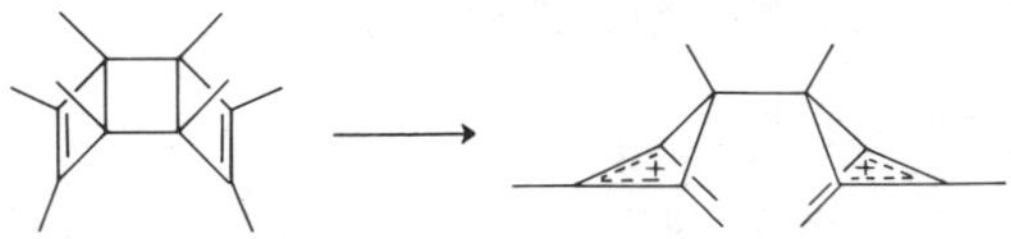

Figure 10.21

Attempts to observe the bishomocyclopropenium cation **49** have been unsuccessful, the allyl cation being preferred, but some evidence for the occurrence of **49** in the reactions of 3-cyclopentenylidene (**50**) has been adduced.

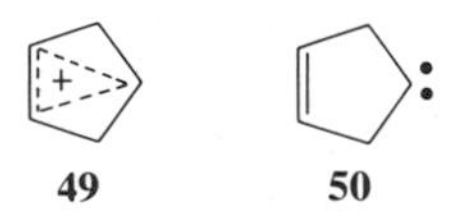

Figure 10.22

The trishomocyclopropenium cation **52** has been obtained by the ionization of the chloride **51** in superacids at −120°C. The ^{1}H and ^{13}C NMR spectra indicate that it is a delocalized system. Evidence, however, suggests that the bicyclic cation **53** is more stable than **52**, and consequently **52** cannot be considered aromatic in the classic sense. Coates' cation **55** is, however, highly delocalized and stabilized, and this stabilization cannot be due to relief of strain.

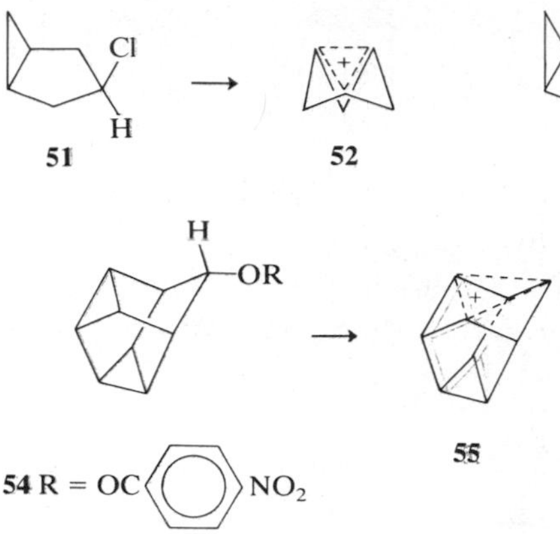

Figure 10.23

In the 10 π-electron series, it has been claimed that treatment of bicyclo[6.1.0]nonatriene (**24**) in THF with a potassium mirror at $-80°C$ gave the monohomocyclooctatetraenyl dianion **56** as the potassium salt. Subsequent experiments have thrown doubt on some of the original spectral data, but **56** can be prepared either by reduction of **24** or by deprotonation of **57**. The ^1H NMR spectrum of **56** shows signals at δ 4.8 (H-4,5), 4.7 (H-2,7), 3.5 (H-1,3,6,8), 1.4 (H^b), and -1.2 (H^a), and the ^{13}C NMR spectrum shows four of the expected five signals. Protons H^a and H^b are strongly coupled ($J = -12.9$ Hz), and H^a is also coupled to H-1,8. The homoaromatic dianion **56** is much more basic and much less stable than the cyclooctatetraenyl dianion, reacting with water more exothermically per electron than potassium!

The bishomocyclooctatetraenyl dianion **58** has been prepared, and both the ^1H and ^{13}C NMR spectra indicate that it is a delocalized system.

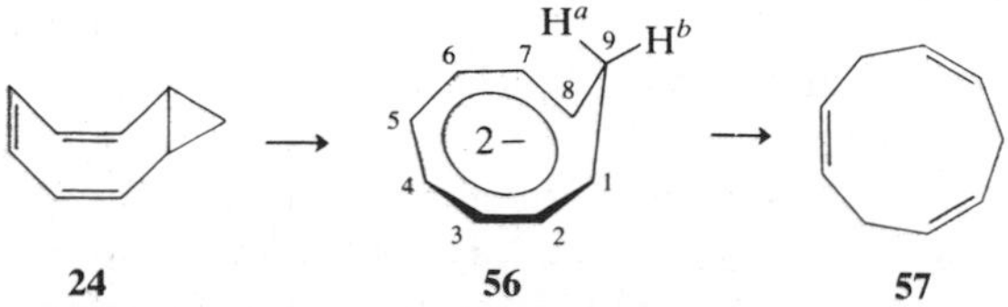

Figure 10.24

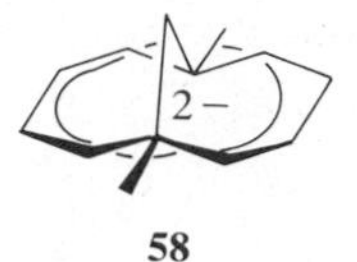

Figure 10.25

The 10 π-electron bridged monohomoaromatic cyclononatetraenyl anion **60** has been prepared by treatment of 1,6-methano[10]annulene (**59**) with sodium methylsulfinyl carbanion in dimethylsulfoxide. The ^{1}H NMR spectrum supports the delocalized symmetric structure.

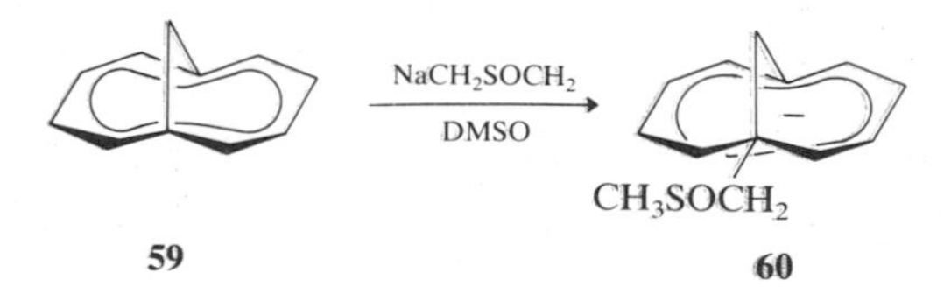

Figure 10.26

There thus appears to be a range of ions containing 2, 6, or 10 π electrons in which cyclic delocalization is interrupted by one or more methylene groups but which are best represented as delocalized systems. Several of the cations are more stable than the isomeric cation which is less delocalized and are thus aromatic in the classic sense, but the thermodynamic stability that accrues through homoaromatic delocalization is small.

There have been extensive discussions as to the possibility of observing neutral homoaromatic systems and even the possibility of stabilizing anions homoaromatically has been questioned. The initial interaction of the filled π orbitals is destabilizing, and only after considerable distortion of the σ framework does the HOMO–LUMO gap decrease sufficiently for stabilization to occur. Are there examples of such effects? This is still an area of contention, but it does appear that 1,6-methano and 1,5-methan[10]annulene are best described as homonaphthalene and homoazulene, respectively, as discussed in Chapter 4. There is a considerable transannular interaction even though the transannular gap is large, and a simple peripheral description does not adequately explain the spectral data. In both compounds, the σ framework has distorted the interacting π bonds, and, in the case of 1,5-methano[10]annulene, the two contributing systems prefer to be of opposite charge, a factor that should further assist HOMO–

LUMO interaction. The 1,6-methano[10]annuleno biphenylene analogue described in Chapter 9 (p. 236) is best represented as a homobenzene connected to a dimethylenebenzocyclobutene, and the 1,6-dimethylenecycloheptatriene part structure appears to be thermodynamically preferred in a range of annelated systems.

Examples of neutral homoaromatic compounds with more than one discontinuity should be even more difficult to find, although the elassovalenes (e.g., **61**) do appear to have some transannular interaction at the central junction. In triquinacene (**62**) the destabilizing effect of the filled shells appears to predominate and the double bonds are repelled from each other.

61 **62**

Figure 10.27

A second group of potentially aromatic compounds involving two non-coplanar π-electron systems was originally postulated by Goldstein, who termed the compounds *bicycloaromatic.* A set of rules was given based on simple MO theory, but these were later modified by Goldstein and Hoffmann, who treated the topology of potentially aromatic systems in a more general manner. *Longicyclic* systems contain three or more conjugated "ribbons" joined by common insulating atoms; the concept is thus related to homoaromaticity, which, together with bicycloaromaticity, is subsumed in their general classification. Examples of longicyclic systems are the pairs of ions **63** and **64**, and **65** and **66**.

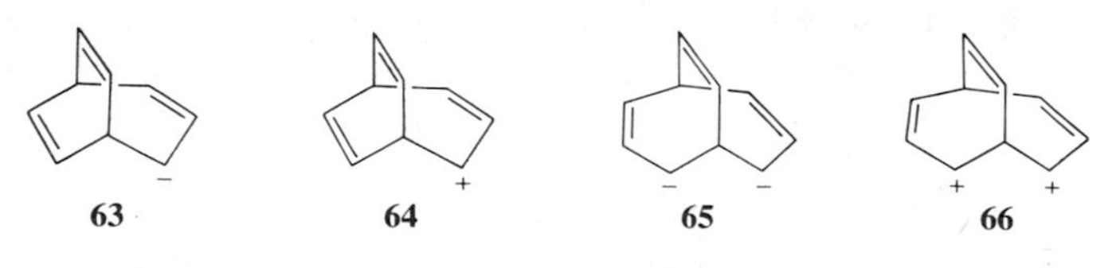

Figure 10.28

The anion **63** contains 8 π electrons in three ribbons, two of which contain $(4n + 2)$ π electrons and the third of which contains $4n$ π electrons; **63** is predicted to be aromatic. The cation **64** contains 6 π electrons in three ribbons, all of which contain $(4n + 2)$ π electrons, and is predicted to be antiaromatic. The dianion **65** contains 10 π electrons in three ribbons, two with $4n$ and one with $(4n + 2)$ π electrons, and is predicted to be aromatic. The dication **66** has 6 π electrons, all three ribbons having $(4n + 2)$ π electrons, and is predicted to be antiaromatic. The rules for ions with three ribbons are that if the total number of π electrons is $4n$, then one ribbon must be $4n$ and the others $(4n + 2)$; whereas if the total number of π electrons is $(4n + 2)$, then two ribbons must be $4n$ and the other $(4n + 2)$. These rules are intuitively satisfying as the individual $(4n + 4n + 2)$ interactions are homoaromatic (although one is intuitively neglecting $4n$, $4n$ and $(4n + 2)$, $(4n + 2)$ interactions!).

The 7-norbornyl cation (**67**) was a known example of a longicyclic [4π-2,2,0] system, which is predicted to be stabilized and was known to be stable. The anion **63** [8π-4,2,2] and its 3-methyl derivative have been prepared, and both appear to be stabilized. The anion **63** has only four resonance signals in the ^{1}H NMR spectrum at δ 5.24 (H-3), 4.98 (H-6,7,8,9), 3.05 (H-2,4), and 2.29 (H-1,5). The ^{13}C NMR spectrum showed four signals at δ 119.0 (C-3), 110.4 (C-6,7,8,9), 58.7 (C-2,4) and 35.1 (C-1,5).

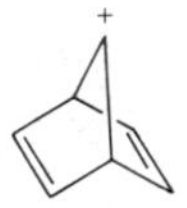

+

67

Figure 10.29

The cation **69** is a longicyclic aromatic [8π-4,2,2] system, but attempts to prepare it led to rearranged products. When the antichloride **68** was hydrolyzed, however, the product alcohol **70** retained its structure and configuration and no rearrangement occurred. The rate of hydrolysis is high, and the cation **69** does appear to be stabilized although other rearranged cations are more stable where accessible.

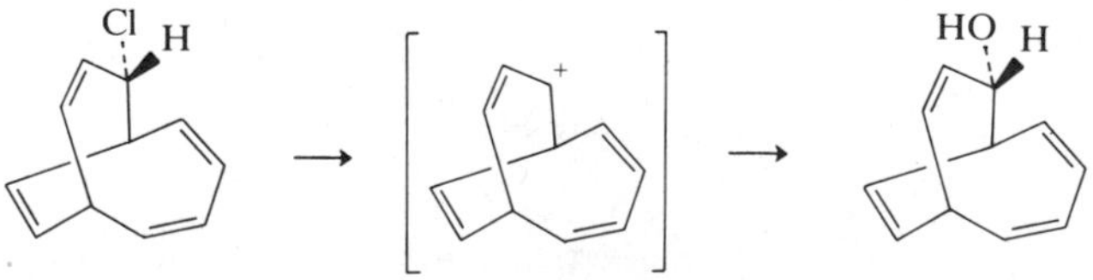

Cl H → + → HO H

68 **69** **70**

Figure 10.30

The reduction of bullvalene (**71**) with sodium-potassium alloy gave the [10π-4,4,2] bicyclo[3.3.2]decatrienyl dianion (**65**) as a yellow-green crystalline solid. Quenching the anion with water gave a mixture of the bicyclic trienes **72**, and oxidation with iodine gave bullvalene. The lithium salt was prepared by metathesis with LiBr, and its ^{1}H NMR spectrum showed absorptions at δ 6.14 (H-3,7), 4.63 (H-9, 10), 3.06 (H-2,4,6,8), and 2.31 (H-1,5) and the ^{13}C NMR spectrum absorptions at δ 131.5 (C-3,7), 106.3 (C-9,10), 75.3 (C-2,4,6,8) and 36.3 (C-1,5). Dihydrobullvalene does not give a similar dianion with sodium-potassium alloy but undergoes cleavage to a mixture of bicyclic dienes.

71 **65** **72**

Figure 10.31

Although still contentious, there does now appear to be some justification in suggesting that there is a further set of aromatic systems that is stabilized by the interaction of more than two noncoplanar arrays of π electrons. The species in which such effects are manifest are all charged, and the evidence for this effect in neutral systems does, at best, rely on negative evidence.

A further type of nonplanar interaction that was predicted to be stabilizing is exemplified by spiro[4.4]nonatetraene (**73**) and was termed *spiroconjugation.* Compound **73** and its dihydroderivative **74** were synthesized by Semmelhack and co-workers. The theoretical prediction was that the mixing of the two π systems would not result in any great increase in bonding energy but would lower the HOMO–LUMO gap. Comparison of the electronic spectra of **73** and **74** shows the former to have its absorption maximum at longer wavelength (**73**, 276 nm; **74**, 254 nm) and with a lower extinction coefficient (**73**, ε 1120; **74**, ε 2750) as predicted. The related bisindene spiroderivatives also show evidence for spiroconjugation.

A number of other topologic arrays has been suggested to be stabilized and other, more recherché forms of stabilization have also been considered,

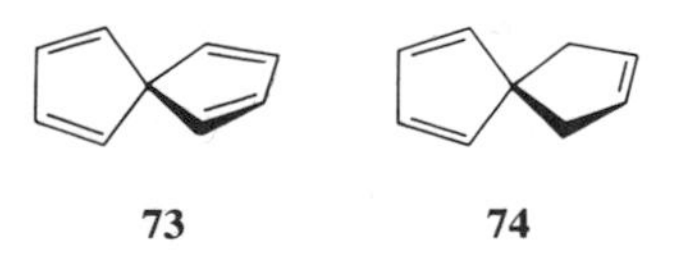

73 **74**

Figure 10.32

for example *trefoil aromatics.* Experimentalists will continue to be stimulated by such theoretical suggestions and the disputes that result from them, and attempts to prepare novel systems will continue. The interested reader is referred to the references in the Further Reading section.

FURTHER READING

For reviews on homoaromaticity, see S. Winstein, *Chem. Soc. Special Publication*, 1967, No. 21, 5; *Quart Rev.*, 1969, **23**, 141; P. M. Warner, in T. Nozoe, R. Breslow, K. Hafner, Shô Itô, and I. Murata (Eds.), *Topics in Nonbenzenoid Aromatic Chemistry*, Vol. 2, Hirokawa, Tokyo, 1977, p. 283; L. A. Paquette, *Angew. Chem. Int. Ed. Engl.*, 1978, **17**, 106.

For a recent account of some of the difficulties of relating proton chemical shifts to homoaromaticity, see R. F. Childs, M. J. McGlinchey, and A. Varadarajan, *J. Am. Chem. Soc.*, 1984, **106**, 5974.

For the x-ray structures of a number of homoaromatic systems and some observations on stabilization, see R. F. Childs *et al.*, *Pure Appl. Chem.*, 1986, **58**, 111.

For a discussion of recent work on homocyclopropenium cations and related systems, see Q. B. Broxterman, H. Hogeveen, and R. F. Kingma, *Pure Appl. Chem.*, 1986, **58**, 89.

For theoretical discussions concerning homoaromaticity, see K. N. Houk, R. W. Gandow, R. W. Strozier, N. G. Radan, and L. A. Paquette, *J. Am. Chem. Soc.*, 1979, **101**, 6797; E. Kaufman, H. Mayr, J. Chandrasekhar, and P. von R. Schleyer, *ibid.*, 1981, **103**, 1375; J. M. Brown, R. J. Elliot, and W. G. Richards, *J. Chem. Soc., Perkin 2*, 1982, 485; R. C. Haddon and K. Ragharchari, *J. Am. Chem. Soc.*, 1983, **105**, 118; D. Cremer, E. Kraka, T. S. Lee, R. F. W. Bader, C. D. H. Lau, T. J. Nguyen-Dang, and P. J. MacDougall, *ibid.*, 1983, **105**, 5069.

For a discussion of homoconjugation in neutral systems, see L. T. Scott, *Pure Appl. Chem.*, 1986, **58**, 105.

For a theoretical discussion of topologic aromaticity, see M. Goldstein and R. Hoffmann, *J. Am. Chem. Soc.*, 1973, **91**, 6193; and for the earlier account of spiroconjugation, see H. E. Simmons and T. Fukunaga, *ibid.*, 1967, **89**, 5208, and R. Hoffmann, A. Imamara, and G. D. Zeiss, *ibid.*, 1967, **89**, 5215.

For trefoil aromatics, see T. Fukunaga, H. E. Simmons, J. J. Wendoloski, and M. D. Gordon, *J. Am. Chem. Soc.*, 1983, **105**, 2729.

11

AROMATIC TRANSITION STATES AND CRITERIA FOR AROMATICITY

11.1. AROMATIC TRANSITION STATES

As may be seen from the discussion in the foregoing chapters, the concept of aromaticity pervades many areas of organic chemistry. The delocalization of π electrons over a number of atoms in a cycle has been shown to lower the energy, *providing* that there are sufficient bonding molecular orbitals to accommodate the electrons in a closed shell. As we have seen, the advantages of delocalization are emphasized in charged systems, since delocalization has the added effect of spreading the charge over a number of atoms. It might therefore be expected that a similar delocalization of π electrons occurring in the transition state of a reaction should lower the energy of that state and allow the reaction to proceed more easily. The concept of aromatic stabilization in the transition state has recently been shown to be extremely useful and may be applied to a variety of reactions. As Dewar has pointed out, the first suggestion of such a stabilization was put forward by Evans to account for the facility with which the Diels–Alder reaction occurs. In its simplest form, this reaction can be exemplified by the addition of butadiene (**1**) to ethylene (**2**) to give cyclohexene (**3**).

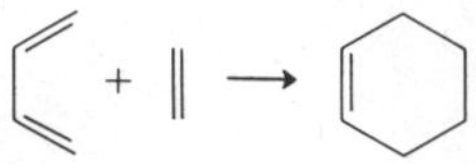

Figure 11.1

The transition state of this reaction may be envisaged to be formed by the interaction of the four $2p$ π-atomic orbitals on the diene **1** with the two $2p$ π-atomic orbitals on the dieneophile **2** as shown in Figure 11.4(a). The set of six atomic orbitals can then form six molecular orbitals, three of which are bonding and can accommodate the 6 π electrons. The problem can thus be seen to be very similar to that of benzene discussed in Chapter 1, except that the interaction between the terminal atoms of the two reactants resembles a homoaromatic rather than a pure π interaction.

The interaction of ethylene (**2**) with itself to give cyclobutane can now be seen to involve a transition state made up of four $2p$ π atomic orbitals as shown in Figure 11.4(b).

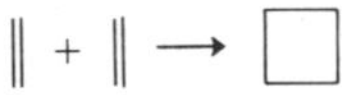

Figure 11.2

This transition state resembles cyclobutadiene, and Evans pointed out that this was nonaromatic, whereas we would now consider it to be antiaromatic. There are four molecular orbitals, one bonding, two nonbonding, and one antibonding, and two electrons will have to enter nonbonding orbitals. Similarly, the dimerization of butadiene to give cycloocta-1,5-diene (**5**) will involve a transition state with eight $2p$ π-atomic orbitals as shown in Figure 11.4(c). This orbital arrangement resembles planar, antiaromatic cyclooctatetraene with 2 π electrons again having to enter NBMOs.

There are a myriad of known examples of the Diels–Alder (4 + 2) π-electron addition, but examples of (2 + 2) or (4 + 4) π-electron cyclo-

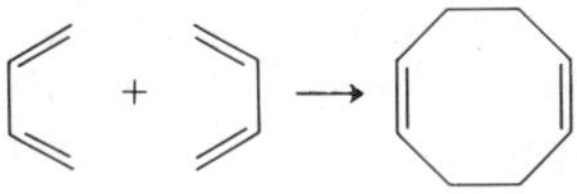

Figure 11.3

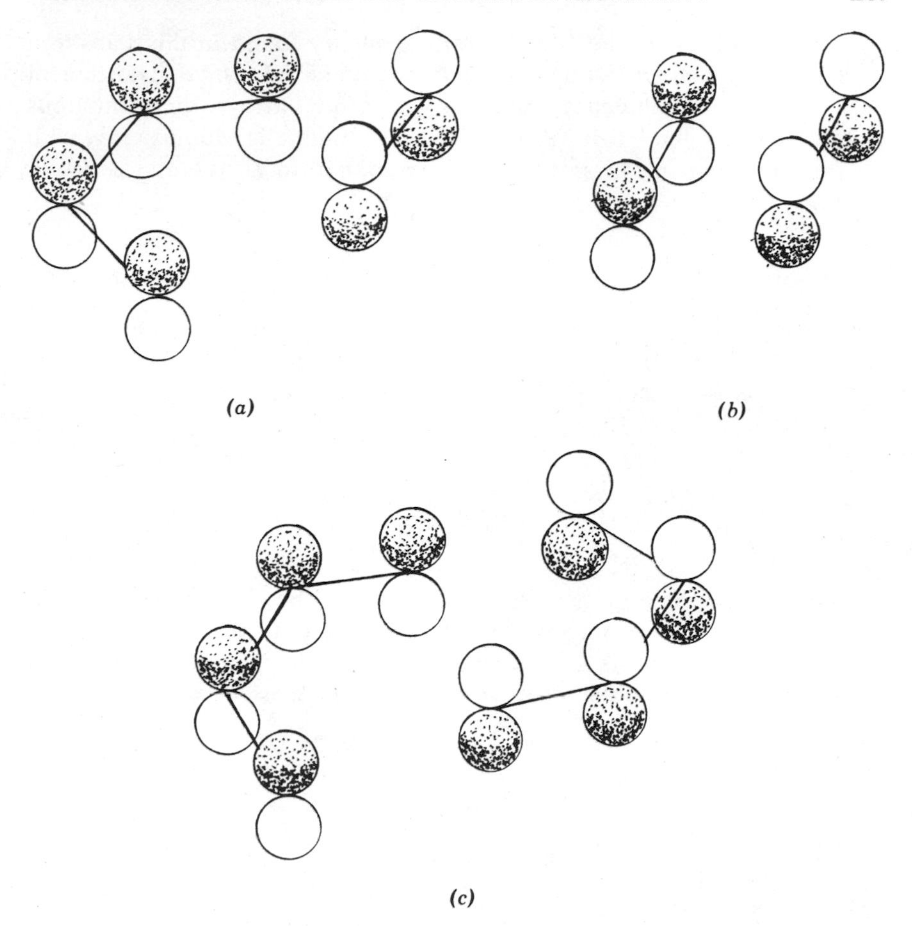

Figure 11.4. (*a*) p_π—Atomic orbitals involved in a 4 + 2 cycloaddition; (*b*) p_π—atomic orbitals involved in a 2 + 2 cycloaddition; (*c*) p_π—atomic orbitals involved in a 4 + 4 cycloaddition (shading indicates phase of orbital).

additions are much rarer and involve radical or other nonconcerted mechanisms. *Suprafacial* addition, to use the terminology of Woodward and Hoffmann, will only take place in a concerted fashion if the transition state contains $(4n + 2)$ π electrons. *Antarafacial* addition, in which orbital interaction occurs on opposite sides of the ring plane, requires a $4n$ array of electrons but is stereochemically impossible to effect in smaller systems. In the (14 + 2) π-electron cycloaddition of tetracyanoethylene (**6**) and heptafulvalene (**7**) the reaction does proceed through a 16 π-electron transition state to give **8** with the *trans*-ring-junction expected from antarafacial addition. The Möbius strip geometry found in this transition state was predicted by Heilbronner to be the preferred orbital arrangement of delocalized $4n$ annulenes, and these two orbital arrays, the continuous Hückel array for $4n + 2$ systems and the discontinuous Möbius array with its sign inversion for $4n$ systems, can be applied to all thermal *pericyclic* reactions.

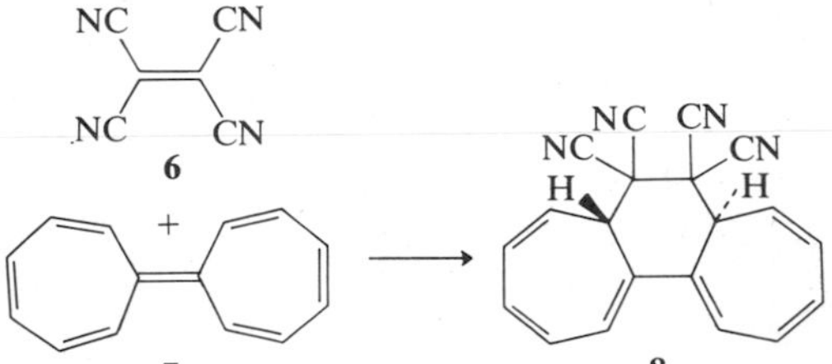

Figure 11.5

Another pericyclic reaction is exemplified by the Cope rearrangement of hexa-1,3,5-triene (**9**) into cyclohexa-1,3-diene (**10**), the transition state of which involves the six $2p$ π orbitals as shown in Figure 11.7. This is a *sigmatropic* reaction in the Woodward–Hoffmann classification.

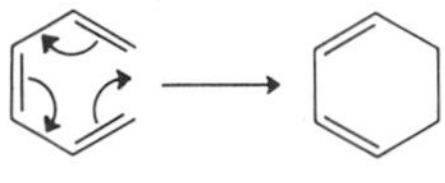

Figure 11.6

The ring closure could occur in two ways: either by rotation of the 1,6-carbon atoms in opposite directions [Fig. 11.7(a)] or in the same direction [Fig. 11.7(b)]. Rotation of atoms in opposite directions was termed *disrotatory* by Woodward and Hoffmann, whereas rotation in the same direction was termed *conrotatory*. As can be seen in Figure 11.7, in the Cope reaction disrotation leads to a Hückel array of the 6 π orbitals, whereas conrotation leads to a Möbius array of the orbitals. Consequently, disrotation should be the predicted course of this electrocyclic reaction. If we now turn to the electrocyclic ring closure of butadiene (**1**) to cyclobutene (**11**), the disrotatory mode gives a Hückel orbital array and conrotation a Möbius array (Fig. 11.9). There are now four electrons in the transition state so that the Möbius, conrotatory mode is preferred.

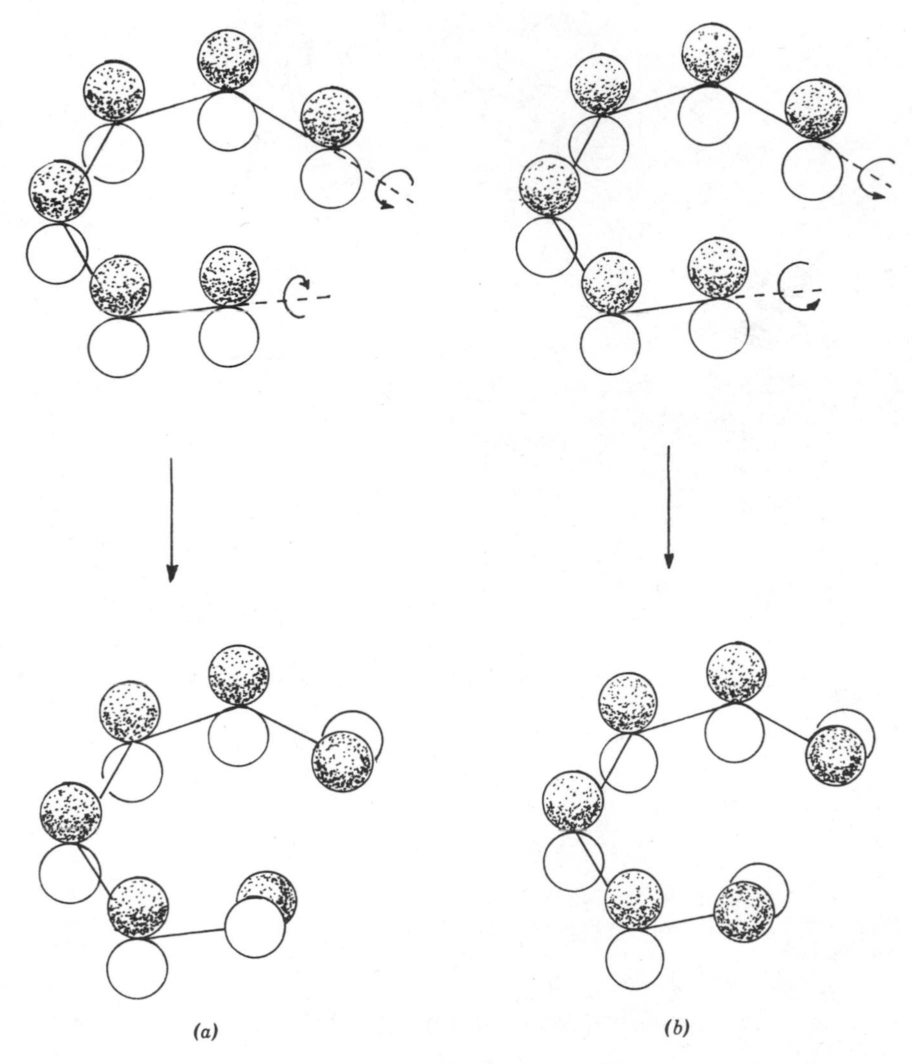

Figure 11.7. (*a*) Disrotatory ring closure in hexatriene; (*b*) conrotatory ring closure in hexatriene.

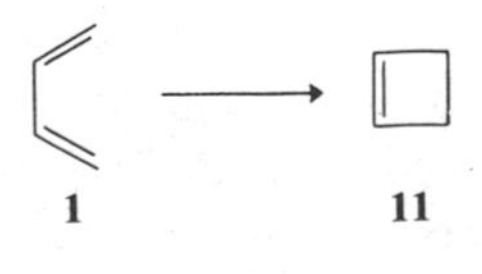

Figure 11.8

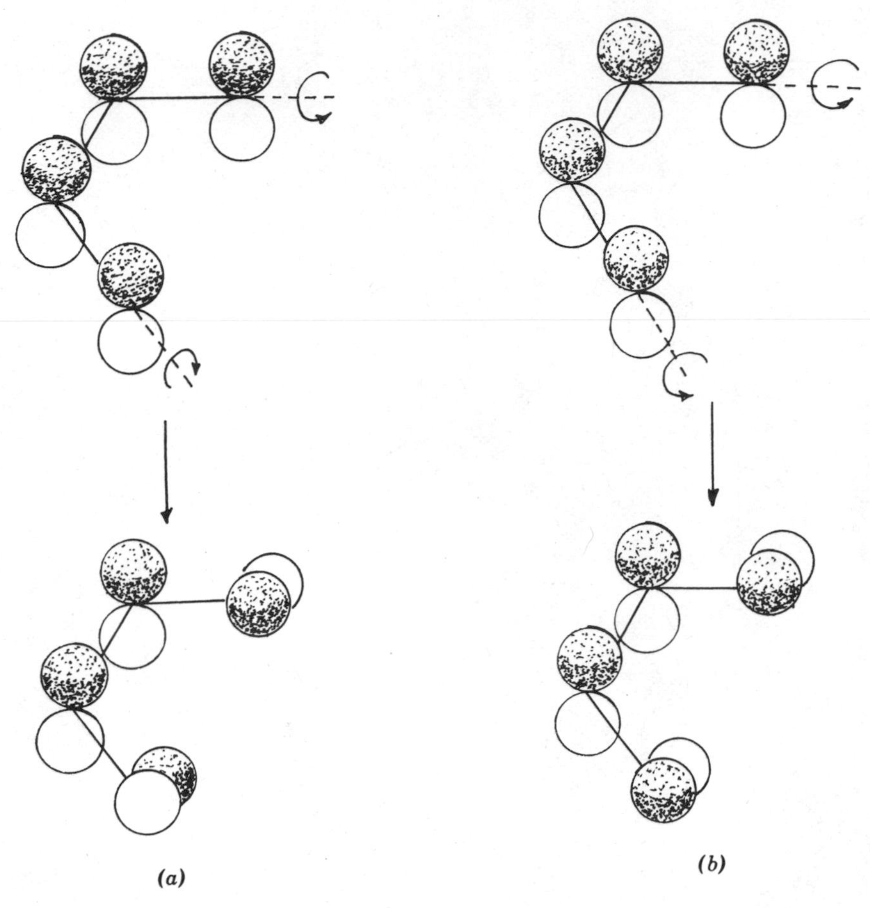

Figure 11.9. (*a*) Disrotatory ring closure in butadiene; (*b*) controtatory ring closure in butadiene.

A body of experimental work confirms the predictions made by this orbital analysis. Thus, thermolysis of the isomeric *cis* (**12**) and *trans* (**14**) 3,4-dimethylcyclobutenes gave stereospecifically the isomeric E,Z- (**13**) and E,E-hexa-2,4-dienes (**15**).

The Z,Z-isomer is also an allowed product from **14**, but the transition state would be more crowded. These reactions are the reverse of the electrocyclic processes described above, but the principle of microscopic reversibility will hold. In the same way, the retro-Diels–Alder reaction will be an allowed concerted process with defined stereochemical consequences.

All the reactions we have so far described have been thermal ground-state processes. Photochemical reactions, involving excited states, may also be

12 13 14 15

Figure 11.10

orbital-symmetry controlled. In such cases the requirement of the transition state will be the *opposite* to that in the ground state. Thus $4n + 2$ arrays of orbitals will prefer Möbius excited transition states, whereas $4n$ orbital arrays will prefer Hückel excited transition states. This can be illustrated by the photochemically induced conversion of hexa-2,4-dienes to 3,4-dimethylcyclobutenes. The E,Z-isomer (**13**) obtained by conrotatory thermal isomerization of *cis*-3,4-dimethylcyclobutene (**12**) on photoirradiation undergoes disrotatory ring closure to *trans*-3,4-dimethylcyclobutene (**14**), whereas **15** photochemically closes to **12**.

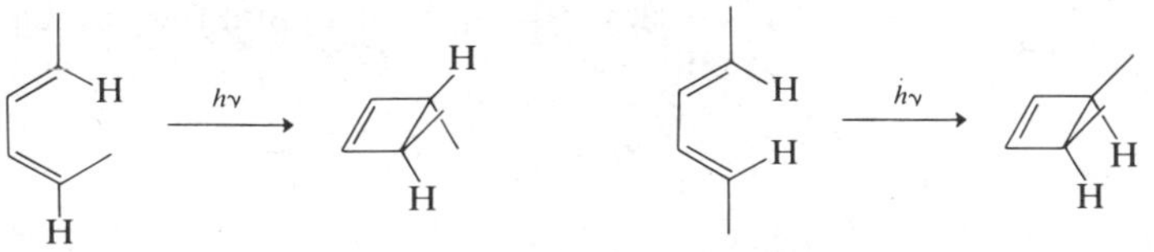

13 14 15 12

Figure 11.11

An advantage of this method of inspecting the aromaticity (or antiaromaticity) of the transition state over other approaches to orbital symmetry correlations is that only the number of interacting atomic orbitals needs to be known for the prediction of the probable stereochemistry and pathway of lowest energy for these types of concerted reactions. Both π and σ orbitals can be included, as is illustrated above for the retro modes of both the electrocyclic and Diels–Alder reactions.

It will have been noted that the arrangement of atomic orbitals in Figures 11.4, 11.7, and 11.9 closely resembles the arrangement of orbitals of a homoaromatic interaction (see Fig. 10.5). Consequently, it is not surprising that homoaromatic transition states should have been invoked in a number of reactions involving charged species. Thus, the acetolysis of *cis*-bicyclo[3.1.0]hex-3-yl *p*-toluenesulfonate (**16**) gives the corresponding *cis*-

acetate **18** as the sole product with retention of configuration. The acetolysis proceeds faster than that of the corresponding *trans*-epimer, and the increase in rate and stereospecificity of the product has been attributed to the intervention of the trishomocyclopropenium cation **17**. Deuteration studies support the intervention of a symmetric transition state.

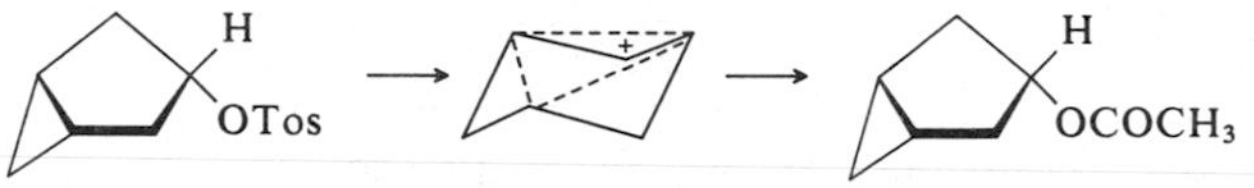

H OTos → + → H $OCOCH_3$

16 **17** **18**

Tos = $pCH_3C_6H_4SO_2$

Figure 11.12

Similarly, a bishomocyclopropenium cation **20** has been invoked as the transition state for the solvolysis of *anti*-7-norbornenyl *p*-toluenesulphonate (**19**). In this case, the cation has been generated in SbF_5–FSO_3H at −50°C, and the ^1H NMR spectrum at low temperature is consistent with the assigned structure. The proton at C-7 is at considerably higher field (δ 3.3) than the protons at C-2, C-3 (δ 7.07), which tends to indicate that the homoaromatic character of this system is small.

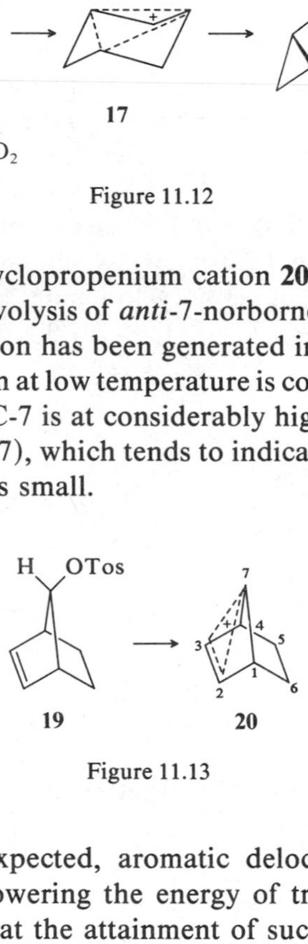

H OTos → 7 + 4 5 3 1 2 6

19 **20**

Figure 11.13

Thus, as would be expected, aromatic delocalization appears to be favored as a means of lowering the energy of transition states, and it is perhaps not surprising that the attainment of such a delocalized state has a profound effect on the course of chemical reactions and the nature of the products.

11.2. CRITERIA FOR AROMATICITY

Throughout the preceding chapters an attempt has been made to distinguish between "aromatic" and "nonaromatic" compounds. Such a distinction requires the use of some criteria in order that a selection can be carried out, and in Chapter 2 some of these criteria were examined. Of course it can, and has, been argued that distinguishing compounds in this way is not a useful exercise and that the concept of aromaticity is so vague as to serve no useful purpose. Recognizing the problems involved, I nevertheless believe that the idea that some compounds have unexpected stability arising from a common property has been historically valuable for the advancement of chemistry and was the first area in organic chemistry in which a fruitful interaction between theoretical and experimental chemists occurred. Although many of the original ideas have been modified, many have been generally accepted, for example Hückel's Rule, and the interplay of experimental observation and theoretical deduction still continues to be fruitful in this area, leading to the synthesis of new nonnatural compounds and the generation of new theoretical concepts.

As was indicated in Chapter 2, two approaches to the classification of compounds as aromatic or nonaromatic can be envisaged. In the first, some theoretical criterion could be proposed, which, if the compound met, would classify it as aromatic. The second would be to advance some experimental parameter, which, given a particular value, would again allow a decision between aromatic and nonaromatic to be made. Both approaches have advantages and disadvantages; the first, choosing a theoretical parameter, provides a firm basis for the classification but it can bring together compounds of dissimilar experimental properties while separating those with similar properties. Also, the more critical the parameter the fewer the number of compounds included, as the quotation on the flyleaf delightfully illustrates. The second, although providing something that the experimentalist can measure, may collect together compounds that have no theoretical common base and perhaps few other common properties. As a resolution between the experimental and theoretical criteria is achieved, we then approach the best of all possible worlds.

As was suggested in Chapter 2, the properties that are principally dependent upon the ground state of the system and that are most readily observed are those involving the magnetic behavior of the molecule. In aromatic systems, the enhanced diamagnetic susceptibility appears to be the most easily investigated and can be observed in three ways: (1) by measurement of the anisotropy of a single crystal, (2) by the determination of the diamagnetic susceptibility exaltation, and (3) by its effect upon the ^{1}H NMR spectrum. Each measurement has both advantages and disadvantages over the others. The diamagnetic anisotropy measurement demands the preparation of a single crystal and the correlation of the magnetic axes of the crystal with the axes of the constituent molecules. However, the measurement of the anisotropy due to the circulation of electrons over the cycle does appear to be the most fundamental observation of the consequences of aromaticity of the three types of measurement. The diamagnetic exaltation (Λ) is more readily obtained and does not require the formation of a crystal nor, consequently, a knowledge of the orientation of the molecules. It does, however, require fairly large samples and, more importantly, it depends on knowledge of the calculated diamagnetic anisotropy of a theoretical but *unreal* molecule. The ^{1}H NMR spectrum is the easiest to observe and requires the least amount of material, and the orientation of the sample is unimportant. As the preceding chapters testify, this is the parameter that has been most commonly obtained. The chemical shifts of a proton in a ^{1}H NMR spectrum are, however, not simply dependent on one factor, and even the chemical shift positions of hydrogens in classically aromatic compounds are widely variable.

Attempts to relate these phenomena to the degree of aromaticity are fraught with difficulty. Values of Λ per unit area from the diamagnetic exaltation or $K_3/(K_1 + K_2)/2$ from the diamagnetic anisotropy are likely to be only crude indicators of the degree of aromaticity. Haddon's success in relating the chemical shift to the ring current by his ring current geometric factor (see Chapter 2) is unexpected and encouraging. It would be of interest to see this criterion extended to a greater variety of compounds, and the observation of Boekelheide that geometric parameters obtained from models rather than from x-ray crystallographic data can be used may encourage experimentalists to carry out such determinations.

A variety of theoretical parameters has been suggested as suitable criteria for determining the aromaticity of a compound. Perhaps the oldest is the resonance energy. This can be a pseudoexperimental measurement if the enthalpy of combustion or of hydrogenation is measured and compared against that of a model system. More recently, however, attention has been focussed on theoretically determining the enthalpy of atomization and comparing that with the theoretical enthalpy of atomization of the corresponding acyclic polyene, as described in Chapter 2. This has the advantage that neither compound need be known, and the criterion, a theoretical resonance energy, is a means of classification independent of observation. Thus, determining an REPE in this way can provide both a basis for classification and a means of comparing the relative aromaticity of compounds. The experimentalist would then want to see how such a classification would mirror the observable properties of known systems.

Besides the comparison with theory of experimental enthalpies of combustion or hydrogenation, two other correlations have been made. Haddon and Aihara have independently shown that the ring current calculated from the observed proton chemical shift by Haddon's method can be related to the resonance energy as described in Chapter 2. Schaad and Hess and, independently, Herndon have shown that a reaction rate measured by Sondheimer and co-workers can be related to the REPE of the product system. Sondheimer's group had synthesized a series of dehydroannulenes annelated by furan rings (e.g., **21**), and found that the rate of Diels–Alder addition to the furan ring depended on the nature of the dehydroannulene. Thus, when the resulting adduct had a $(4n + 2)$ dehydroannulene component (e.g., **22b**), the rate of addition to the furan was greater than when the dehydroannulene component was a $4n$ system (e.g., **22a**). They further

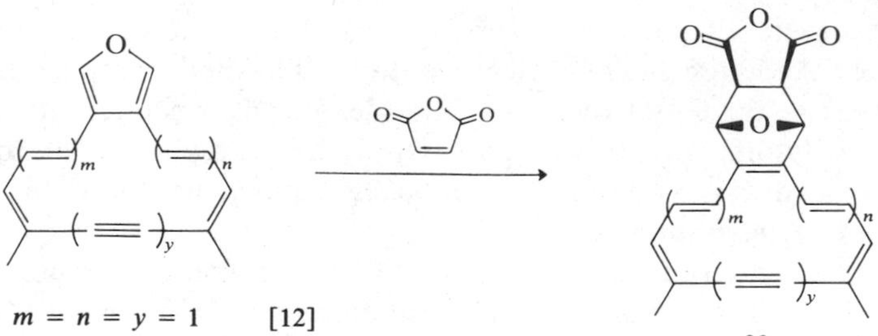

21a $m = n = y = 1$ [12]
21b $m = n = 1, y = 2$ [14]
21c $m = 1, n = y = 2$ [16]
21d $m = n = y = 2$ [18]

22

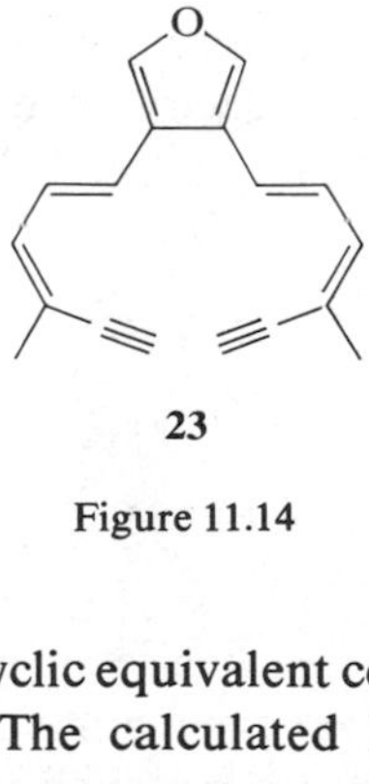

23

Figure 11.14

found that the rate with an acyclic equivalent component (**23**) was intermediate between the two rates. The calculated REPE for the aromatic and antiaromatic dehydroannulenes gave an excellent correlation with the observed reaction rates. Earlier, correlations of RE with the rate of Diels–Alder addition had been obtained for polycyclic benzenoid compounds.

The correlation with the theoretical REPE of such disparate phenomena as the reaction rate of Diels–Alder addition and the observed proton chemical shifts of a variety of annulenes begins to provide some evidence that a theoretical index is being established for a common phenomenon that we have described as aromaticity.

Other theoretical and experimental criteria could be used to determine aromaticity. A second experimental ground-state property that could be used as a criterion is the C—C bond length of aromatic systems, and such a criterion would be excellent for the annulenes. If nonaromatic or antiaromatic, an alternation in bond length would be expected, whereas in aromatic annulenes the bond lengths should all be the same and approach the 139 pm bond length of the benzene C—C bond. This criterion would be largely met by the annulenes, although it must be observed that in [18]annulene the C—C bonds are not all equal and that we also have to be aware that the x-ray data need interpretation. When the bond length criterion is applied to heterocyclic or polycyclic compounds, wide variations in bond length are likely. Although these problems could be alleviated by using other model systems (e.g., the C—N and C—C bond lengths of pyridine), the technique is still fairly difficult and lengthy to apply and a high level of confidence in the x-ray data is required. This probably requires the studies to be carried out at low temperatures.

Using the criterion of the absence of first- or second-order bond fixation may, it has been suggested, severely limit the number of aromatic compounds (one!). This method does, however, give valuable insight into the properties of a variety of nonbenzenoid aromatic compounds. Thus, pentalene and heptalene both exhibit second-order bond fixation, and the two forms (e.g., **24a** and **24b** for pentalene) are in tautomeric equilibrium rather than resonance. The interconversion of the two rectangular forms of cyclobutadiene is another example of second-order bond fixation (Chapter 2).

24a **24b**

Figure 11.15

Clearly, compromises must be made in using any of these criteria of aromaticity. The concept may be vague, but this is not necessarily a disadvantage. Certain types of aromatic systems, such as the annulenes and annulenyl ions, can now be defined by criteria that are generally acceptable. Applying the same definitions to more complex systems is more difficult. Nevertheless, I hope that the reader of this book will have found a continuous thread connecting the various compounds that have been discussed. They may or may not consider that the more esoteric topologic interactions are too far removed from those occurring in benzene to be usefully grouped with them. Time will no doubt justify or modify their view.

FURTHER READING

For the authors' own account of the Woodward–Hoffmann Rules, see R. B. Woodward and R. Hoffmann, *The Conservation of Orbital Symmetry*, Verlag Chemie GmbH, Weinheim, W. Germany, 1970.

For the correlation of the rules with aromatic and antiaromatic transition states, see M. J. S. Dewar, *The Molecular Orbital Theory of Organic Chemistry*, McGraw-Hill, New York, 1969, and *Tetrahedron*, Suppl. 8, 1966, 75. For the Möbius strip as a chemical concept, see E. Heilbronner, *Tetrahedron Lett.*, 1964, 1923; H. E. Zimmerman, *J. Am. Chem. Soc.*, 1966, **88**, 1564, 1566.

For general reviews of orbital symmetry, see T. L. Gilchrist and R. C. Storr, *Organic Reactions and Orbital Symmetry*, C.U.P., Cambridge, 1972; G. B. Gill and M. R. Willis, *Pericyclic Reactions*, Chapman-Hall, London, 1974. See also, H. E. Simmons and J. F. Bunnett (Eds.), *Orbital Symmetry Papers*, A.C.S., Washington, D.C., 1974.

For examples of homoaromatic transition states see, S. Winstein, *Chem. Soc. Special Publ.*, No. 21, 1967, 5; R. E. Leone and P. von R. Schleyer, *Angew. Chem. Int. Ed. Eng.*, 1970, **9**, 860.

For the correlation between REPE and Sondheimer's rate data, see A. B. Hess and L. J. Schaad, *J. Chem. Soc. Chem. Commun.*, 1977, 243; and also W. C. Herndon, *ibid.*, 1977, 817.

For a discussion of double-bond fixation, see G. Binsch, *Jerusalem Symposia on Quantum Chemistry and Biochemistry*, 1971, **3**, 25.

See also the text by Lewis and Peters (Further Reading, Chapter 1).

It has recently been suggested that benzene is a symmetric hexagon because of the σ framework, the π electrons, which would otherwise prefer to be localized, being forced to delocalize by this symmetry requirement of the σ electrons: P. C. Hiberty, S. S. Shaik, J-M. Lefour, and G. Ohanessian, *J. Org. Chem.*, 1985, **50**, 4659. However, the authors do appear to have chosen benzene length bonds for their hexagon, and it would be interesting to know what size hexagon the σ framework would prefer.

INDEX

Some modern methods of organic synthesis

Cambridge Texts in Chemistry and Biochemistry